Topics in Current Physics

34

Topics in Current Physics   Founded by Helmut K. V. Lotsch

1 **Beam-Foil Spectroscopy**
Editor: S. Bashkin

2 **Modern Three-Hadron Physics**
Editor: A. W. Thomas

3 **Dynamics of Solids and Liquids by Neutron Scattering**
Editors: S. W. Lovesey and T. Springer

4 **Electron Spectroscopy for Surface Analysis**
Editor: H. Ibach

5 **Structure and Collisions of Ions and Atoms**
Editor: I. A. Sellin

6 **Neutron Diffraction**
Editor: H. Dachs

7 **Monte Carlo Methods** in Statistical Physics
Editor: K. Binder

8 **Ocean Acoustics**
Editor: J. A. DeSanto

9 **Inverse Source Problems** in Optics
Editor: H. P. Baltes

10 **Synchrotron Radiation**
Techniques and Applications
Editor: C. Kunz

11 **Raman Spectroscopy** of Gases and Liquids
Editor: A. Weber

12 **Positrons in Solids**
Editor: P. Hautojärvi

13 **Computer Processing of Electron Microscope Images**
Editor: P. W. Hawkes

14 **Excitons** Editor: K. Cho

15 Physics of **Superionic Conductors**
Editor: M. B. Salamon

16 **Aerosol Microphysics I**
Particle Interactions
Editor: W. H. Marlow

17 **Solitons**
Editors: R. Bullough, P. Caudrey

18 **Magnetic Electron Lenses**
Editor: P. W. Hawkes

19 Theory of **Chemisorption**
Editor: J. R. Smith

20 **Inverse Scattering Problems** in Optics
Editor: H. P. Baltes

21 **Coherent Nonlinear Optics**
Recent Advances
Editors: M. S. Feld and V. S. Letokhov

22 **Electromagnetic Theory of Gratings**
Editor: R. Petit

23 **Structural Phase Transitions I**
Editors: K. A. Müller and H. Thomas

24 **Amorphous Solids**
Low-Temperature Properties
Editor: W. A. Phillips

25 **Mössbauer Spectroscopy II**
The Exotic Side of the Method
Editor: U. Gonser

26 **Crystal Cohesion and Conformational Energies** Editor: R. M. Metzger

27 **Dissipative Systems in Quantum Optics** Resonance Fluorescence
Optical Bistability, Superfluorescence
Editor: R. Bonifacio

28 **The Stratospheric Aerosol Layer**
Editor: R. C. Whitten

29 **Aerosol Microphysics II**
Chemical Physics of Microparticles
Editor: W. H. Marlow

30 **Real-Space Renormalization**
Editors: T. W. Burkhardt and J. M. J. van Leeuwen

31 **Hyperfine Interactions of Radioactive Nuclei**
Editor: J. Christiansen

32 **Superconductivity in Ternary Compounds I**
Structural, Electronic, and Lattice Properties
Editors: Ø. Fischer and M. B. Maple

33 **Molecular Collision Dynamics**
Editor: J. M. Bowman

34 **Superconductivity in Ternary Compounds II**
Superconductivity and Magnetism
Editors: M. B. Maple and Ø. Fischer

Superconductivity in Ternary Compounds II

Superconductivity and Magnetism

Edited by M.B. Maple and Ø. Fischer

With Contributions by: H. F. Braun M. Decroux
B. D. Dunlap Ø. Fischer F. Y. Fradin A. J. Freeman
P. Fulde H. C. Hamaker M. Ishikawa T. Jarlborg
D. C. Johnston J. Keller C. W. Kimball J. W. Lynn
M. B. Maple D. E. Moncton J. Muller G. K. Shenoy
G. Shirane W. Thomlinson L. D. Woolf

With 136 Figures

Springer-Verlag Berlin Heidelberg GmbH 1982

Prof. M. Brian Maple

Department of Physics and Institute for Pure and Applied Physical Sciences, University of California, San Diego, La Jolla, CA 92093, USA

Prof. Øystein Fischer

Département de Physique de la Matière Condensée, Université de Genéve, 24, Quai Ernest Ansermet, CH-1211 Genéve 4

ISBN 978-0-387-11814-7 ISBN 978-1-4899-3768-1 (eBook)
DOI 10.1007/978-1-4899-3768-1

Library of Congress Cataloging in Publication Data. Main entry under title: Superconductivity in ternary compounds II. (Topics in current physics ; 34) Bibliography: p. Includes index. 1. Superconductivity. 2. Superconductors–Magnetic properties. 3. Rare earth metal compounds–Magnetic properties. I. Maple, M. Brian, 1939– . II. Fischer, Ø. (Øystein), 1942– . III. Braun, H. (Hans F.) IV. Series. QC612.S8S833 1982 537.6'23 82-10647

Originally published by Springer-Verlag Berlin Heidelberg New York in 1982
Softcover reprint of the hardcover 1st edition 1982

Offset printing and bookbinding: Konrad Triltsch, Graphischer Betrieb, Würzburg.
2153/3130-543210

Dedicated to the memory of
Bernd T. Matthias

Preface

This Topics in Current Physics (TCP) Volume 34 is concerned primarily with superconductivity and magnetism, and the mutual interaction of these two phenomena in ternary rare earth compounds. It is the companion of TCP Volume 32 - Superconductivity in Ternary Compounds: Structural, Electronic and Lattice Properties.

The interplay between superconductivity and magnetism has intrigued theoreticians and experimentalists alike for more than two decades. V.L. Ginzburg first addressed the question of whether or not superconductivity and ferromagnetism could coexist in 1957, and B.T. Matthias and coworkers carried out the first experimental investigations on this problem in 1959. The early experiments were made on systems that consisted of a superconducting element or compound into which small concentrations of rare earth impurities with partially-filled 4f electron shells had been introduced. These dilute impurity systems were chosen because the scattering of conduction electrons by paramagnetic rare earth impurity ions usually has a strong destructive "pair breaking" effect on superconductivity, typically driving the superconducting transition temperature to zero at impurity concentrations of only a few atomic percent. Unfortunately, analysis of these early experiments was complicated by clustering and/or the formation of short range or "glassy" types of magnetic order so that definitive conclusions regarding the coexistence of superconductivity and magnetism could not be reached.

The ternary rare earth (RE) compounds $Re_xMo_6X_8$ and $RERh_4B_4$ had a great impact on the recent developments on this subject which experienced a revival around 1976-1977. This can be attributed to two special properties of these particular compounds: (1) The strength of the exchange interaction between the conduction electron spins and the rare earth magnetic moments is very weak (~ 0.01 eV-atom), and (2) the rare earth ions form an ordered sublattice. The small exchange interaction, which can be traced to the cluster character of these compounds, allows them to remain superconducting, even in the presence of relatively large concentrations of rare earth ions. It also results in magnetic ordering temperatures that are comparable to the superconducting transition temperatures. The ordered sublattice of rare earth ions leads to long-range magnetic order, with sharp magnetic ordering temperatures that are delineated by well defined features in the physical properties. This has greatly facilitated the interpretation of the experiments on the interaction of superconductivity and magnetism in ternary rare earth compounds and accounts for the rapid progress that has been made in this field during the last five years or so.

TCP Volume 34 is organized in the following way. Chapter 1 gives a brief introduction to the problem of the coexistence of superconductivity and magnetism and provides some of the key references. A discussion of the systematics of superconductivity in ternary compounds as well as many useful tables of data can be found in Chapter 2. Chapter 3 concentrates on the critical fields and critical currents of ternary molybdenum chalcogenides. Superconductivity, magnetism and their mutual interaction are the subjects of Chapters 4 and 5, Chapter 4 being devoted to the rare earth rhodium borides and rare earth transition metal stannides, while Chapter 5 considers the rare earth molybdenum chalcogenides. The band structure and magnetism of the ternary molybdenum chalcogenides and rhodium borides are discussed in Chapter 6. In Chapter 7, microscopic measurements such as NMR and Mössbauer effect on ternary compounds are reviewed, while in Chapter 8, neutron scattering investigations of magnetic ordering in ternary superconductors are recounted. Finally, the theory of magnetic superconductors is treated in Chapter 9.

It is hoped that the information contained within these two TCP volumes will enable the reader to assess the current status of this exciting and rapidly developing field in condensed matter science.

June 1982

M. Brian Maple
Øystein Fischer

Contents

List of Contributors

Braun, Hans F.
Département de Physique de la Matière Condensée, Université de Genève,
32, Bd. d'Yvoy, 1211 Genève 4, Switzerland

Decroux, Michel
Département de Physique de la Matière Condensée, Université de Genève,
24, Quai Ernest-Ansermet, 1211 Genève 4, Switzerland

Dunlap, Bob D.
Argonne National Laboratory, Argonne, IL 60439, USA

Fischer, Øystein
Département de Physique de la Matière Condensée, Université de Genève,
24, Quai Ernest Ansermet, 1211 Genève 4, Switzerland

Fradin, Frank Y.
Argonne National Laboratory, Argonne, IL 60439, USA

Freeman, Arthur J.
Department of Physics and Astronomy and Materials Research Center,
Northwestern University, Evanston, Il 60201, USA

Fulde, Peter
Max-Planck-Institut für Festkörperforschung
7000 Stuttgart 80, Fed. Rep. of Germany

Hamaker, H. Christopher
Argonne National Laboratory, Rm. D-226, Bldg. 212, 9700 South Cass Avenue,
Argonne, IL 60439, USA

Ishikawa, Masayasu
Département de Physique de la Matière Condensée, Université de Genève,
32, Bd. d'Yvoy, 1211 Genève 4, Switzerland

Jarlborg, Thomas
Department of Physics and Astronomy and Materials Research Center,
Northwestern University, Evanston, IL 60201, USA

Johnston, David C.
Corporate Research-Science-Laboratories, Exxon Research and Engineering
Company, P.O. Box 45, Linden, NJ 07036, USA

Keller, Joachim
Fakultät für Physik, Universität Regensburg
8400 Regensburg, Fed. Rep. of Germany

Kimball, Clyde W.
Northern Illinois University, DeKalb, Il 60115, USA

Lynn, Jeffrey W.
Department of Physics, University of Maryland, College Park, MD 20742, USA

Maple, M. Brian
Department of Pyhsics and Institute for Pure and Applied Physical Sciences
University of California, San Diego, La Jolla, CA 92093, USA

Moncton, David E.
Department of Physics, Brookhaven National Laboratory, Upton, NY 11973, USA

Muller, Jean
Département de Physique de la Matière Condensée, Université de Genève,
32, Bd. d'Yvoy, 1211 Genève 4, Switzerland

Shenoy, Gopol K.
Argonne National Laboratory, Argonne, Il 60439, USA

Shirane, Gen
Department of Physics, Brookhaven National Laboratory, Upton, NY 11973, USA

Thomlinson, William
Department of Physics, Brookhaven National Laboratory, Upton, NY 11973, USA

Woolf, Lawrence D.
Exxon Research and Engineering Company, P.O. Box 45, Linden, NJ 07036, USA

1. Magnetic Superconductors

M. B. Maple and Ø. Fischer

1.1 Introduction

A considerable amount of experimental and theoretical research on the interplay of superconductivity and magnetism has followed the discovery of the superconducting ternary rare earth (RE) compounds $REMo_6X_8$ (X = S, Se) [1.1,2] and $RERh_4B_4$ [1.3]. Thus, this second volume on "Superconductivity in Ternary Compounds" is largely devoted to that problem which has fascinated experimentalists and theoreticians alike for more than two decades. The purpose of this chapter is to provide the reader with a brief introduction to the subject and some of the key references. It is by no means complete, and we refer the reader to the following chapters for a detailed description of the phenomena that have been discovered and are presently under investigation.

1.2 How the Problem of Magnetic Superconductors Originated

The suppression of superconductivity by applied magnetic fields implies that ferromagnetism and superconductivity are two very different cooperative phenomena that are unlikely to occur simultaneously within the same material. The first person to point out that the two phenomena were mutually exclusive appears to have been GINZBURG [1.4] in 1957. Shortly thereafter, in 1959, MATTHIAS and coworkers [1.5] carried out the initial experiments in this field. What they found was of great importance: superconductivity is not only destroyed by ferromagnetic ordering, but also by rather small amounts (~ 1 at.%) of paramagnetic impurities. Moreover, by measuring the depression of the superconducting critical temperature T_c of lanthanum that resulted from the addition of 1 at.% of RE ions throughout the entire RE series, they found that the rate of depression of T_c correlated with the total spin, rather than the magnetic moment, of the paramagnetic RE impurity ions. This implicated the exchange interaction between the spins of the impurity ions and conduction electrons as the cause for that profound effect on superconductivity [1.6]. Since then, many of the phenomena in this field have been described by starting from an interaction of the type[1] (footnote see next page)

$$\mathcal{H}_{ex} = \frac{1}{N}\sum_i \Gamma(\underline{\sigma}\cdot\underline{S}_i) \tag{1.1}$$

where Γ is the exchange interaction parameter, $\underline{S}_i$ is the spin of the ith localized moment, $\underline{\sigma}$ is the conduction electron spin, and N is the total number of atoms in the system.

The theory for the effect of paramagnetic impurities on superconductivity was formulated to second order in Γ by ABRIKOSOV and GOR'KOV (hereafter denoted by the short hand AG) [1.7] in 1961, and subsequently confirmed by experiment [1.8]. Deviations of the depression of T_c with paramagnetic impurity concentration from the behavior predicted by the AG theory have been observed; they arise from the effects of interactions among the impurities, crystalline electric fields and Kondo scattering which can be described by appropriate theories that are beyond the scope of the original AG theory [1.9].

1.3 Pairbreaking Effects — Impurities and External Fields

The dramatic effect that magnetic impurities have on a superconductor can be qualitatively understood in terms of "pairbreaking." One essential condition for constructing a BCS-type state, according to ANDERSON's prescription of pairing time reversed states [1.10], is that the two states to be paired are degenerate in energy, i.e., that the time reversal operator commutes with the Hamiltonian. This latter condition is not fulfilled if the exchange interaction between an impurity spin and the conduction electron spins is included in the Hamiltonian. Thus this type of interaction breaks up the superconducting electron "Cooper" pairs by affecting the two electrons differently. The AG theory leads to the following well-known expression for T_c versus the concentration x of paramagnetic impurities

$$\ln\left(\frac{T_c}{T_{c_0}}\right) = \psi\left(\frac{1}{2}\right) - \psi\left(\frac{1}{2} + \frac{\rho}{2T_c/T_{c_0}}\right)$$

where (1.2)

$$\rho = (x/4k_BT_{c_0})N(E_F)\Gamma^2S(S+1) \quad ,$$

T_{c_0} is the transition temperature when $x = 0$, and ψ is the digamma function. When typical values for the exchange interaction parameter Γ and density of states at the Fermi level $N(E_F)$ are substituted into this equation, it gives a critical concentration, x_{crit}, of the order of 1 at.%.

1 In some chapters throughout these two volumes, the Hamiltonian represented by (1.1) is written as $\mathcal{H}_{int} = -2\mathcal{J}\underline{s}\cdot\underline{S}$ where $-2\mathcal{J} = \Gamma$, $\underline{S} = \underline{S}_i$ and $\underline{s} = \underline{\sigma}$.

Following this initial theoretical work, it soon became clear that the effect of a magnetic field on a superconductor could also be described as a pairbreaking effect. It is therefore not surprising that the upper critical magnetic field of a type II superconductor $H_{c2}(T)$, as a function of temperature T, which is determined by the interaction of the conduction electron orbits with the external magnetic field, is also given by (1.2), with

$$\rho = (v_F^2 e\tau / 3\pi T_{c_0})H \tag{1.3}$$

where v_F is the Fermi velocity, e is the electron charge, τ is the transport relaxation time, and T_c has been replaced by T [1.11,12]. Another mechanism that determines the upper critical magnetic field is associated with the molecular field which acts on the spins of the conduction electrons. This leads to a first-order transition in a BCS superconductor at T = 0 [1.13-15], but gives a second-order transition when spin orbit scattering is present and can, in the limit of strong spin orbit scattering, be described by (1.2) [1.16] with $\rho \sim H^2$. A very interesting and useful result is that when several pairbreaking mechanisms are operating simultaneously, the system can still be described by (1.2) in the strong spin orbit scattering limit when the different pairbreaking parameters are added together [1.16,17]. The different aspects of pairbreaking in ternary superconductors are dealt with in Chaps.3,4,5 and 9.

1.4 Impact of Ternary Rare Earth Compounds

The strong pairbreaking effects of magnetic ions in a superconductor have, in the past, been an obstacle to studies of the interplay of superconductivity and long range magnetic order. In dense magnetic compounds likely to show magnetic order, one anticipates that superconductivity will be destroyed via conduction electron-impurity spin exchange scattering effects even in the absence of magnetic ordering, due to the effects we have just discussed. An obvious way of avoiding this difficulty is to search for materials where the exchange interaction is particularly weak. MATTHIAS and coworkers found several systems that appeared to satisfy this criterion such as $Ce_{1-x}Gd_xRu_2$ and $Y_{1-x}Gd_xOs_2$ [1.18,19]. After these initial discoveries, interest focused on pseudobinary rare earth systems of this type where the curves of T_c and the magnetic ordering temperature T_M versus x intersected with the expectation that some kind of coexistence would occur in the region of overlap [1.20]. However, the magnetic order that was observed in these relatively dilute alloy systems was a short range or "spin glass" type, rather than true long range magnetic order. In addition, it often proved very difficult to avoid problems associated with chemical clustering. Thus, although considerable progress was made in understanding pairbreaking effects during this period, the questions relating directly to the mutual effect of long-range magnetic order and superconductivity were left more or less unresolved.

The ternary rare earth compounds changed this situation completely. The series of $REMo_6X_8$ and $RERh_4B_4$ compounds [1.21], as well as other materials to be described in the next chapter, differ from the pseudo-binary compounds in that they contain an ordered sublattice of magnetic RE ions, but nevertheless are superconducting. The occurrence of superconductivity even in the presence of relatively large concentrations of RE ions can be attributed to the crystal structures of these materials which are built up from RE ions and transition metal clusters. The superconductivity appears to be associated primarily with the transition metal 4d electrons which interact only weakly with the RE ions. During the last five years a considerable amount of effort has been expended in investigating the interaction between superconductivity and magnetism in ternary RE compounds, and most of this volume is devoted to research on this problem.

1.5 Antiferromagnetic Superconductors

Many of the $REMo_6X_8$ and $RERh_4B_4$ compounds exhibit the coexistence of superconductivity and long-range antiferromagnetic order [1.22-32]. In retrospect, this may not seem too surprising in view of the fact that over the scale of the superconducting coherence length, the exchange field of an antiferromagnet averages to zero. Nevertheless, antiferromagnetic order can effect superconductivity by means of several different mechanisms as evidenced by the behavior of H_{c2} as a function of temperature. For some antiferromagnetic superconductors such as $ErMo_6S_8$ [1.22] and $SmRh_4B_4$ [1.29], H_{c2} increases below the Néel temperature T_N, while for others, such as $REMo_6S_8$ for RE = Tb, Dy and Gd [1.22,24,32], as well as $NdRh_4B_4$ [1.28], H_{c2} decreases rather abruptly below T_N. Analysis of the H_{c2} vs T data for the $REMo_6S_8$ antiferromagnetic superconductors in terms of multiple pair breaking theory revealed an additional pairbreaking parameter whose temperature dependence is similar to that of the antiferromagnetic order parameter (sublattice magnetization) [1.30]. With the exception of $GdMo_6S_8$ [1.32], no evidence for fluctuation effects in the pairbreaking has been found.

The interrelation between superconductivity and antiferromagnetism was first addressed theoretically in 1963 by BALTENSPERGER and STRÄSSLER [1.33] who concluded that the two phenomena were not incompatible. Numerous theories have been advanced recently to explain the physical properties of antiferromagnetic superconductors, particularly the anomalous decrease of H_{c2} in the vicinity of T_N that certain of them display [1.34-41]. Several mechanisms by means of which superconductivity is modified by antiferromagnetic order have been considered such as: 1) the reduction in pairbreaking due to the decrease of the mean magnetization and, in turn, the conduction electron spin polarization below T_N; 2) the increase of pairbreaking due to magnetic moment fluctuations in the vicinity of T_N; 3) the decrease of the attractive phonon mediated electron-electron pairing interaction by antiferromagnetic magnons; 4) the reduction of available phase space for virtual pair scattering due to the introduction of gaps in the one electron excitation spectrum E(k) by the change in lattice periodicity associated with antiferromagnetic order, and 5) the pairing of electrons with finite momentum. The pairing of electrons with finite momentum, which was, in fact, originally proposed by BALTENSPERGER and STRÄSSLER [1.33], has been incorporated in several recent theories [1.35,37], including the calculations by ZWICKNAGL and FULDE [1.37] of the curves of H_{c2} vs T for the $REMo_6S_8$ compounds [1.22]. However, there seems to be some controversy as to whether such a new pairing state actually exists [1.36]. A detailed account of experiments on antiferromagnetic superconductors is given in Chaps.4,5 and 7 and the theoretical aspects are treated in Chap.9.

1.6 Ferromagnetism and Superconductivity

Two ternary RE compounds, $HoMo_6S_8$ [1.42,43] and $ErRh_4B_4$ [1.44,45], exhibit reentrant superconductive behavior due to the onset of long-range ferromagnetic ordering of their RE magnetic moments. These two materials, which become superconducting at an upper critical temperature T_{c1}, lose their superconductivity at a lower critical temperature $T_{c2} \sim T_M$, where T_M is the Curie temperature. A first-order transition from the superconducting to the ferromagnetic normal state occurs at T_{c2}, as revealed by thermal hysteresis in various physical properties and a feature in the specific heat near T_{c2} [1.46,47].

It is not too surprising that it is the ferromagnetism that survives the struggle between these two phenomena. Relative to the paramagnetic normal state, the decrease in free energy of the ferromagnetic state is approximately $cN(k_BT_M)$ where c is the atomic fraction of magnetic ions and N is the total number of atoms, whereas the decrease in free energy of the superconducting state is roughly given by $(k_BT_c/E_F)N(k_BT_c)$, the factor k_BT_c/E_F being the proportion of conduction electrons that participate in the superconductivity. Usually, $c \gg k_BT_c/E_F$, so that the ferromagnetic state will be favored over the superconducting state. This argument does not, of course, exclude the possibility that the two phenomena coexist, and a great deal of attention has therefore been focused on what actually happens in the vicinity of the Curie temperature.

Careful inspection of experimental data, has, in fact, suggested that there might be a narrow coexistence region for temperatures just below T_M [1.48,49], although this has not been established with any real certainty. Rather extensive neutron scattering experiments in the vicinity of T_M have been carried out on both $HoMo_6S_8$ [1.50,51] and $ErRh_4B_4$ [1.52,53] and have proven to be very informative. The experiments on $ErRh_4B_4$ have yielded evidence for a spatially inhomogeneous state in a narrow temperature interval above T_{c2} in which superconducting regions coexist with normal ferromagnetic regions. Moreover, for both $HoMo_6S_8$ and $ErRh_4B_4$ within the superconducting regions, the RE magnetic moments are in a sinusoidally modulated magnetic state with a wavelength of the order of a few hundred Angstroms. Thus, microscopic coexistence of superconductivity and magnetism occurs in these ferromagnetic superconductors as well, but the interaction between these two phenomena transforms the ferromagnetic state within the superconducting regions into an oscillatory magnetic state in which the net magnetization is zero.

A number of possible explanations have been advanced for oscillatory magnetic states in superconductors. The earliest of these, which actually preceded the experimental observation of an oscillatory state, was proposed by ANDERSON and SUHL in 1959 [1.54]. This so-called cryptoferromagnetic state was based on the assumption that the magnetic ions interact via the RKKY indirect exchange mechanism; the oscillatory character of the magnetization resulted from the change in the q-dependent magnetic suceptibility caused by the superconductivity.

Recently, BLOUNT and VARMA [1.55] suggested that the electromagnetic interaction between the magnetization of the localized ions and the momenta of the conduction electrons could lead to a spiral magnetization state in a ferromagnetic superconductor. The essence of the argument is as follows: If a spontaneous magnetization appears in a superconductor, the superconductor will respond with screening currents (Meissner effect) to compensate the average magnetization. In order to avoid this energetically costly process, the magnetic moments tilt slightly so as to form a long-wavelength oscillatory state. If this wavelength is shorter than the superconducting penetration depth, no Meissner effect will result.

Several other theoretical investigations of periodic magnetic structures with wavelengths $\sim$100 Å that coexist with superconductivity in ferromagnetic superconductors above T_{c2} have recently appeared in the literature [1.56-61]. Many of them are based on the electromagnetic interaction since it is widely believed to be stronger than the exchange interaction in the $REMo_6X_8$ and $RERh_4B_4$ compounds. However, some estimates suggest that the electromagnetic and exchange interactions are of comparable strength in these materials [1.62,63], so that this question does not seem to be completely settled. On the other hand, it has also been suggested that the periodic magnetic structure above T_{c2} may be due to the development of a spontaneous flux line lattice [1.64,65] when $M > H_{c1}$, where M is the magnetization and H_{c1} is the lower critical magnetic field. The formation of such a self-induced vortex state in a magnetic superconductor appears to have been considered first by KREY about a decade ago [1.66]. The measurements that are described in Chap.8 seem to exclude this possibility, although the precise nature of the oscillatory magnetization state has yet to be determined.

Recent neutron diffraction measurements on a single-crystal specimen of $ErRh_4B_4$ by SINHA et al. [1.53] indicate that the sinusoidally modulated state is linearly polarized, rather than spiral. GREENSIDE et al. [1.61] have shown that when the magnetic anisotropy is sufficiently strong, such a linearly polarized sinusoidal state can be preferred over both spiral and vortex states. Other alternatives have recently been examined such as a laminar structure [1.67], stabilized by the rare earth magnetization in a self-consistent manner, and combined spiral magnetic and spontaneous vortex states [1.68].

Re-entrant superconductivity due to the onset of ferromagnetic order was actually predicted some time ago by GOR'KOV and RUSINOV [1.64]. In their theory, the destruction of superconductivity is caused by pairbreaking that results from the average polarization of the conduction electron spins due to the ferromagnetically aligned magnetic moments. This effect can be reduced by both spin-orbit and exchange scattering, permitting the coexistence of superconductivity and ferromagnetism. However, prior to this, JACCARINO and PETER [1.70] pointed out that this average polarization can be reduced to zero in a strong external field if the exchange interaction between the conduction electrons and the localized magnetic moments were negative (antiferromagnetic). It was suggested that this could even allow a ferromagnet to become superconducting (if it was predisposed to become superconducting in the absence of ferromagnetism), although this is not a consideration for $HoMo_6S_8$ and $ErRh_4B_4$ since the exchange interaction should be positive for both of these compounds. In addition, a detailed analysis of this effect shows that it can only be observed in materials that are potentially high-field superconductors [1.71]. The Jaccarino-Peter compensation effect has been observed in superconductors containing paramagnetic rare earth ions where the rare earth spins are aligned by the application of an external magnetic field. In the system $Sn_{1-x}Eu_xMo_6S_8$, this effect leads to an increase in H_{c2} as x is increased as well as pronounced anomalies in the curves of H_{c2} versus T [1.72]. The negative exchange interaction in this system is consistent with NMR [1.73], Mössbauer effect [1.74] and EPR measurements [1.75], and with band structure calculations [1.76]. Similar behavior has recently been found in the $La_{1-x}Eu_xMo_6S_8$ [1.77] system. More details about this effect can be found in Chaps.3,6 and 7.

1.7 Superconductivity and Competing Magnetic Interactions

A number of experiments have been carried out on pseudoternary $RERh_4B_4$ and $REMo_6X_8$ systems. Two types of pseudoternaries have been formed, one in which a second RE element is substituted at the RE sites, and, in the case of the $RERh_4B_4$ compounds, another in which a different transition metal is substituted at the Rh sites. Not only is this an alternative method for studying the interaction between superconductivity and long-range magnetic order, but it also allows the effects of competing types of magnetic moment anisotropy and/or magnetic order to be explored.

Investigations on a variety of pseudoternary systems have revealed several complex and interesting temperature vs composition phase diagrams for both the $RERh_4B_4$ [1.78-82] and $REMo_6S_8$ [1.83,84] systems that have been the subject of several recent theories [1.85-87]. In addition, evidence has been found for various types of multicritical points [1.78,81,85-87], the suppression of the Curie temperature by superconductivity [1.88], the coexistence of superconductivity and antiferromagnetism with $T_N > T_c$ [1.89,90], and possibly, even the coexistence of superconductivity and ferromagnetism [1.83]. More complete descriptions of the experiments on pseudoternary rare earth compounds appear in Chaps.4 and 5.

1.8 Concluding Remarks

In this chapter, we have tried to provide the reader with a brief introduction to the general subject of interacting superconducting and magnetic order parameters. The ideas discussed here have stimulated many experimental investigations of magnetic superconductors and the development of theoretical models to explain their remarkable physical properties. The current status of this interesting and timely topic is reviewed throughout this volume.

References

1.1 Ø. Fischer, A. Treyvaud, R. Chevrel, M. Sergent: Solid State Commun. *17*, 721 (1975)
1.2 R.N. Shelton, R.W. McCallum, H. Adrian: Phys. Lett. *56*A, 213 (1976)
1.3 B.T. Matthias, E. Corenzwit, J.M. Vandenberg, H. Barz: Proc. Natl. Acad. Sci. USA *74*, 1334 (1977)
1.4 V.L. Ginzburg: Sov. Phys. JETP *4*, 153 (1957)
1.5 B.T. Matthias, H. Suhl, E. Corenzwit: Phys. Rev. Lett. *1*, 92 (1959)
1.6 H. Suhl, B.T. Matthias: Phys. Rev. *114*, 977 (1959)
1.7 A.A. Abrikosov, L.P. Gor'kov: Sov. Phys. JETP *12*, 1243 (1961)
1.8 M.B. Maple: Phys. Lett. A*26*, 513 (1968)
1.9 For a review, see M.B. Maple: Appl. Phys. *9*, 179 (1976)
1.10 P.W. Anderson: J. Phys. Chem. Solids *11*, 26 (1959)
1.11 P.G. deGennes: Phys. Cond. Mat. *3*, 79 (1964)
1.12 K. Maki: Physics *1*, 21 (1964)
1.13 B.S. Chandrasekhar: Appl. Phys. Lett. *1*, 7 (1962)
1.14 A.M. Clogston: Phys. Rev. Lett. *9*, 266 (1962)
1.15 G. Sarma: J. Phys. Chem. Solids *24*, 1029 (1963)
1.16 P. Fulde, K. Maki: Phys. Rev. *141*, 275 (1966)
1.17 R.D. Parks: In *Superconductivity*, Vol.2, ed. by P.R. Wallace (Gordon and Breach, New York 1969) pp.625-690
1.18 B.T. Matthias, H. Suhl, E. Corenzwit: Phys. Rev. Lett. *1*, 449 (1958)
1.19 H. Suhl, B.T. Matthias, E. Corenzwit: J. Phys. Chem. Solids *19*, 346 (1959)
1.20 For a review, see Ø. Fischer, M. Peter: In *Magnetism*, Vol.5, ed. by H. Suhl (Academic, New York 1973) pp.327-352;
S. Roth: Appl. Phys. *15*, 1 (1978)

1.21 Ø. Fischer, M.B. Maple (eds.): *Superconductivity in Ternary Compounds I*, Topics in Current Physics, Vol.32 (Springer, Berlin, Heidelberg, New York 1982)
1.22 M. Ishikawa, Ø. Fischer: Solid State Commun. *24*, 747 (1977)
1.23 D.E. Moncton, G. Shirane, W. Thomlinson, M. Ishikawa, Ø. Fischer: Phys. Rev. Lett. *41*, 1133 (1978)
1.24 C.F. Majkrzak, G. Shirane, W. Thomlinson, M. Ishikawa, Ø. Fischer, D.E. Moncton: Solid State Commun. *31*, 773 (1979)
1.25 W. Thomlinson, G. Shirane, D.E. Moncton, M. Ishikawa, Ø. Fischer: Phys. Rev. B*23*, 4455 (1981)
1.26 R.W. McCallum, D.C. Johnston, R.N. Shelton, M.B. Maple: Solid State Commun. *24*, 391 (1977)
1.27 M.B. Maple, L.D. Woolf, C.F. Majkrzak, G. Shirane, W. Thomlinson, D.E. Moncton: Phys. Lett. *77*A, 487 (1980)
1.28 H.C. Hamaker, L.D. Woolf, H.B. MacKay, Z. Fisk, M.B. Maple: Solid State Commun. *31*, 139 (1979)
1.29 H.C. Hamaker, L.D. Woolf, H.B. MacKay, Z. Fisk, M.B. Maple: Solid State Commun. *32*, 289 (1979)
1.30 Ø. Fischer, M. Ishikawa, M. Pelizzone, A. Treyvaud: J. Phys. *40*, C5-89 (1979)
1.31 M.B. Maple: J. Phys. *39*, C6-1374 (1978)
1.32 M. Ishikawa, Ø. Fischer, J. Muller: J. Phys. *39*, C6-1379 (1978)
1.33 W. Baltensperger, S. Strässler: Phys. Cond. Materie *1*, 20 (1963)
1.34 K. Machida: J. Low Temp. Phys. *37*, 583 (1979)
1.35 K. Machida, K. Nokura, T. Matsubara: Phys. Rev. B*22*, 2307 (1980)
1.36 M.J. Nass, K. Levin, G.S. Grest: Phys. Rev. Lett. *46*, 614 (1981)
1.37 G. Zwicknagl, P. Fulde: Z. Physik B*43*, 23 (1981)
1.38 T.V. Ramakrishnan, C.M. Varma: Phys. Rev. B*24*, 137 (1981)
1.39 O. Sakai, M. Tachiki, T. Koyama, H. Matsumoto, H. Umezawa: Phys. Rev. B*24*, 3830 (1981)
1.40 S.D. Mahanti, K.P. Sinha, M.V. Atre: J. Low Temp. Phys. *44*, 1 (1981)
1.41 Y. Suzumura, A.D.S. Nagi: Solid State Commun. *41*, 413 (1982)
1.42 M. Ishikawa, Ø. Fischer: Solid State Commun. *23*, 37 (1977)
1.43 J.W. Lynn, D.E. Moncton, W. Thomlinson, G. Shirane, R.N. Shelton: Solid State Commun. *26*, 493 (1978)
1.44 W.A. Fertig, D.C. Johnston, L.E. DeLong, R.W. McCallum, M.B. Maple, B.T. Matthias: Phys. Rev. Lett. *38*, 387 (1977)
1.45 D.E. Moncton, D.B. McWhan, J. Eckert, G. Shirane, W. Thomlinson: Phys. Rev. Lett. *39*, 1164 (1977)
1.46 L.D. Woolf, D.C. Johnston, H.B. MacKay, R.W. McCallum, M.B. Maple: J. Low Temp. Phys. *35*, 651 (1979)
1.47 L.D. Woolf, M. Tovar, H.C. Hamaker, M.B. Maple: Phys. Lett. *71*A, 137 (1979); Phys. Lett. *72*A, 481 (1979)
1.48 H.R. Ott, W.A. Fertig, D.C. Johnston, M.B. Maple, B.T. Matthias: J. Low Temp. Phys. *33*, 159 (1978)
1.49 G. Cort, R.D. Taylor, J.D. Willis: Physica *108*B, 809 (1981)
1.50 J.W. Lynn, J.L. Ragazzoni, R. Pynn, J. Joffrin: J. Phys. Lett. *42*, 368 (1981)
1.51 J.W. Lynn, G. Shirane, W. Thomlinson, R.N. Shelton, D.E. Moncton: Phys. Rev. B*24*, 3817 (1981)
1.52 D.E. Moncton, D.B. McWhan, P.H. Schmidt, G. Shirane, W. Thomlinson, M.B. Maple, H.B. MacKay, L.D. Woolf, Z. Fisk, D.C. Johnston: Phys. Rev. Lett. *45*, 2060 (1980)
1.53 S.K. Sinha, G.W. Crabtree, D.G. Hinks, H.A. Mook: Phys. Rev. Lett. *48*, 950 (1982)
1.54 P.W. Anderson, H. Suhl: Phys. Rev. *116*, 898 (1959)
1.55 E.I. Blount, C.M. Varma: Phys. Rev. Lett. *42*, 1079 (1979)
1.56 H. Suhl: J. Less-Common Metals *62*, 225 (1978)
1.57 R.A. Ferrell, J.K. Bhattacharjee, A. Bagchi: Phys. Rev. Lett. *43*, 154 (1979)
1.58 L.N. Bulaevski, A.I. Rusinov, M. Kulič: Solid State Commun. *30*, 59 (1979)
1.59 H. Matsumoto, H. Umezawa, M. Tachiki: Solid State Commun. *31*, 157 (1979)
1.60 K. Machida, T. Matsubara: Solid State Commun. *31*, 791 (1979)
1.61 H.S. Greenside, E.I. Blount, C.M. Varma: Phys. Rev. Lett. *46*, 49 (1981)
1.62 H.B. MacKay, L.D. Woolf, M.B. Maple, D.C. Johnston: Phys. Rev. Lett. *42*, 918 (1979); Phys. Rev. Lett. *43*, 89 (1979)

1.63 M. Kulić: Phys. Lett. *81*A, 359 (1981)
1.64 M. Tachiki, H. Matsumoto, T. Koyama, H. Umezawa: Solid State Commun. *34*, 19 (1980)
1.65 C.G. Kuper, M. Revzen, A. Ron: Phys. Rev. Lett. *44*, 1545 (1980)
1.66 U. Krey: Int. J. Magnetism *3*, 65 (1972); Int. J. Magnetism *4*, 153 (1973)
1.67 M. Tachiki: Proc. 16th Int. Conf. Low Temp. Phys., Part III, to appear in Physica B + C, 1982
1.68 C.R. Hu, T.E. Ham: Physica *108*B, 1041 (1981)
1.69 L.P. Gor'kov, A.I. Rusinov: Sov. Phys. JETP *19*, 922 (1964)
1.70 V. Jaccarino, M. Peter: Phys. Rev. Lett. *9*, 280 (1962)
1.71 Ø. Fischer: Helv. Phys. Acta *45*, 332 (1972)
1.72 Ø. Fischer, M. Decroux, S. Roth, R. Chevrel, M. Sergent: J. Phys. C*8*, L474 (1975)
1.73 C.P. Slichter: *Principles of Magnetic Resonance*, 2nd. ed., Springer Series in Solid-State Sciences, Vol.1 (Springer, Berlin, Heidelberg, New York 1980)
1.74 F.Y. Fradin, G.K. Shenoy, B.D. Dunlap, A.T. Aldred, C.W. Kimball: Phys. Rev. Lett. *38*, 719 (1977)
1.75 R. Odermatt: Helv. Phys. Acta *54*, 1 (1981)
1.76 T. Jarlborg, A.J. Freeman: Phys. Rev. Lett. *44*, 178 (1980); J. Magn. Magn. Mat. *15-18*, 1579 (1980)
1.77 M.S. Torikachvili, M.B. Maple: Solid State Commun. *40*, 1 (1981)
1.78 D.C. Johnston, W.A. Fertig, M.B. Maple, B.T. Matthias: Solid State Commun. *26*, 141 (1978)
1.79 R.H. Wang, R.J. Laskowski, C.Y. Huang, J.L. Smith, C.W. Chu: J. Appl. Phys. *49*, 1392 (1978);
S. Kohn, R.H. Wang, J.L. Smith, C.Y. Huang: J. Appl. Phys. *50*, 1862 (1979)
1.80 H.C. Ku, F. Acker, B.T. Matthias: Phys. Lett. *76*A, 399 (1980)
1.81 L.D. Woolf, M.B. Maple: In *Ternary Superconductors*, ed. by G.K. Shenoy, B.D. Dunlap, F.Y. Fradin (North-Holland, Amsterdam 1981) pp.181-184
1.82 H.C. Hamaker, M.B. Maple: In *Ternary Superconductors*, ed. by G.K. Shenoy, B.D. Dunlap, F.Y. Fradin (North-Holland, Amsterdam 1981) pp.201-204
1.83 M. Ishikawa, M. Sergent, Ø. Fischer: Phys. Lett. *82*A, 30 (1981)
1.84 C.F. Majkrzak, G. Shirane, M.B. Maple, M.S. Torikachvili: Bull. Am. Phys. Soc. *27*, 320 (1982);
C.F. Majkrzak, G. Shirane, M.B. Maple, M.S. Torikachvili, L. Brossard: To be published
1.85 R.M. Hornreich, H.G. Schuster: Phys. Lett. *70*A, 143 (1979)
1.86 C. Balseiro, L.M. Falicov: Phys. Rev. B*19*, 2548 (1979)
1.87 B. Schuh, N. Grewe: Solid State Commun. *37*, 145 (1981)
1.88 L.D. Woolf, D.C. Johnston, H.A. Mook, W.C. Koehler, M.B. Maple, Z. Fisk: Proc. 16th Int. Conf. Low Temp. Phys,, Part III, to appear in Physica B + C, 1982
1.89 L.D. Woolf, S.E. Lambert, M.B. Maple, H.C. Ku, W. Odoni, H.R. Ott: Physica *108*B, 761 (1981)
1.90 H.C. Hamaker, H.C. Ku, M.B. Maple, H.A. Mook: Bull. Am. Phys. Soc. *27*, 247 (1982), and to be published

2. Systematics of Superconductivity in Ternary Compounds

D.C. Johnston and H.F. Braun

With 20 Figures

The discoveries of superconductivity in two particular classes of ternary compounds by MATTHIAS et al. [2.1,2] catalyzed the recent intense research on ternary superconductors [2.3]. These classes are: (i) rhombohedral or triclinic Chevrel phases $M_xMo_6X_8$, where X is a chalcogen and M can be any of a large number of metal atoms [2.4-6], including the rare earths (RE) [2.7,8], and (ii) primitive tetragonal borides MRh_4B_4, where M can be Y, Th or one of nine RE atoms [2.2,9]. The former class of compounds exhibits the highest upper critical magnetic fields ever observed (e.g., $H_{c2} \approx 600$ kG for $PbMo_6S_8$ [2.10-12]), whereas those of the latter class are quite low. In addition, when M is a magnetic RE element, superconductivity occurs in both classes of compounds even though the RE concentrations are high (7 at.% and 11 at.%, respectively) [2.2,7,8]. Since the RE atoms form a spatially ordered array in each of the two structures, these materials have provided the first opportunities to explore the interaction between superconductivity and *long-range* magnetic order [2.13-16]. The unique superconducting behavior of these materials also provided the impetus for work in the synthesis of other and diverse ternary systems; this in turn led to the discovery of a large number of new ternary superconductors, many of which also contain spatially ordered arrays of RE atoms. Our primary purpose in this chapter is to survey the systematics of superconductivity in ternary compounds with an emphasis on those recently discovered.

A great number of metallic binary compounds containing spatially ordered arrays of local magnetic moments are known; however, not one of them has been found to exhibit superconductivity. In view of the occurrence of superconductivity in many ternary compounds containing spatially ordered arrays of magnetic ions, the absence of superconductivity in all such binary compounds seems phenomenal in retrospect.

The essential characteristic of a ternary compound is that it contains three elements, each occupying a distinct set or sets of crystallographic sites; this crystallographic definition clearly distinguishes ternary compounds from binary or pseudobinary phases in which only two sets of lattice sites are occupied or which contain only two species of atoms. Hence, the unique superconducting and magnetic behavior of the former compounds must necessarily derive from their detailed crystal structures and stoichiometries in a fundamental way. In this chapter, we will therefore consider superconductivity in ternary compounds primarily from a structural and chemical point of view. We will limit our discussion to those classes of materials exhibiting superconducting behavior which may serve to elucidate the interaction between superconductivity and magnetism. The Chevrel phases and related compounds will only be touched on here since they are the subjects of two excellent recent review articles [2.5,6] as well as other chapters in these volumes.

The present chapter is organized as follows. The dependences of the superconducting critical temperature on structural, chemical and physical parameters are considered in Sect.2.1 according to ternary structure class. Tables containing lists of crystallographic, superconducting and magnetic data for these structure classes are placed at the end of the chapter in order not to interrupt the continuity of the text. These tables are numbered according to their citation in the text, however. The chapter is concluded in Sect.2.2 with some observations regarding superconductivity in ternary compounds and some suggestions and speculations on future developments in the area of ternary superconductivity.

2.1 Review of Superconductivity in Ternary Compounds by Structure Class

2.1.1 Compounds with MT_4B_4 Stoichiometry

a) *Primitive Tetragonal* $CeCo_4B_4$ *Structure*

The crystal structure of the primitive tetragonal MRh_4B_4 compounds was solved by VANDENBERG and MATTHIAS [2.9] from X ray powder data obtained for the Y member, and confirmed by YVON and GRÜTTNER [2.17] by single crystal structure analysis. The structure was found to be of the $CeCo_4B_4$ type [2.18]. The M atoms completely occupy a slightly distorted face-centered-cubic (fcc) array, the Rh atoms form separated tetrahedra, and the boron atoms form B_2 dimers. The M atoms and centers of the Rh tetrahedra form a slightly distorted NaCl lattice [2.19], as shown in Fig.2.1. That the M, Rh and B atoms each completely occupy a distinct set of lattice sites is crucial to the superconducting and magnetic properties observed for the MRh_4B_4 compounds containing magnetic RE atoms. In particular, the spatial ordering of the M atom array is a factor which allows the occurrence of long-range magnetic order of the RE moments. The crystallographic data and superconducting and magnetic critical temperatures (T_c and T_m) for $CeCo_4B_4$-type compounds are listed in Table 2.1 [2.2,9, 13,15,18,20-25], and the occurrence of this structure type among MT_4B_4 systems is summarized in Table 2.2 (see Appendix 2A).

The Rh tetrahedra in the MRh_4B_4 compounds form sheets in the a-b plane, where the Rh-Rh interatomic distances within a plane (2.6-2.8 Å) are approximately the same as in Rh metal (2.69 Å) but substantially smaller than between planes (3.1 Å) [2.9,17]. Two orientations of Rh tetrahedra occur; the Rh network can be viewed as consisting of two infinite quasi two-dimensional extended clusters of Rh_4 tetrahedra, where each extended cluster contains Rh_4 tetrahedra of the same orientation and

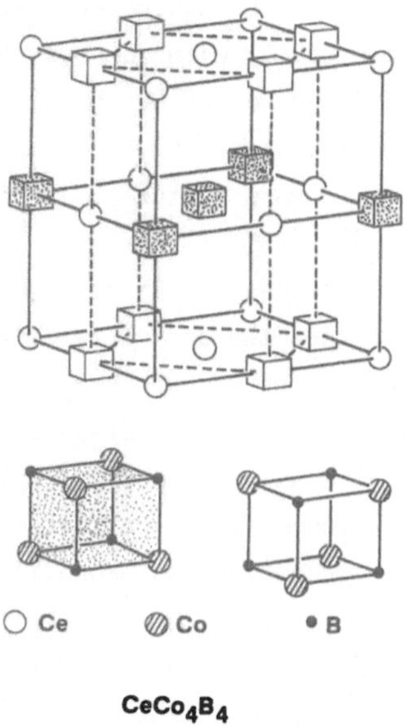

Fig.2.1. Idealized crystal structure of $CeCo_4B_4$-type compounds. The dashed lines outline the primitive tetragonal unit cell; for clarity, the cubes representing the Rh_4B_4 units are not drawn to scale (after [2.22])

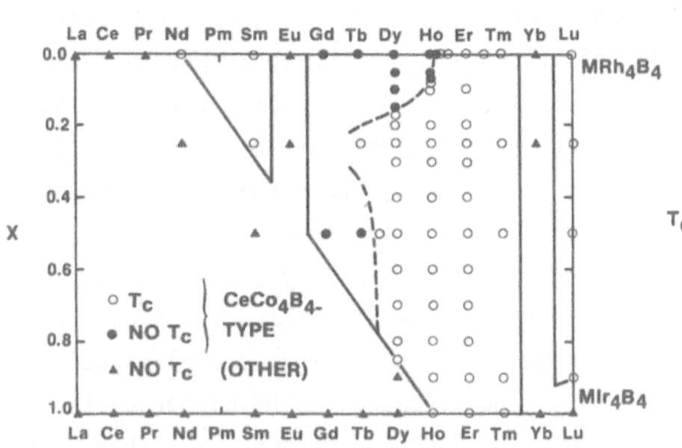

Fig.2.2. Illustration of the distribution of superconducting compounds of $CeCo_4B_4$-type in pseudoternary $RE(Rh_{1-x}Ir_x)_4B_4$ alloys (after KU and BARZ [2.30])

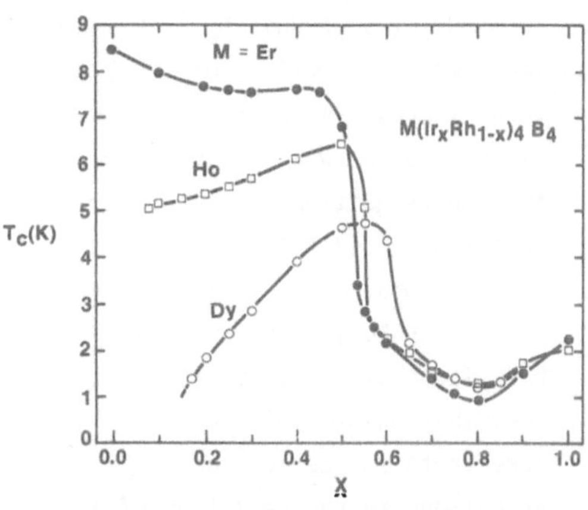

Fig.2.3. Variation of superconducting transition temperature (T_c) with composition in the $CeCo_4B_4$-type systems $RE(Rh_{1-x}Ir_x)_4B_4$ (after KU and ACKER [2.29])

where the two extended clusters alternate along the c axis as illustrated in Fig. 2.1. Clustering of transition metal atoms has been found to be a feature common to other binary and ternary high T_c superconductors, and is therefore probably a contributing factor to the high critical temperatures observed [2.26].

Experimental evidence that the degree of crystallographic order is a parameter which is very important to the superconducting and magnetic interactions in the $CeCo_4B_4$-type compounds was recently obtained on several fronts. KU [2.27] and KU et al. [2.24,28-30] have found that isoelectronic series of pseudoternary $RE(Rh_{1-x}Ir_x)_4B_4$ compounds can be formed with the $CeCo_4B_4$-type structure, as shown in Fig.2.2. The composition dependences of T_c for the Dy, Ho and Er pseudoternaries are shown in Fig.2.3, where it is seen that T_c drops precipitously with increasing x at x = 0.5-0.6 in each case. Since similar behavior was observed for $Lu(Rh_{1-x}Ir_x)_4B_4$, this effect is not of magnetic origin and may reflect ordering of the Rh and Ir atoms within the transition metal clusters or a fundamental change in the electronic structure [2.24,27-30]. From Figs.2.2 and 2.3, substitution of 25% of the Rh by Ir induces superconductivity in the Tb, Dy and Ho members, where the pure Rh compounds show only ferromagnetism; this occurs in spite of the fact that the spatially ordered RE sublattice in each case is not disturbed. Finally, the T_c's of several MRh_4B_4 compounds have been found to be unusually sensitive to α-particle damage [2.31-33], a feature which is common to superconducting transition metal cluster compounds [2.31]. Each of the above effects is a rather spectacular illustration of the importance of the degree of crystallographic order to the observed superconducting and magnetic interactions in the $CeCo_4B_4$-type MRh_4B_4 compounds.

In contrast to the highly nonlinear composition dependences of T_c found for the $RE(Rh_{1-x}Ir_x)_4B_4$ compounds, pseudoternary alloys of the type $(M^{(1)}_{1-x}M^{(2)}_x)Rh_4B_4$ usually show approximately linear variations of the T_c with composition if $M^{(1)}$ and $M^{(2)}$ both form MRh_4B_4 compounds by themselves, irrespective of whether $M^{(1)}$ and/or $M^{(2)}$ carry local magnetic moments or not [2.34-42]. This shows that the RE ion-conduction electron exchange interaction is unusually weak. Further, this behavior emphasizes the differences between the roles played in the superconductivity by the M sublattice and the transition metal clusters, consistent with results from band structure calculations [2.43-45] which show that the electronic states at the Fermi level in the MRh_4B_4 compounds arise mainly from the Rh atoms and are primarily of d-character.

When analyzing the variation of T_c with RE in a series of compounds containing rare earth elements, it is important to separate the magnetic from the nonmagnetic contributions to this variation. Unfortunately, one nonmagnetic end member of the rare-earth series (La) does not form a $CeCo_4B_4$-type $RERh_4B_4$ compound, and these contributions to the variation of T_c with RE in this system must therefore be assessed indirectly. Shown in Fig.2.4 is a compilation of all reported T_c data obtained from ac magnetic susceptibility measurements for $CeCo_4B_4$-type $RERh_4B_4$ compounds [2.2] and pseudoternary mixtures of them [2.34,35,37,40,42] plotted versus the atomic number of the RE (or average atomic number in the case of pseudoternaries). This figure clearly shows that the atomic number itself is not the primary parameter which scales the T_c's.

For the heavy RE members, the T_c of the $RERh_4B_4$ compounds and the rate of depression of T_c of $LuRh_4B_4$ upon dilute substitution for Lu by these RE's both vary approximately linearly with the deGennes Factor (dGF), $dGF \equiv (g-1)^2 J(J+1)$ [2.46], where g is the Landé g factor and J is the total angular momentum of the RE^{3+} Hund's Rule ground state. This proportionality is the dilute limit of the more generally applicable prediction of the Abrikosov-Gor'kov (AG) theory that the exchange inter-

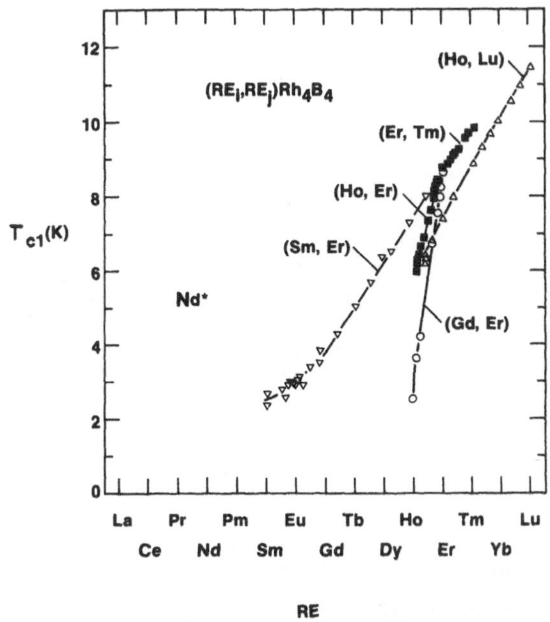

Fig.2.4. Variation of the (upper) superconducting critical temperature for ternary and pseudoternary $CeCo_4B_4$-type $RERh_4B_4$ compounds compiled from literature data, plotted versus the atomic number of the RE atom (or average atomic number in the cases of pseudoternaries)

action between the conduction electrons with spin $\underline{s}$ and local magnetic moments with spin $\underline{S}$ ($H = -2I\underline{S}\cdot\underline{s}$) results in a reduction of T_c below the value (T_{co}) it would have had in the absence of exchange scattering [2.47]. Shown in Fig.2.5 is a plot of T_c vs dGF for the compounds illustrated in Fig.2.4 where the average dGF has been used for the pseudoternary mixtures.

All of the data in Fig.2.5 for compounds containing RE = Gd to Lu and for dGF < 5 follow a common universal behavior; in particular, the universal curve predicted by the AG theory could be scaled to fit these data using the parameters T_{co} = 11.4 K and initial slope $dT_c/d(dGF)$ = -1.09 K, as shown by the solid curve in Fig.2.5 labeled "AG". For dGF < 5, the curve fits the data for RE = Gd to Lu to within about ± 0.2K, typical of systematic differences for the T_c of a given nominal composition which can arise from variations in the details of sample preparation. This agreement suggests that T_{co} and the product $N(E_F)I^2$ are independent of RE for these compounds, where $N(E_F)$ is the electronic density of states at the Fermi level and I was defined above. However, significant negative deviations do occur from the solid curve for $(Er_{1-x}Gd_x)Rh_4B_4$ alloys with Gd concentration higher than x = 0.2 (dGF $\geq$ 5) and for compounds containing the light RE elements Nd and Sm; i.e., the discrepancy increases as the average size of the RE increases past that of Ho [2.48].

The progressively larger negative deviations of the data in Fig.2.5 with increasing size of the RE atom could be due to corresponding increases in the parameters I^2 and/or $N(E_F)$ or to a breakdown of the AG theory in this regime. An alternative possibility is that these deviations occur due to decreasing values of T_{co}, i.e., from nonmagnetic effects. Although the correct explanation is not yet clear, the latter alternative is supported by several recent experimental results. A theoretical fit to upper critical magnetic field data for $SmRh_4B_4$ [2.21] suggested that

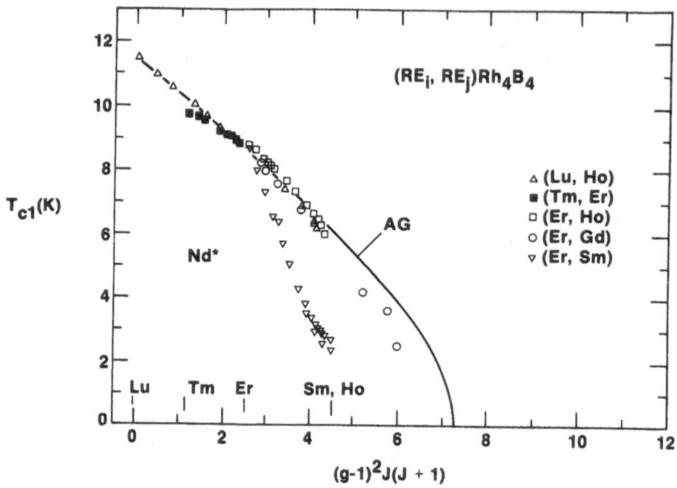

Fig.2.5. Superconducting transition temperature (T_{c1}) versus deGennes factor, $(g-1)^2J(J+1)$, for the $CeCo_4B_4$-type compounds of Fig.2.4. The solid curve labeled "AG" is a fit to the data using the Abrikosov-Gor'kov theory (see text)

$T_{co} \sim 5$ K in this compound, and T_c measurements on dilute alloys (x = 0.01-0.06) in the $(Lu_{1-x}La_x)Rh_4B_4$ system [2.46] suggested a similar value of T_{co} for the nonmagnetic hypothetical compound $LaRh_4B_4$. In addition, from [2.21] and NMR data on $SmRh_4B_4$ [2.49], the exchange coupling does not appear to be any stronger in this compound than in the heavy $RERh_4B_4$ members. Strong variations in T_{co} have been concluded to occur with RE in $RE(Rh_{1-x}Ru_x)_4B_4$ compounds which have a very similar structure (see Sect.2.1.1b).

The influence of hydrostatic pressure on the superconducting and/or magnetic critical temperatures has been determined for all of the MRh_4B_4 compounds forming the $CeCo_4B_4$ structure [2.41,50-52] as well as for a substantial number of $(M^{(1)}_{1-x}M^{(2)}_x)Rh_4B_4$ pseudoternary compounds [2.36,41,51-53]. The rich variety of pressure effects observed mirrors the unusual behavior of the zero-pressure T_c's themselves, as illustrated in Fig.2.6 [2.50,51]. Discontinuities in $T_c(p)$ and in the slope of $T_c(p)$ have been observed for $LuRh_4B_4$ [2.50,52] and YRh_4B_4 [2.41], respectively, possibly due to pressure-induced electronic or lattice transformations; the data for $LuRh_4B_4$ are shown in Fig.2.7. The factors contributing to these incipient instabilities are not yet understood, but these may also be important to the superconducting and magnetic interactions in the MRh_4B_4 compounds at zero pressure. In this context, X ray diffraction measurements at 4 K failed to reveal any evidence for a low-temperature lattice distortion at zero pressure in $LuRh_4B_4$ [2.54].

b) *Body-Centered-Tetragonal $LuRu_4B_4$ Structure*

Two high temperature or metastable modifications of the $CeCo_4B_4$-type MRh_4B_4 compounds have been found to exist. One of these is the body-centered-tetragonal (bct) $LuRu_4B_4$-type and the other has the orthorhombic $LuRh_4B_4$ structure (Sect.2.1.1c). The $LuRu_4B_4$-type modification is found in arc-melted ingots slightly deficient in

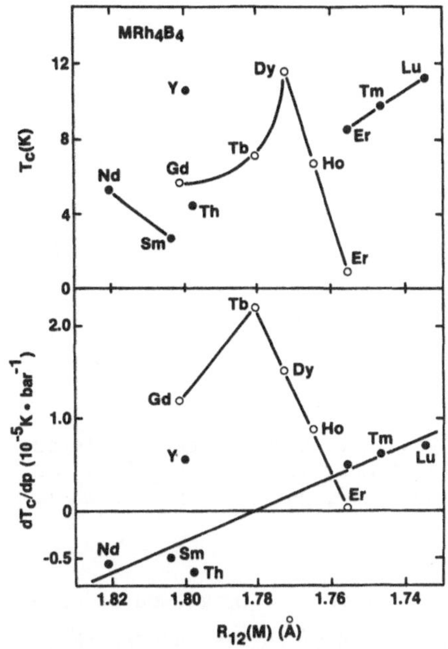

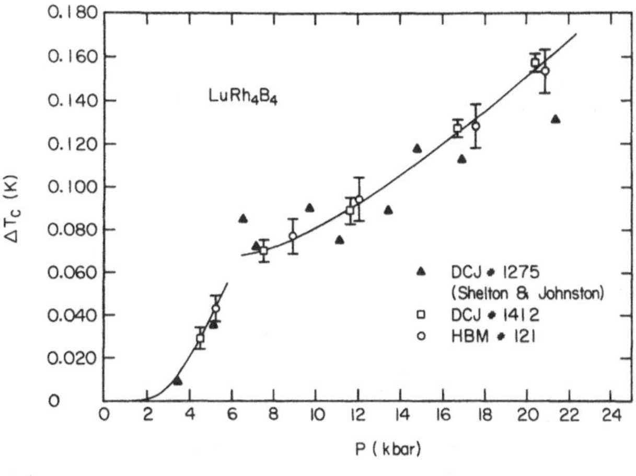

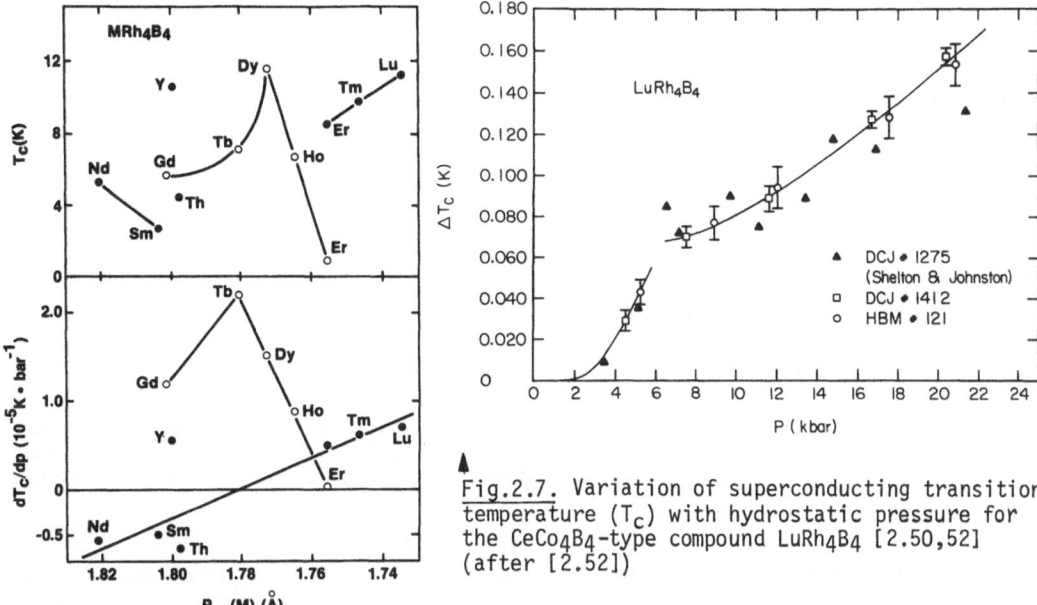

Fig.2.7. Variation of superconducting transition temperature (T_c) with hydrostatic pressure for the $CeCo_4B_4$-type compound $LuRh_4B_4$ [2.50,52] (after [2.52])

Fig.2.6. Superconducting (solid circles) and magnetic (open circles) transition temperatures and their pressure derivatives for $CeCo_4B_4$-type MRh_4B_4 compounds (after [2.50,51])

boron with respect to the composition MRh_4B_4 [2.55]. This phase can be stabilized at the nominal stoichiometry MT_4B_4 by substituting a few percent of the Rh by Ru, and in fact, many MRu_4B_4 compounds also crystallize in this structure [2.55-57]. The structure was solved by powder X ray methods for the $LuRu_4B_4$ member and was found to be a new structure type [2.55]. The structure was confirmed by single crystal X ray studies of YRu_4B_4 and of $Y(Rh_{1-x}Ru_x)_4B_4$ pseudoternary compounds [2.17]. The basic building blocks of the bct structure were found to be the same as in the $CeCo_4B_4$-type phases, i.e., a slightly distorted fcc array of isolated M atoms, separated Ru_4 tetrahedra with two different orientations and B_2 dimers. Not surprisingly, the unit cell dimensions of the two structures bear simple relationships to each other: $\sqrt{2}\, a_{LuRh_4B_4} \approx c_{LuRh_4B_4} \approx (1/2) c_{LuRu_4B_4}$. In both structure types, the T_4 tetrahedra and M atoms are arranged in a NaCl-type array [2.19]. However, whereas the T_4 tetrahedra of the same orientation in the $CeCo_4B_4$-type phases are joined into sheets perpendicular to the c-axis (Fig.2.1), they are distributed equally in each plane in an ordered way in the $LuRu_4B_4$ structure, as shown in Fig. 2.8. The smallest distance between Ru atoms belonging to different tetrahedra (2.77 Å) occurs between tetrahedra of the same orientation in adjacent planes, rather than within a plane; this distance is comparable to those (2.71 Å, 2.78 Å) between Ru atoms in the same tetrahedron, as well as to the Ru-Ru interatomic distance in Ru metal (2.71 Å). The distances between Ru atoms in differently oriented tetrahedra are significantly larger (2.98 Å, 3.10 Å) and the like-oriented tetra-

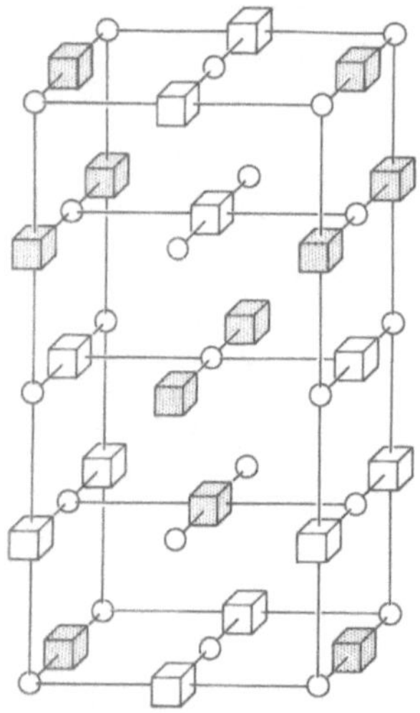

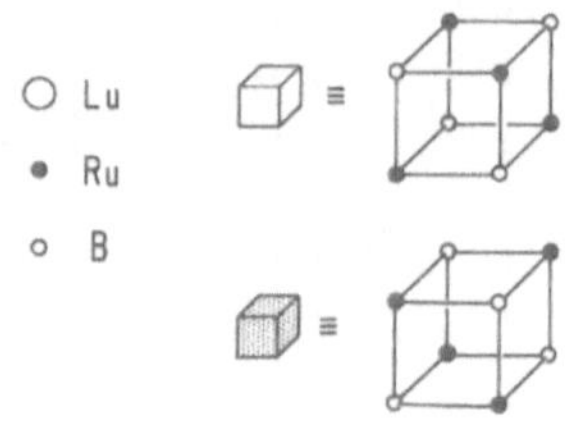

Fig.2.8. Idealized crystal structure of body-centered-tetragonal LuRu4B4-type compounds; as in Fig.2.1, the cubes representing the Ru4B4 units are not drawn to scale for clarity

hedra form a three-dimensional net. The structure can therefore be viewed as containing two interpenetrating extended clusters consisting of Ru_4 tetrahedra (Fig. 2.8), where each extended cluster contains only Ru_4 tetrahedra of the same orientation [2.58].

The various MT_4B_4 compounds reported to crystallize in the $LuRu_4B_4$ structure so far are shown in Table 2.2, and crystallographic data and critical temperatures are listed in Table 2.3 [2.27,55-57,59-61]. The large enhancement of the T_c's and depression of the T_m's for the $M(Rh_{0.85}Ru_{0.15})_4B_4$ compounds in Table 2.3 relative to those of the corresponding MRu_4B_4 ternaries is striking. The interactions between the magnetic RE atoms not only appear to change in magnitude, but also to change in sign in some cases as the Ru concentration is increased [2.60,62,63]. This effect has been documented for the isomorphous bct $Gd(Rh_{0.85}Ru_{0.15})_4B_4$ and $GdRu_4B_4$ compounds [2.62], where the magnetic interactions were found to be antiferromagnetic (θ = -5.6 K) and ferromagnetic (θ = 3.3 K), respectively. The same sign change in the Weiss temperature with composition was also recently inferred for the bct $Dy(Rh_{1-x}Ru_x)_4B_4$ system [2.60,63]. The differences noted above in the superconducting and magnetic interactions between the Rh-rich and Ru-rich $M(Rh_{1-x}Ru_x)_4B_4$ compounds are each associated with a change in the sign of the deviations of the c/a ratios from the value of 2; from Table 2.3, all of the former compounds have c/a ratios slightly less than two, whereas all of the latter exhibit ratios slightly greater than 2 [2.55]. It is therefore likely that c/a ≡ 2 is a unique ratio at which the Fermi surface changes character with changing electron concentration. This hypothesis is supported by measurements described below.

Detailed studies of the composition dependences of the critical temperatures and lattice parameters for bct $M(Rh_{1-x}Ru_x)_4B_4$ systems have been carried out for both magnetic (Dy [2.60,63], Er [2.64]) and nonmagnetic (Y [2.17,55,62]) M atoms. The

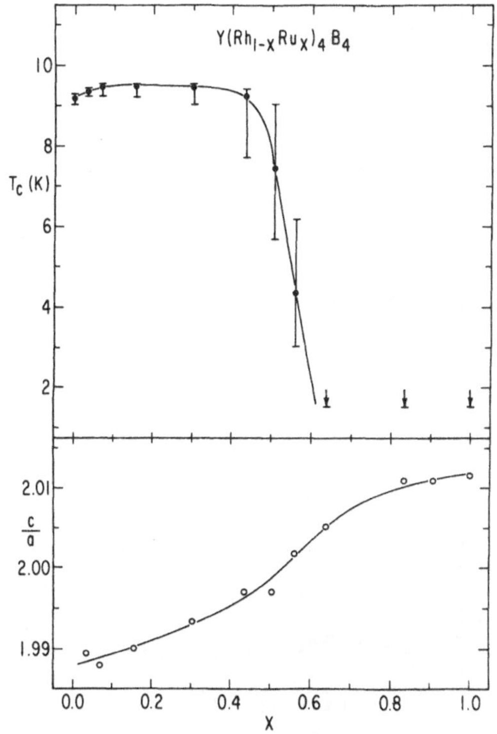

Fig.2.9. Superconducting transition temperature (T_c) and crystallographic c/a ratio for the $LuRu_4B_4$-type pseudo-ternary system $Y(Rh_{1-x}Ru_x)_4B_4$ [2.62], plotted versus composition (x)

variation of T_c with composition for the illustrative $Y(Rh_{1-x}Ru_x)_4B_4$ system is shown in Fig.2.9 [2.55,62], where T_c is nearly constant at ≈9 K for x < 0.4 but drops to less than 1.5 K between x = 0.4 and x = 0.6 and remains below 1.5 K at larger x values. The sharp drop in T_c occurs just as the c/a ratio increases through the value of 2.000 at x = 0.5, a composition where the c/a variation also shows an inflection point. A similar variation of T_c with x was reported for the Er [2.64] and Dy [2.60,63] systems and, as cited in the last section, for $CeCo_4B_4$-type $M(Rh_{1-x}Ir_x)_4B_4$ systems. These strong similarities suggest a common origin for the anomalous T_c behavior, even though the latter systems are isoelectronic whereas the former are not.

The nonlinear variations of T_c and the lattice parameters with composition in the $Y(Rh_{1-x}Ru_x)_4B_4$ system near x = 0.5 are both correlated with anomalous behavior of the B-B interatomic distance [2.62]. The composition dependence of the B-B interatomic distance computed [2.62] from the composition-dependent atomic positions reported by YVON and GRÜTTNER [2.17] is shown in Fig.2.10a. With increasing electron concentration (decreasing x), d(B-B) decreases monotonically from its value [1.78(5) Å] at x = 1 to the short distance of 1.45(10) Å at x = 0.5, then abruptly increases by 30% back to 1.88(11) Å by x = 0.4 and remains approximately constant at lower x values. These data suggest that the B_2 dimers are acting as electron sinks for the electrons donated by the Rh (d^7) atoms upon substitution for

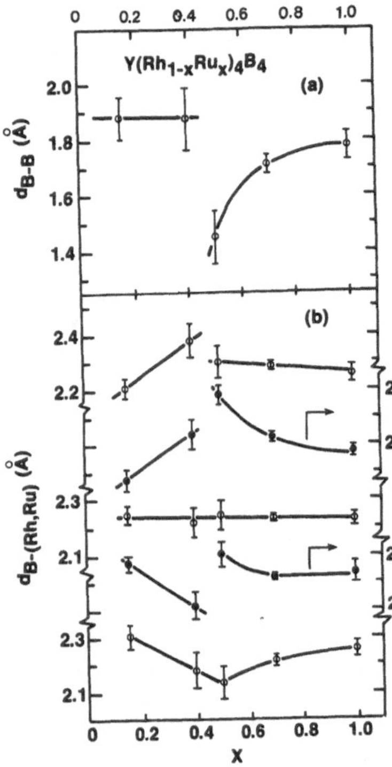

Fig.2.10. Composition dependence of the B-B and B-(Rh,Ru) interatomic distances in the $LuRu_4B_4$-type $Y(Rh_{1-x}Ru_x)_4B_4$ system [2.62]

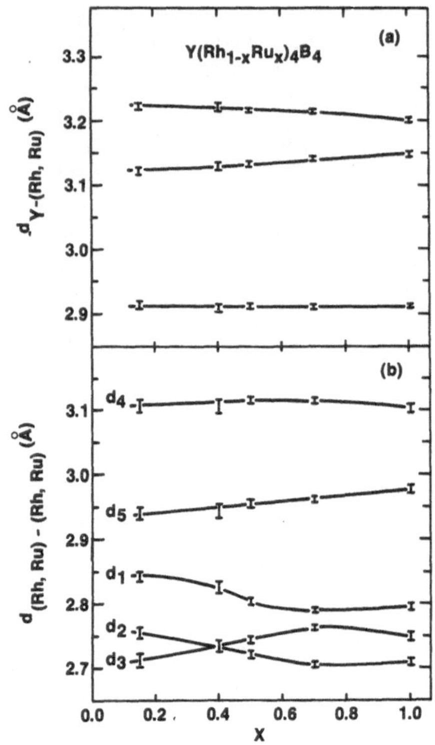

Fig.2.11. Composition dependence of the metal-metal interatomic distances Y-(Rh,Ru) and (Rh,Ru)-(Rh,Ru) in the $LuRu_4B_4$-type $Y(Rh_{1-x}Ru_x)_4B_4$ system [2.17,62]

Ru (d^6) atoms until x decreases to ≈0.5, after which the "trapped" electrons are nearly discontinuously released; it is apparently at this point that T_c abruptly increases and at which c/a decreases through the value of 2 and exhibits an inflection point in its composition dependence [2.62].

The large variations in the B-B interatomic distance in Fig.2.10 are accompanied by large and complex variations in the distances between each B atom and its other five neighbors to which it is bonded, the (Rh,Ru) atoms, as shown in Fig.2.10b [2.62]. The sensitivity to composition of the interatomic distances between the metal atoms, shown in Fig.2.11 [2.17,62], is an order of magnitude smaller than seen in Fig.2.10. In addition, significant nonlinearities with composition and the largest overall changes in Fig.2.11 occur only for the small intracluster (Rh,Ru)-(Rh,Ru) distances. Therefore, it appears that the anomalous superconducting behavior in Fig.2.9 arises from an abrupt composition-induced redistribution of electron density between the B_2 dimers and the (Rh,Ru) extended clusters [2.62].

The similarities between the T_c data in Figs.2.9,3 therefore suggest that the abrupt T_c variations in the latter figure could also arise from the mechanism sug-

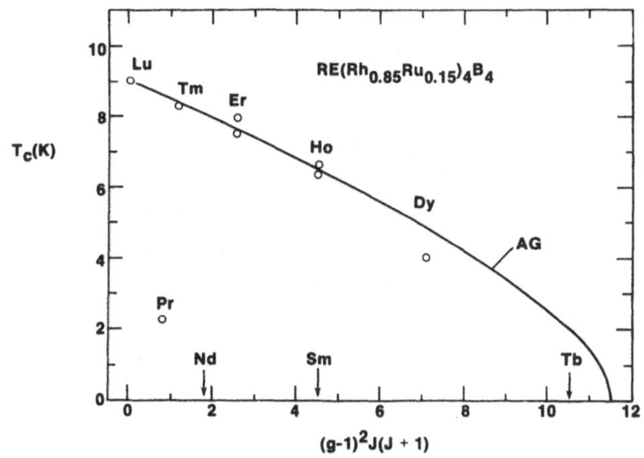

▲ Fig.2.12. Superconducting transition temperature T_c versus deGennes factor for the $LuRu_4B_4$-type compounds $RE(Rh_{0.85}Ru_{0.15})_4B_4$; the notation and comments are as in the caption to Fig.2.5 [2.65]

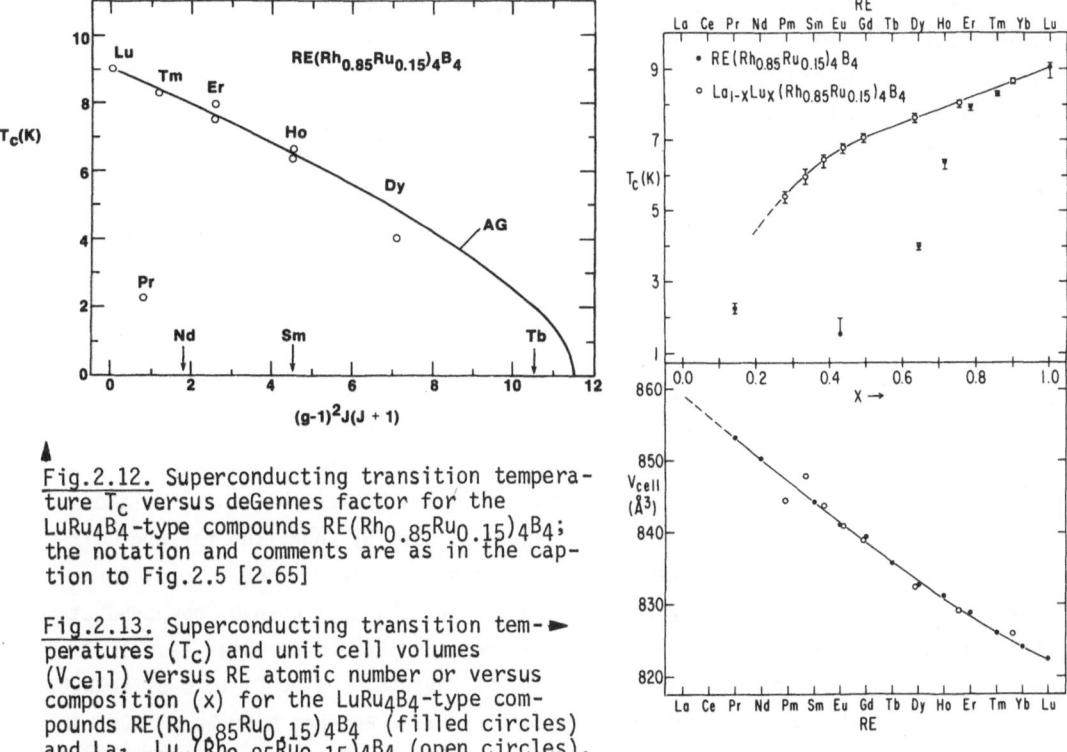

Fig.2.13. Superconducting transition temperatures (T_c) and unit cell volumes (V_{cell}) versus RE atomic number or versus composition (x) for the $LuRu_4B_4$-type compounds $RE(Rh_{0.85}Ru_{0.15})_4B_4$ (filled circles) and $La_{1-x}Lu_x(Rh_{0.85}Ru_{0.15})_4B_4$ (open circles), respectively [2.65] ►

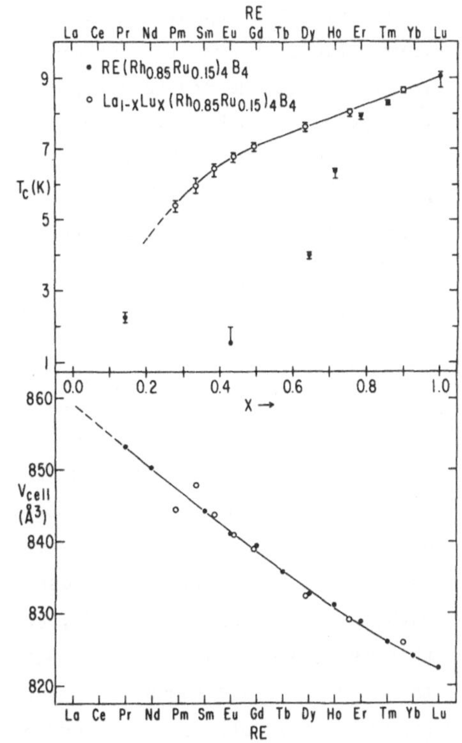

gested above; band structure calculations [2.43-45] have shown that the electronic states at the Fermi level in the $CeCo_4B_4$-type MRh_4B_4 compounds do contain a significant contribution deriving from the boron atoms, consistent with this suggestion.

The relative contributions of magnetic and nonmagnetic factors in establishing the observed variation of T_c with RE in bct $RE(Rh_{0.85}Ru_{0.15})_4B_4$ compounds has been assessed [2.65]. The T_c data were analyzed as outlined in the last section. Shown in Fig.2.12 is a plot of T_c versus the deGennes factor (dGF). As in Fig.2.5, a good fit of the AG theory to the data was obtained for RE = Ho to Lu (solid curve in Fig. 2.12); however, for larger RE ions, negative deviations from the AG curve become increasingly apparent. The AG curve in Fig.2.12 corresponds to T_{co} = 9.05 K, initial slope $dT_c/d(dGF)$ = -0.54 K and total exchange coupling strength $nI^2N(E_F) = 0.94\times10^{-5}$ eV. The latter two values are about one-half as large as those cited in the last section for the $CeCo_4B_4$-type $RERh_4B_4$ compounds, indicating an appreciably smaller $N(E_F)$ and/or I^2 than is present in the heavy RE members of the latter structure class. Measurements of T_c and unit cell volume for nonmagnetic $(La_{1-x}Lu_x)(Rh_{0.85}Ru_{0.15})_4B_4$ alloys (open circles) are compared with those of the $RE(Rh_{0.85}Ru_{0.15})_4B_4$ compounds (filled circles) in Fig.2.13 [2.65]. These data indicate that the negative deviations of the data in Fig.2.12 from the AG curve largely arise from decreases in T_{co}, i.e., from nonmagnetic effects [2.65].

The pressure dependences of the superconducting or magnetic transition temperatures have been reported for thirteen compounds with the $LuRu_4B_4$ structure [2.66]. In all cases, the pressure dependence was found to be linear, in contrast to results for several $CeCo_4B_4$-type compounds cited in the last section. Other qualitative differences in the behavior of the two structural classes were also found. For example, whereas the T_c's of the Er, Tm and Lu members of the $CeCo_4B_4$-type MRh_4B_4 phases all increase with pressure, those of the corresponding and nearly isoelectronic bct $M(Rh_{0.85}Ru_{0.15})_4B_4$ compounds each decrease with pressure, even though the T_c's within each pair of compounds are quite comparable.

c) *Orthorhombic $LuRh_4B_4$ Structure*

Studies of the superconducting properties and phase relationships in the Lu-Rh-B system near the composition $LuRh_4B_4$ led to the discovery of a third superconducting polytype of MT_4B_4 stoichiometry [2.58,67,68]. The latter phase was isolated in nearly single phase form at the composition $LuRh_{4.05}B_4$ by annealing the arc-melted ingot at ≈1200 C followed by quenching. This phase was found to be in equilibrium with the compounds RhB, $LuRhB_4$, $LuRh_3B_2$ and the $CeCo_4B_4$-type $LuRh_4B_4$ compound at this temperature; at somewhat lower temperatures the phase decomposes [2.67]. The third polytype was found to be the stable low temperature (1150°C) form of $YbRh_4B_4$ (arc-melted ingots have the $LuRu_4B_4$ structure [2.67]), and has also been found in annealed (1250°C) ingots of $RERh_{4+x}B_4$ ($x \approx 0.1$) for RE = Tm, Er and Ho [2.58,67,68]; the radius range for the RE atoms that form the phase appears to be severely limited compared with the other MT_4B_4 polytypes (cf. Table 2.2). With the exception of the Yb member, each of the compounds exhibits superconductivity [2.67,68]. A selection of T_c data is shown in Table 2.4; the T_c's are lower than those obtained for the corresponding $CeCo_4B_4$-type or Ru-stabilized $LuRu_4B_4$-type polymorphs of $RERh_4B_4$ compounds (Tables 2.1,3). The phases in Table 2.4 exhibit X ray powder patterns very similar to those of the $CeCo_4B_4$-type MRh_4B_4 compounds, and for this reason their presence as impurity phases in the latter materials is very difficult to detect from X ray powder data [2.67].

The structure of the $LuRh_4B_4$ polytype was solved through a single crystal X ray diffraction study [2.58]. The unit cell symmetry was found to be orthorhombic (space group Ccca), which shows that the tetragonal cell proposed in [2.67] is a pseudocell of the true cell. The orthorhombic lattice parameters are shown in Table 2.4, along with those of the Er, Tm and Yb members [2.58]. As in the $LuRu_4B_4$ structure, orthorhombic $LuRh_4B_4$ contains B_2 dimers and two three-dimensional interpenetrating extended clusters of Rh_4 tetrahedra; these extended clusters are qualitatively different than those in $CeCo_4B_4$-type MRh_4B_4 compounds, since in the latter phases they are quasi two-dimensional and do not interpenetrate [2.58]. The dimensionality of the extended clusters seems to be uniquely associated with the conduction electron-RE ion exchange interaction strength: from a plot of T_c versus deGennes factor for the compounds in Table 2.4, $nI^2N(E_F)$ is found to be nearly the same as in the $LuRu_4B_4$-type $RE(Rh_{0.85}Ru_{0.15})_4B_4$ compounds, and this value is a factor of two smaller than found in the $RERh_4B_4$ phases with the $CeCo_4B_4$-type structure.

d) *Other MT_4B_4 Structure Types*

The fourth and last polytype with MT_4B_4 stoichiometry for which the structure is known is typified by $NdCo_4B_4$, which crystallizes in a primitive tetragonal structure with space group $P4_2/n$ [2.69]. The members of this structure class, shown in Table 2.2 [2.19,25,57,69-71], evidently do not exhibit superconductivity above 1 K [2.30], in contrast to the other three polytypes. Qualitative differences in the structure from the other three [2.19,69] evidently give rise to changes in the electronic and/or phonon properties which are unfavorable to the occurrence of superconductivity.

Further tetragonal variants with MT_4B_4 stoichiometry have recently been reported, typified by $CeRe_4B_4$ ($a \approx 7.4$ Å, $c \approx 10.6$ Å) [2.72] and $CeFe_4B_4$ ($a \approx 7.1$ Å, $c \approx 27.4$ Å, space group P4/ncc) [2.73], but the details of the crystal structures are not yet known. ROGL [2.70] has reported the occurrence of MT_4B_4 compounds in the eight systems (Y, Gd, Tb, Dy, Ho, Er, Tm, Yb)-Os-B. In addition, T_c's were sometimes observed in the Lu-Rh-B system near the composition $LuRh_4B_4$ which could not be attributed to any of the known phases in the system [2.67]; this suggests the existence of yet another polytype of $LuRh_4B_4$.

2.1.2 Orthorhombic $LuRuB_2$ Structure

Orthorhombic $LuRuB_2$ and $YRuB_2$ exhibit relatively high T_c's of 10.0 and 7.8 K, respectively, [2.74] which among ternary borides are second only to those of the MT_4B_4 phases. These T_c's are comparable to the value reported for orthorhombic NbPS (T_c = 12.5 K) [2.75] which also has some similar structural features as will be noted below. The MTB_2 compounds crystallizing in the $LuRuB_2$ structure are shown in Table 2.5, along with their lattice parameters and superconducting or magnetic critical temperatures [2.27,74,76]. So far, only compounds containing trivalent M atoms and T = Os or Ru have been reported to form $LuRuB_2$-type phases [2.76].

The $LuRuB_2$ structure was solved through a single crystal X ray structural study [2.76], and a projection of the structure on the a-c plane is shown in Fig.2.14 [2.77]. The Lu and Ru atoms form planar nets perpendicular to the b-axis; the boron atoms are dimerized (d[B-B] = 1.74 Å) as in the MT_4B_4 compounds and are located between the Lu-Ru planes. The dimer axis is perpendicular to these planes and the B_2 dimers are weakly coupled (d[B-B] = 1.92 Å) to form zig-zag chains running perpendicular to the metal planes (i.e., parallel to the b-axis); this is in contrast to the MT_4B_4 compounds, in which the B_2 dimers are isolated from each other. An important feature of the structure is that the Lu-Lu distance within a plane is very short (3.10 Å) [2.76] in comparison to the normal CN12 metallic interatomic distance

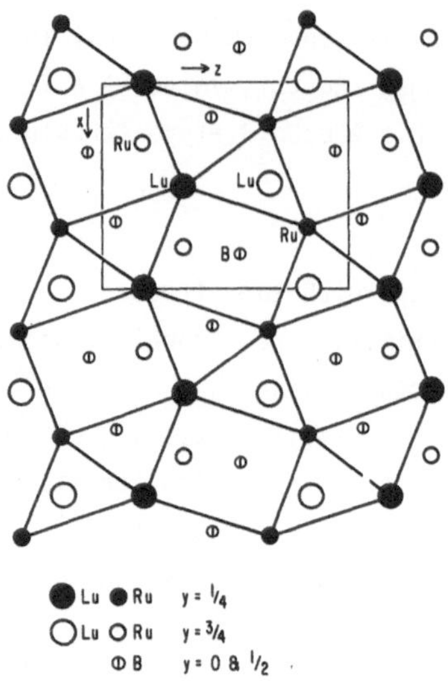

Fig.2.14. Projection of the orthorhombic $LuRuB_2$-type structure onto the a-c plane; the heavy solid lines illustrate the nature of the Lu-Ru planar nets (after [2.77]). The boron y parameters shown are approximations of the actual values

of 3.47 Å [2.78], whereas within the layer d[Ru-Ru] = 3.03 Å is about 12% longer than in Ru metal (2.71 Å). The interlayer Lu-Lu distance (3.34 Å) is much longer than within a plane and there are no Ru-Ru interlayer contacts (d[Ru-Ru] = 4.07 Å (2X) and 4.16 Å (4X)). Thus, the Lu atoms form zig-zag chain clusters lying in the a-c plane and running in the a direction (Fig.2.14); these chains are held together within the metal layers by Lu-Ru bonds for which the average bondlength (d[Lu-Ru] = 3.05 Å) is very close to that calculated (3.07 Å) from the CN12 metallic radii. $LuRuB_2$ is a cluster compound of a kind qualitatively different than the superconducting MT_4B_4 or Chevrel phase compounds. In the latter materials, the transition metal atoms form clusters and the M atoms are isolated; in $LuRuB_2$, these roles of the M and T atoms in the clustering behavior are reversed. As noted above, the B_2 clusters are weakly bonded together to form chains in $LuRuB_2$, whereas they are isolated from each other in the MT_4B_4 compounds. In this connection, $LuRuB_2$ is similar to NbPS in which the phosphorus atoms form P_2 dimers which are weakly coupled together to form chains [2.75].

Only those compounds within the $LuRuB_2$ structure class which contain nonmagnetic M atoms exhibit superconductivity (Table 2.5), and the T_c peaks at M = Lu for both the $MRuB_2$ and $MOsB_2$ series of compounds. The T_c's of the Os compounds are about a factor of four lower than the T_c's of the respective isoelectronic Ru members. From Table 2.5, substitution of Os for Ru does not systematically affect the unit cell volume, the b-axis parameter (i.e., the metal interlayer distance) or the area of the unit cell in the a-c plane. However, systematic upward shifts of the c-parameters and downward shifts of the a-parameters ($\approx$0.02 Å) do occur upon replacing Ru by Os; this suggests that the values of the (variable) atomic coordinates and interatomic distances within the unit cell may be changing significantly with this substitution and may be responsible for the deleterious effect on the superconducting transition temperatures. Further structural studies would be of value in this regard.

The magnetic ordering temperatures in Table 2.5 are seen to approach 50 K, which is a factor of four larger than the maximum value (12 K) observed in the $RERh_4B_4$ compounds. These enhanced T_m values are probably due to both the shorter RE-RE interatomic distances and the higher RE concentration in the $RERuB_2$ phases relative to the other compounds. The interaction between magnetic ordering and superconductivity was investigated in the $(Lu_{1-x}Tm_x)RuB_2$ system where re-entrant superconductivity was observed for $0.52 \le x \le 0.62$ [2.77].

2.1.3 $CeCo_3B_2$ and Related Structures

An isotypic series of compounds with the stoichiometry MCo_3B_2 was reported in 1969 by KUZ'MA et al. [2.79]. The $CeCo_3B_2$ structure was found [2.79] to be an ordered structure derived from the structure [2.80] of the binary compound $CaCu_5$, with Co and B, respectively, replacing the Cu atoms on two crystallographically distinct lattice sites. Approximately ninety compounds have now been found to form at the stoichiometry $M_{\approx 1}T_{\approx 3}B_2$ [2.25,27,55,79,81-90] or $M_{\approx 1}T_{\approx 3}Si_2$ [2.91,92] which crystallize in the $CeCo_3B_2$ structure or in a very closely related structure, where T = Fe, Ru, Os, Co, Rh, Ir, Ni or Pt. $ZrIr_{\approx 3}B_{\approx 2}$ and $HfIr_{\approx 3}B_{\approx 2}$ have been found to form compounds, but their unit cell symmetry and structure are not yet known [2.93]; the corresponding Co compounds have a rhombohedral structure [2.94]. The distribution of structures among the MT_3B_2 phases reported so far is shown Table 2.6. In the following, we will be primarily concerned with those phases for which superconductivity or magnetic data [2.27,55,84-87,91,92,95] have been reported.

a) *Hexagonal $CeCo_3B_2$ Structure*

About half of the compounds in Table 2.6 crystallize in the hexagonal $CeCo_3B_2$ structure (labeled "A" in Table 2.6). The lattice parameters and superconducting or magnetic critical temperatures reported for the $CeCo_3B_2$-type MT_3B_2 compounds with T = (Fe, Ru, Os) are listed in Table 2.7 and for T = (Co, Rh, Ir) in Table 2.8.

The $CeCo_3B_2$ structure is shown in Fig.2.15. It consists of an alternating stacking sequence of two types of layers along the c-axis. One layer contains the Ce and boron atoms, while the other contains only Co. The Ce atoms form a simple hexagonal lattice. The Co atoms form trigonal prisms, each of which has a boron atom at its body center. These prisms are joined into infinite columns in the c-direction by sharing of the triangular faces and the columns are edgeshared to form a three-dimensional array. A significant feature of the $CeCo_3B_2$ structure which is qualitatively different from those of the ternary borides discussed above and of the Chevrel phases is that all atoms occupy special positions which are fixed with respect to the cell axes, i.e., the atomic positions cannot take variable x, y, or z coordinate values.

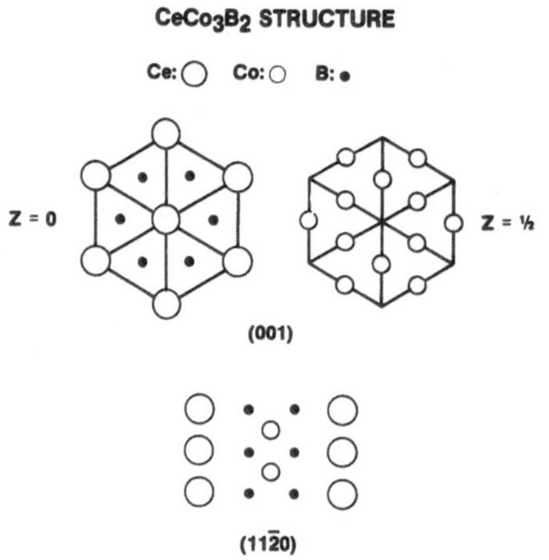

Fig.2.15. Illustration of the planar nets present in the structure of hexagonal $CeCo_3B_2$

Thus, the interatomic distances between the atoms in each compound are constrained to bear specific relationships to each other, and these distances are fixed by the lattice parameters alone.

In the $CeCo_3B_2$-type compounds, the B-B distance d[B-B] ≈ 3.0 Å both within an M-B layer and between two such layers and the boron atoms are therefore isolated from each other, in contrast to the chains or B_2 dimers in the other borides discussed above. The M atoms form chains along the c-axis (d[M-M] ≈ 3.0 Å) and these chains are isolated from each other (d[M-M] = 5.0-5.5 Å); the intrachain M-M distance is very short and nearly the same as in the $LuRuB_2$-type compounds. The T atoms form two-dimensional nets with d[T-T] = 2.5-2.75 Å, values which are close to the values in the respective T metals; between T layers, d[T-T] is more than 10% longer. Therefore, in the MT_3B_2 compounds we have a type of clustering not seen yet in our review: the M and T atoms both respectively form clusters and these clusters interpenetrate, whereas the B atoms are isolated from each other.

From Tables 2.7,8, a remarkable phenomenon is seen: the c-axis parameters of all the compounds are nearly identical, irrespective of the nature of the M and T atoms; these atoms exert their influence primarily on the length of the a-axis. The inescapable conclusion is that the c-axis is somehow determined primarily by the boron atoms. Consistent with this, it will be seen below that substituting Si for B does increase the parameter corresponding to c by ≈15% (0.6 Å), whereas the a value is essentially unchanged. The influence of the boron atoms on the crystallography of the MT_3B_2 compounds is not presently understood; however, this is again a spectacular illustration that the boron atoms are not electronically inert in the materials of interest in this chapter.

b) *Monoclinic $ErIr_3B_2$ Structure*

The $ErIr_3B_2$ structure is a slight monoclinic distortion of the hexagonal $CeCo_3B_2$ structure within the basal plane [2.27,87]. The magnetic ordering temperatures obtained from ac magnetic susceptibility measurements and the lattice parameters for the $ErIr_3B_2$-type compounds are listed in Table 2.9; no superconductivity was observed above 1.2 K for any of these materials [2.27,87]. The T_m values attain maxima of 45 K (Tb) for the MRh_3B_2 compounds and 30 K (Gd) within the MIr_3B_2 series; these values are comparable to those of the $CeCo_3B_2$-type compounds. The interatomic distances are also nearly identical to those found for the $CeCo_3B_2$-type series and the crystal-chemical comments made in the last section are therefore applicable here.

c) *Hexagonal $Ba_{2/3}Pt_3B_2$ and $LaRu_3Si_2$ Structures*

The structures of the title compounds were determined from X ray powder diffraction data by SHELTON [2.84], and by VANDENBERG and BARZ [2.92], respectively. Both structures may be considered to be distorted derivatives of the $CeCo_3B_2$ structure. The structure of $LaRu_3Si_2$ is shown in Fig.2.16 where the distortions of the Ru sublattice have been amplified by a factor of five to illustrate the resultant structural features. The La atoms form a simple hexagonal lattice as in the $CeCo_3B_2$ structure. The T metal array of Fig.2.15 becomes distorted in such a way that larger and smaller triangular clusters are formed which are twisted by about 4° with respect to their orientation in the $CeCo_3B_2$ structure; the twist angle and T_3 triangle size reverse and alternate, respectively, from one T layer to the next giving rise to a 2c superlattice. The Si atoms center the twisted and deformed trigonal prisms formed by the large and small Ru triangular faces in adjacent metal atom layers. The Ru-Ru distance within a small triangle is 2.76 Å, whereas the distance between these is 2.93 Å. The interlayer Ru-Ru distance (3.57 Å) is no longer a metallic distance [2.92]. Thus, the Ru sublattice forms a two-dimensional cluster lattice as pointed out in [2.92], and this clustering is much more pronounced than in the $CeCo_3B_2$-type or $ErIr_3B_2$-type compounds. On the other hand, the one-dimensional clustering of the M atoms becomes much less pronounced; the La-La distance is now 3.58 Å, far greater than that (3.0-3.1 Å) in the $CeCo_3B_2$-type phases. Whereas the Ru atoms in Fig.2.16 remain coplanar, the Si atoms do not; the nature of the puckering is shown in Fig.2.16. The magnitude of the Si sublattice distortion is not yet known [2.92].

The lattice parameters and T_c or T_m values from ac magnetic susceptibility measurements for $LaRu_3Si_2$-type MRu_3Si_2 compounds [2.91,92] are listed in Table 2.10. The nonmagnetic Y, La and Th members each exhibit superconductivity with the maximum T_c occurring for M = La (7.6 K). Similar to the $CeCo_3B_2$-type and $ErIr_3B_2$-type compounds, magnetic ordering was found for most of the remaining $LaRu_3Si_2$-type phases containing magnetic RE atoms at temperatures sometimes exceeding 25 K.

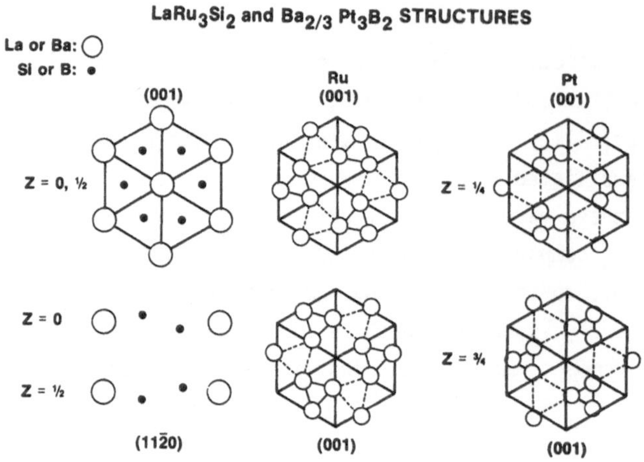

Fig.2.16. Relationships between the hexagonal $LaRu_3Si_2$ and $Ba_{2/3}Pt_3B_2$ structures. Both structures are derived from the $CeCo_3B_2$ structure (compare with Fig.2.15)

The $Ba_{2/3}Pt_3B_2$ structure [2.84] is closely related to that of $LaRu_3Si_2$, the two structures differing only in the nature of the distortion in the transition metal layers as shown in Fig.2.16; here again, the Pt sublattice distortion has been magnified by a factor of about five to illustrate the structural features. In $Ba_{2/3}Pt_3B_2$, the triangular faces of the distorted Pt trigonal prisms are not twisted with respect to each other. Only two-thirds of the M positions are occupied by Ba [2.84] and it is possible that the structure in Fig.2.16 is therefore a substructure of the true structure [2.84]. The interatomic distances are vastly different in the two structures of Fig.2.16. Within a Pt layer, the Pt-Pt distances are 2.86 Å within a small Pt triangle and 3.30 Å between them. In contrast, the interlayer Pt-Pt distance is only 2.65 Å, 4.3% shorter than in Pt metal. The nature of the transition metal sublattice in $Ba_{2/3}Pt_3B_2$ is therefore qualitatively different from that in the $CeCo_3B_2$, $ErIr_3B_2$ and $LaRu_3Si_2$-types in which two-dimensional clusters are formed. The Pt atoms in $Ba_{2/3}Pt_3B_2$ can be viewed as forming zig-zag chains (d[Pt-Pt] = 2.65 Å) running in the c direction, where the interchain distances are d[Pt-Pt] = 2.86 Å (2X) and 3.30 Å (2X). This type of clustering apparently occurs due to the high concentration of vacancies on the Ba sites which allows the c-axis parameter to collapse. However, electronic factors are probably also important here because the c-axis parameter appears to be determined primarily by the boron atoms in the other MT_3B_2 phases.

The lattice parameters and superconducting critical temperatures of the three known members of the $Ba_{2/3}Pt_3B_2$ structure class are shown in Table 2.11 [2.84]. The T_c is seen to increase monotonically with the unit cell volume, attaining a value of 5.6 K for the Ba member.

d) *Other MT_3B_2 Variants*

Eleven MT_3B_2 compounds in Table 2.6, typified by YOs_3B_2, have been found to crystallize in a structure different from those discussed in previous sections [2.27]. The powder diffraction data for these compounds could be indexed on an orthorhombic lattice, but the structure has not yet been solved [2.27]. Superconductivity was observed in this class of material for YOs_3B_2, (T_c = 6 K), $LaRu_{2.75}B_2$ (4 K) and $ThOs_3B_2$ (3 K); except for $YbOs_3B_2$, the remainder order magnetically at temperatures up to and exceeding 45 K [2.27]. The (Zr, Hf)Ir_3B_2 compounds [2.93] do not exhibit superconductivity above 1.2 K [2.96]. Superconducting transition temperatures have not been reported for the remaining compounds in Table 2.6.

Finally, we note that compounds of stoichiometry MT_6B_4 exist, i.e., $M_{0.5}T_3B_2$. First reported in 1977 (YRh_6B_4 [2.55]), the X ray patterns bear a resemblance to those of the $CeCo_3B_2$-type phases. The structure of the MT_6B_4 phases is not yet known although powder X ray data have been indexed on a hexagonal lattice with parameters (a = 5.65 Å, c = 6 × 2.85 Å [2.97]) which bear a close relationship to those of the $CeCo_3B_2$-type compounds. The compounds YRh_6B_4 and $LuRh_6B_4$ do not become superconducting above 1.5 K [2.55,62]. Itinerant electron ferromagnetism has been reported for $LaRh_6B_4$ and $EuRh_6B_4$ below T_m = 6 K and 19 K, respectively [2.97], a property shared by $CeRh_3B_2$ [2.95]; moreover, YRh_6B_4 and $LuRh_6B_4$ exhibit strongly enhanced Pauli paramagnetism [2.97]. In view of these results and the probable similarities between the MT_3B_2 and MT_6B_4 structures, measurements of the magnetic properties of the "nonmagnetic" members of the former class would be of great interest with respect to understanding the systematics of superconductivity among the MT_3B_2 materials. Data for such MRu_3B_2 compounds [2.86] do show enhancements in the volume susceptibility by factors of 5-15 over that [2.22] for $CeCo_4B_4$-type $LuRh_4B_4$.

e) *Hexagonal ZrRuSi Structure*

A system of ternary transition metal phosphides typified by ZrRuP was recently reported which exhibits superconducting critical temperatures (T_c^{onset} = 13.0 K for ZrRuP) approaching the highest values known among ternary compounds [2.98]. The lattice parameters and T_c's reported for these hexagonal phosphides are listed in Table 2.12 [2.98].

The structure of ZrRuP was found from powder X ray diffraction data [2.98] to be that of hexagonal ZrRuSi [2.99], which is an ordered Fe_2P structure [2.100]. The ZrRuSi structure is closely related to that of $CeCo_3B_2$, as illustrated in Fig.2.17. The Zr and Ru atoms lie in different layers and the metal arrangement in each layer can be derived from that of Co in the Co layer of $CeCo_3B_2$. Whereas the Zr configuration is only slightly distorted from that of the Co, the Ru atom array is highly distorted; the Ru atoms form triangular clusters as was seen in Fig.2.16 for the Ru and Pt in $LaRu_3Si_2$ and $Ba_{2/3}Pt_3B_2$. In fact, the Ru_3 array in ZrRuSi becomes

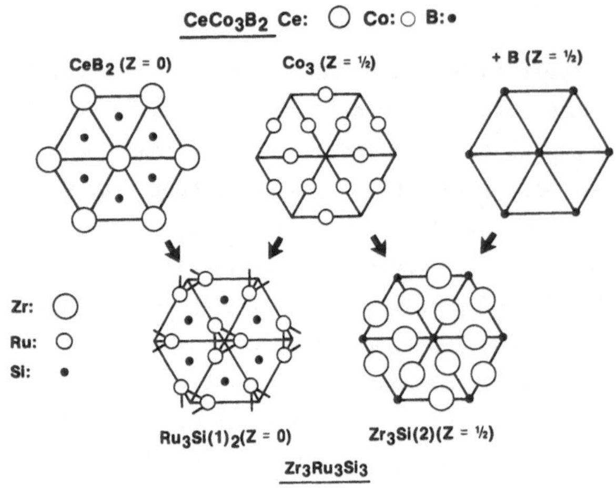

Fig.2.17. Derivation of the hexagonal ZrRuSi structure from that of $CeCo_3B_2$

identical to the Pt_3 array if each Ru_3 triad is rotated by 30° about its 3-fold axis. The Si array in the Ru_3Si_2 layer of Fig.2.17 is identical to the B sublattice in $CeCo_3B_2$ and Si is also inserted as a simple hexagonal sublattice into the Zr_3 layer. Thus, the CeB_2 layers in $CeCo_3B_2$ become Ru_3Si_2 layers in ZrRuSi and the Co_3 layers become Zr_3Si layers; the sum yields the composition $Zr_3Ru_3Si_3 \equiv$ ZrRuSi.

From Table 2.12, the T_c's of the Ti(Ru,Os)P compounds ($\lesssim$ 1.3 K) are an order of magnitude smaller than those (5.5-11.8 K) of the four respective (Zr,Hf)(Ru,Os)P ternaries. This discrepancy also exists between the T_c's of the latter materials and those of most other Fe_2P-type compounds of which we are aware. This paradox appears to arise from differences in the detailed nature of the crystallographic ordering of the three types of atoms in these phases. Most of the compounds known to form an ordered Fe_2P-type $M^{I}M^{II}X$ structure (see, e.g., [2.101,102] and references therein) order in a way that would correspond to switching the Ru and Si atoms in Fig.2.17, a change which apparently seriously degrades the superconducting properties. Inductive T_c measurements on 20 Fe_2P-type compounds within the systems (Y,Zr,Hf,Th,U)-(Ru,Rh,Ir,Pd,Pt,Cu)(Al,Ga,In) [2.103] showed no evidence of superconductivity for any of them above 1.5 K except for ThPtGa (T_c = 2.63-2.53 K) and UPtIn (3.00-2.47 K). In contrast, a URhAl sample exhibited a ferromagnetic cusp in the ac susceptibility at 26.7 K, as did a different sample of UPtIn at 19 K; it is not known which of these superconducting and magnetic effects, if any, are due to impurity phases [2.103]. The Fe_2P-type compounds MPtSn [2.104] exhibit no superconductivity above 1.1 K for M = Y and Er [2.105].

Additional evidence further indicates the strong influence of crystallographic order on the superconducting properties of the Fe_2P-type materials [2.98,106]. On the other hand, electronic factors are also evidently important to the occurrence of superconductivity in the ordered Fe_2P-type phases because ZrRuSi which is isostructural to ZrRuP exhibits no superconductivity above 1.2 K [2.98]. Single crystal structural studies would clearly be of great value in understanding the T_c differences seen among the Fe_2P-type phases; these could more accurately establish the occupancy factors for the various atomic positions in these materials than can powder diffraction methods. Finally, superconductivity has been found in ternary arsenides which are isostructural to ZrRuP [2.98].

2.1.4 Primitive Cubic $Pr_3Rh_4Sn_{13}$ and Related Structures

a) *$Pr_3Rh_4Sn_{13}$ Structure*

Independent discoveries of superconductivity or magnetic ordering were recently reported for what now appear to be isostructural series of primitive cubic ternary transition metal germanides [2.107] and stannides [2.108]. The structure was solved for the Pr-Rh-Sn compound using powder X ray diffraction [2.109]; the ideal stoichiometry found therefrom was $Pr_3Rh_4Sn_{13}$, with two formula units per unit cell ($a = 9.693$ Å). A very precise single crystal X ray study of $Yb_3Rh_4Sn_{13}$ [2.110] confirmed the structure and single crystal studies also suggested isotypy for $Y_3Ru_4Ge_{13}$ and $Lu_3Os_4Ge_{13}$ [2.111] (but see below). About fifty compounds with general formula $M_3T_4X_{13}$ have now been reported [2.105,108,111-114] and are listed as phase I [2.108] in Table 2.13. Some samples of phase I stannides (denoted phase I* in [2.105] and phase IV in [2.112,113]) exhibit line splitting at high Bragg angles in X ray powder diffraction patterns; HODEAU et al. [2.110] have found that this effect can arise from the occurrence of two cubic lattices with slightly different lattice parameters.

The lattice parameters and superconducting or magnetic critical temperatures reported for phase I $M_3T_4X_{13}$ compounds in Table 2.13 are listed in Table 2.14 [2.105, 108-114]. Superconductivity has been observed in these systems only for the nonmagnetic M atoms Ca, Sr, Y, La, Lu, Th and nonmagnetic (?) Yb; RE elements with incomplete 4f shell do not exhibit superconductivity and in some cases order magnetically. A clear correlation exists in Table 2.14 between T_c and the M atom valence: the T_c decreases as this valency increases. The highest T_c's observed (up to 8.7 K) are for the divalent Ca and Sr and divalent (?) Yb M atoms (stannide series); the T_c's of the trivalent Y, La and Lu members of the stannides and germanides are lower ($T_c \lesssim 4K$) and that of $ThRh_xSn_y$ still lower ($T_c \approx 2K$). Even the Ca-Co-Sn compound exhibits a relatively high T_c of 5.9 K [2.112]. The magnetic interactions between magnetic RE atoms in the stannides and germanides appear to be of roughly comparable strength, based on the T_m values for the RE members near Gd in each series.

The $Pr_3Rh_4Sn_{13}$ structure [2.109-111] has a number of interesting features. The Sn atoms occupy two different types of positions: 2 Sn(1) atoms per unit cell form a bcc sublattice and appear to be cationic in character; the remainder [Sn(2)] are anionic in nature [2.110,115] and coordinate the Sn(1), Pr and Rh atoms [2.109-111]. The formula can therefore be properly written $Sn(1)Pr_3Rh_4Sn(2)_{12}$ [2.110]. The $Sn(1)Pr_3Rh_4$ sublattice structure is shown in Fig.2.18. As noted above, Sn(1) forms a bcc sublattice and the Pr and Sn(1) together form the Cr_3Si (A-15) structure in which nonintersecting and mutually perpendicular chains of Pr atoms run across the

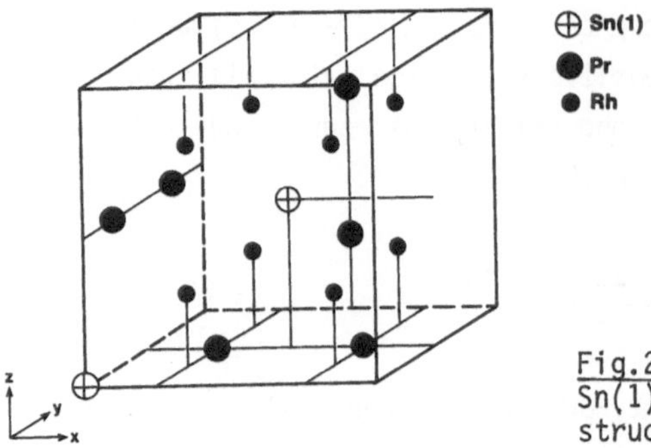

Fig.2.18. $Sn(1)Pr_3Rh_4$ sublattice of the $Sn(1)Pr_3Rh_4Sn(2)_{12}$ structure; this sublattice structure is identical to the structure of $NaPt_3O_4$

cell faces [2.110]. The Rh atoms form an interpenetrating simple cubic array; one Rh_4 cube is situated at the center of each A-15 unit cell as shown in Fig.2.18. This structure of the $Sn(1)Pr_3Rh_4$ sublattice is identical to the structure [2.116] of the ternary oxide $NaPt_3O_4$ with the obvious replacements Sn(1) for Na, Pr for Pt and Rh for O. The Sn(2) sublattice is discussed in detail in [2.110,111]. HODEAU et al. [2.110] have shown that the Yb and Sn atoms in $SnYb_3Rh_4Sn_{12}$ show mutual substitutional disorder on the Sn(1) and Yb sites of the structure (5.7% and 8.5 at.%, respectively). This substitutional disorder could be significantly affecting the superconducting and magnetic properties of phase I stannides.

In the $M_3T_4Ge_{13}$ compounds, the thermal parameters of both Ge(1) and Ge(2) are rather large and, for Ge(2), strongly anisotropic, corresponding to rms displacements of 0.34 Å for Ge(1) and 0.2 Å for Ge(2) [2.111]. This indicates that the structure model for $Y_3Ru_4Ge_{13}$ and $Lu_3Os_4Ge_{13}$ [2.111] is really an average structure; the true structure could therefore have a lower symmetry [2.111].

In contrast to all structures discussed in previous sections, there are no M-M, T-T and only weak M-T contacts in the $M_3T_4X_{13}$ compounds: d[M-2M] = d[T-6T] = 4.5-4.8 Å and d[M-4T] = d[T-3M] = 3.2-3.4 Å; the X(2) atoms effectively isolate each T and M atom from the others. Thus, there are no M-M, M-T or T-T clusters. These materials might, therefore, be expected to exhibit physical properties typical of sp-element superconductors.

b) *Ternary Compounds Related to $Pr_3Rh_4Sn_{13}$-Type Materials*

In addition to the phase I stannides discussed above, two other ternary phases with approximately the same composition have been found in M-T-Sn systems [2.105,108, 112,113,117]. Phase III is face centered cubic with $a \approx 13.7$ Å [2.108,110,118], and phase II is a tetragonal superstructure of phase III with $a \approx 13.7$ Å and $c \approx 27.4$ Å [2.118]. The lattice parameters of phases I, II and III are simply related: $\sqrt{2}\, a^{I} \approx a^{III} \approx a^{II} \approx c^{II}/2$; it is therefore likely that the phase II and phase III structures, not yet reported in detail, have similar features which in turn are similar to those of phase I. The occurrence of phases II and III among M-T-Sn systems is shown in Table 2.13.

Crystallographic, superconducting and magnetic data reported for phase II and phase III stannides are shown in Table 2.15 [2.105,108,112,113,117-120]. Re-entrant superconductivity occurs in the phase II compound $ErRh_{1.1}Sn_{3.6}$ [2.108,115,117,119, 120] and in the phase III compound $ErOs_xSn_y$ [2.105] at zero applied magnetic field and in $TmRh_{1.3}Sn_{4.0}$ in applied fields greater than 1.2 kOe [2.120]. The (upper) T_c's of the phase II and phase III materials ($T_c \lesssim 4.5$ K) are lower than values found in phase I stannides (Table 2.14). Two of the phase III compounds ($ErRu_xSn_y$ and YRu_xSn_y) and pseudoternaries formed from them exhibit a very slight distortion [2.105] and have been denoted "phase III*" in [2.105]. The degree of crystallographic order appears to be important to the observed properties of $ErRh_xSn_y$ [2.119].

Superconductivity or magnetic ordering has recently been reported [2.121] for compounds crystallizing in the body-centered-cubic $LaFe_4P_{12}$-type structure [2.122]. The T_c observed for $LaFe_4P_{12}$ (4.0 K) is high for an iron compound. The T_c is 7.1 K for $LaRu_4P_{12}$ and magnetic ordering transitions were reported below 2K for $NdRu_4P_{12}$, $PrFe_4P_{12}$ and $NdFe_4P_{12}$. The stoichiometry and structure of $LaFe_4P_{12}$ bear close resemblances to those of primitive cubic $SnPr_3Rh_4Sn_{12}$; the $LaFe_4$ sublattice structure is identical to the $Sn(1)Rh_4$ sublattice although the P_{12} and $Sn(2)_{12}$ sublattices are quite different. In $LaFe_4P_{12}$, the Fe atoms are octahedrally coordinated by P, whereas in $SnPr_3Rh_4Sn_{12}$ the coordination of Rh by Sn(2) is trigonal prismatic.

2.1.5 Primitive Tetragonal $Sc_2Fe_3Si_5$ Structure

Iron is an element not commonly found in superconductors. In fact, if certain titanium-iron and zirconium-iron solid solutions are disregarded, only two superconducting binary iron compounds are known above 1 K: Th_7Fe_3 (T_c=1.9 K [2.123]) and U_6Fe (T_c = 3.9 K [2.124]). In other compounds such as the Chevrel-phases [2.5] and A15-phases [2.125], dissolved iron carrying a magnetic moment depresses their superconducting critical temperatures. Thus, the recent discovery of superconductivity in the series $M_2Fe_3Si_5$ [2.126] at temperatures up to 6 K poses some interesting questions about the origin of the superconducting electrons.

The $Sc_2Fe_3Si_5$-type structure (Fig.2.19, after [2.127]) is primitive tetragonal [2.128]. The iron atoms occupy two sets of point positions, forming isolated squares in the basal plane [Fe(1)] and isolated linear chains running perpendicular to this plane [Fe(2)]. The Fe-Fe distances in both sets are approximately equal (2.64 Å and 2.67 Å), while the nearest distance between iron atoms belonging to different sets is considerably larger (4.10 Å). Thus, the transition metal in this class of compounds forms two types of clusters, though not in the most stringent sense [2.26] since the Fe-Fe intracluster distances are larger than in iron metal (2.48 Å).

As shown in Table 2.16 [2.126,127,129,130], the iron based compounds with M = Sc, Y and Lu are superconducting with the highest critical temperatures observed

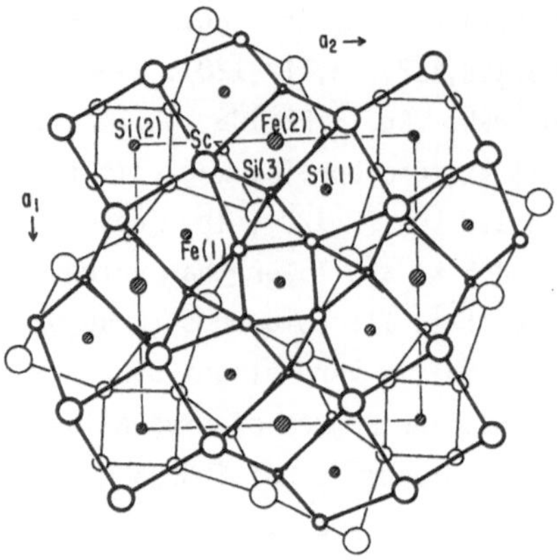

Fig.2.19. Projection onto the basal plane of the primitive tetragonal structure of $Sc_2Fe_3Si_5$; the planar nets at z=1/2 and z=0 are connected by light and heavy lines, respectively; shaded symbols: z=1/4, 3/4 (after [2.127])

for iron compounds. Remarkably, $Tm_2Fe_3Si_5$ also exhibits superconductivity with T_c = 1.3 K but re-enters the normal state at the lower temperature of 1.1 K [2.130]; this is only the third known spatially ordered and stoichiometric compound exhibiting this behavior, the others being $ErRh_4B_4$ and $HoMo_6S_8$. Other $M_2Fe_3Si_5$ compounds containing lanthanide ions with incomplete 4f shells show antiferromagnetic ordering at temperatures below 11 K. At higher temperatures the magnetic susceptibilities of these compounds follow a Curie-Weiss law with an effective moment of the rare-earth ion approximately equal to the free ion moment [2.127,129,131]. Among the other $M_2T_3Si_5$ compounds, only $Y_2Re_3Si_5$ was found to become superconducting above 1 K. Most striking is the absence of superconductivity for $Y_2T_3Si_5$ and $Lu_2T_3Si_5$ in which Fe is replaced by the isoelectronic metals Ru or Os.

Mössbauer studies of $M_2Fe_3Si_5$ compounds [2.131,132] show that the magnetic moment on both Fe(1) and Fe(2) is less than 0.03 μ_B. In compounds containing magnetic rare earths the Fe remains nonmagnetic in zero field, while a small negative induced moment is observed at the iron site in an applied field. The iron 3d electrons may be nonmagnetic because they participate in bonding. The presence of covalent interactions is suggested by the conspicuous shortening of the Fe-Si distances, frequently observed in silicides. Whether the 3d electrons of Fe participate in the superconductivity of $M_2Fe_3Si_5$ compounds is an interesting question; from the absence of superconductivity in the isoelectronic Os and Ru compounds in Table 2.16, these 3d electrons do indeed appear to be crucial to the superconductivity.

In pseudoternary solid solutions $(M_{1-x}M'_x)_2Fe_3Si_5$ between pairs of superconducting compounds with (M,M') = Lu, Sc, or Y, the T_c is rapidly depressed to below 1 K for x values of only 0.1 [2.130]. Preliminary results on the series with M = Lu and M' = Tb to Tm indicate that the depression of T_c depends primarily on the size difference between M and M' [2.130]. The T_c of $Y_2Fe_3Si_5$ almost doubles under hydrostatic pressure and reaches a maximum of 5 K at 18 kbar [2.133]. Both results seem to indicate that the Fermi energy is in the vicinity of a peak in the density of states and can be moved through it by the application of mechanical or "chemical" pressure with drastic consequences to the superconducting properties of these materials.

2.1.6 Primitive Tetragonal $Sc_5Co_4Si_{10}$ Structure

The $LaFe_4P_{12}$, $Pr_3Rh_4Sn_{13}$ and $Sc_2Fe_3Si_5$-type classes discussed above have one or more superconducting members containing a normally magnetic 3d element. Another recently discovered class of this type is typified by tetragonal $Sc_5Co_4Si_{10}$ [2.127,134-137]. As shown in Fig.2.20 (after [2.134]), the Co and Si atoms form planar nets of hexagons and pentagons which are connected along [001] via Co-Si-Co zig-zag chains to form a three-dimensional network; the hexagon-pentagon layers are separated by layers of Sc. As in the $LaFe_4P_{12}$-type and $Pr_3Rh_4Sn_{13}$-type phases, the transition metal (Co) atoms are isolated from each other.

$Sc_5Co_4Si_{10}$-type silicides and germanides of stoichiometry $M_5T_4X_{10}$ form with a range of transition metals and trivalent M atoms. The crystallographic, superconducting and magnetic data for these compounds are shown in Table 2.17 [2.134-136]. All seven compounds not containing a magnetic M element exhibit superconductivity, three of them between 8 and 9 K. Superconductivity is also seen for the prototypic compound $Sc_5Co_4Si_{10}$ with T_c = 5.0-4.8 K; this T_c is high for a Co compound and is also intriguing because the binary compound $CoSi_2$ itself becomes superconducting (T_c = 1.2 K, CaF_2-type [2.138]). Magnetic susceptibility measurements on $Sc_5Co_4Si_{10}$ indicate that the Co atoms do not carry a local magnetic moment [2.135]; here again, this quenching of the moment correlates with conspicuously short (covalent) Co-Si (and Si-Si) interatomic distances [2.134].

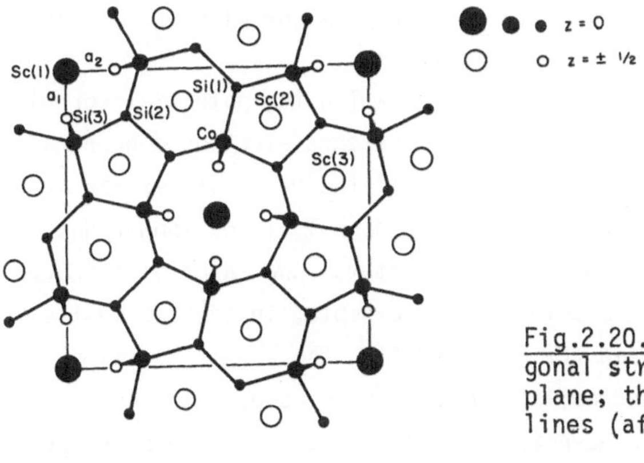

Fig.2.20. Projection of the primitive tetragonal structure of $Sc_5Co_4Si_{10}$ onto the basal plane; the Co-Si net is connected by the heavy lines (after [2.134], modified)

In contrast to the extensive occurrence of superconductivity among the compounds containing nonmagnetic M atoms, all four $RE_5Ir_4Si_{10}$ compounds with magnetic RE atoms (RE = Dy-Tm) order magnetically below 6 K. In these compounds, the effective magnetic moments are close to the free ion values, similar to the $RE_2Fe_3Si_5$-type phases.

2.2 Concluding Remarks

In Sect.2.1 we have indicated where possible the systematic trends associated with superconductivity in each of a number of ternary structure classes. Most classes were shown to exhibit some form of clustering behavior for the M and/or the T atoms in these M-T-X systems, where M is a nontransition metal, T a transition metal and X a post-transition element. First pointed out as an important aspect of unusual superconductors by VANDENBERG and MATTHIAS [2.26], this is one of the most pervasive systematic trends seen in this review of recently discovered ternary superconductors. Ultimately, whether the apparent crystallographic clustering is important physically should be reflected in pronounced anisotropy in the physical properties where the anisotropy must exhibit the symmetry of the cluster network. Indeed, such anisotropy has been documented, for example, for Chevrel phases MMo_6X_8 [2.5] and for the recently discovered pseudo one-dimensional superconductor $Tl_2Mo_6Se_6$ [2.139]; in addition, large anisotropies have been found for directionally solidified samples of $CeCo_4B_4$-type $RERh_4B_4$ compounds via magnetic measurements, although the orientation of the field with respect to the crystallographic (and 2D cluster) axes was not clear [2.29,140].

Physically important clustering behavior has been found to be correlated with the occurrence of significant charge transfer from/to the clusters within the structure. In the case of the Chevrel and related phases, YVON and others have convincingly shown that the charge on the transition metal clusters correlates well with the observed variations in T_c and in other physical properties [2.6,141-146]. In Sect.2.1, we have alluded to work which has suggested that abrupt composition-induced charge transfer between the B_2 dimers and the transition metal sublattice in $M(Rh_{1-x}T_x)_4B_4$ compounds is responsible for the abrupt composition-induced T_c anomalies observed in these systems. Much work remains to be done to further define general relationships between superconductivity and charge transfer to/from the clusters within ternary compounds exhibiting crystallographic clustering.

In contrast to many classes of binary compounds in which the potential for high T_c in a given compound can be predicted from average valence electron concentration rules, this simple parameter is not generally useful in the realm of ternary compounds [2.147,148]. This shows that the potential for high T_c is no longer determined by the electronic properties of the individual atoms. One major difference arises in the crystallography of the structures. For example, in the two most extensively studied binary structure classes (Cr_3Si and NaCl), the atomic positions are fixed with respect to the unit cell edges; the $CeCo_3B_2$-type compounds are the only ones discussed in Sect.2.1 which also have this property and it is therefore perhaps not a coincidence that the T_c's of some MT_3B_2 compounds are comparable to those of the corresponding MT_5 binary compounds [2.87]. In most ternary structure classes exhibiting interesting superconducting properties, the atomic positions are variable with respect to the unit cell axes so that variable and unpredictable (a priori) clustering and electronic and lattice properties can result while retaining the same structure and similar lattice parameters upon change in composition. In addition, the important classes of ternary superconductors usually have noncubic crystal structures, in contrast to the high T_c binary compounds; this gives additional flexibility to the detailed atomic arrangement via the variable ratios of the cell edges as well as the cell angles in some cases.

MATTHIAS [2.149] briefly noted another unique feature of some ternary materials some years ago which is relevant to understanding the origin of their superconducting properties. He noticed that in all of the Chevrel phases $M_xMo_6X_8$, the M and Mo elements are immiscible and do not form compounds by themselves. Formation of ternary compounds from such combinations of elements can therefore be viewed as forcing two desired but immiscible elements M and T to be present in the same compound $M_xT_yX_z$; the third element X therefore acts as a "glue" to hold these elements together. This view would not be tenable if M-T bonds existed in the resulting ternary, but this is rarely found to be the case. In particular, the aversion of RE and Mo for each other in the binary combinations is preserved in the $REMo_6X_8$ compounds since the RE atoms are bonded only to the X atoms. This resulting spatial isolation is one factor which is responsible for the unusual superconducting/magnetic effects observed in these materials. Numerous other examples of the above crystallographic-chemical correlation could be cited. It is of interest in this regard that the compounds discussed in Sect.2.1 do not fit into this classification.

We are left with the intriguing question: "Where does one look next to find new/unusual ternary superconductors?" Although it is not possible to answer this question with precision, relevant comments have been made above and we discuss this question a little more in the following final paragraphs.

In the ternary $M_xT_yX_z$ compounds discussed in these volumes, M is usually a pretransition or rare-earth metal, T a transition metal and X a post transition nonmetal. The nature of these elements gives rise to large electronegativity differences between the two metals and/or between the metals and the nonmetal; these differences therefore appear to be crucial to the occurrence of the structures and properties observed. This also immediately implies the occurrence of charge transfer within the structure, discussed above. These electronegativity differences are favorable to the formation of stoichiometric compounds, as is graphically illustrated in [2.25], in which the potential number of M-T-B ternary boride compounds found for a given combination of M and T is shown to be a rapidly increasing function of the electronegativity difference between M and T. The likelihood of finding new ternary superconductors is therefore also related to these electronegativity differences.

A straight-forward way of finding new ternary compounds is to "design" them from the known structures and chemical compositions of binary compounds. One of the best known examples of this procedure is the intercalation of metal atoms into a binary tunnel or layer structure to form a ternary insertion compound; in this instance, previously empty crystallographic sites within a binary compound are filled with a third element. Formation of the ternary compound $Mo_6S_6Br_2$ [2.150,151] occurs by replacing the S(2) atoms in Mo_6S_8 [$Mo_6S(1)_6S(2)_2$] by Br atoms, and in this case a spectacular increase in T_c results. Here, a subset [S(2)] of atoms (S) occupying more than one set of lattice sites in a binary compound is replaced completely by a third element. This method constitutes a direct approach for synthesizing new ternary compounds which only requires study of the structures and compositions of binary compounds. It can be generalized to higher-component systems; for example, true quaternary compounds $M(1)M(2)_3T_4X_{12}$ with the $Sn(1)Pr_3Rh_4Sn(2)_{12}$ structure could probably be obtained by replacing the atoms in the Sn(1) positions of $Pr_3Rh_4Sn_{13}$-type ternary compounds by a fourth element.

2A. Appendix: Tables of Crystallographic, Superconducting and Magnetic Data for 14 Classes of Ternary Compounds

Table 2.1. Lattice parameters and superconducting or magnetic critical temperatures for $CeCo_4B_4$-type compounds

M	a_o[Å]	c_o[Å]	c/a	T_c[K]	T_m[K]	Reference
			MRh_4B_4			
Y	5.308	7.403	1.395	11.3	-	2.2,9
Nd	5.333	7.468	1.400	5.3	1.31, 0.89	2.2,9,20
Sm	5.312	7.430	1.399	2.7	0.87	2.2,9,21
Gd	5.309	7.417	1.397	-	5.62	2.2,9
Tb	5.303	7.404	1.396	-	7.08	2.2,9
Dy	5.302	7.395	1.395	-	12.03	2.2,9
Ho	5.293	7.379	1.394	-	6.56	2.2,9
Er	5.292	7.374	1.393	8.7	0.9	2.2,9,13,15,22
Tm	5.287	7.359	1.392	9.8	0.4	2.2,9,23
Lu	5.294	7.359	1.390	11.7	-	2.2,9,22
Th	5.356	7.538	1.400	4.3	-	2.2,9
			MIr_4B_4			
Ho	-	-	-	2.0	-	2.24
Er	5.408	7.278	1.346	2.1	-	2.24
Tm	5.404	7.281	1.347	1.6	-	2.24
			MCo_4B_4			
Sc	-	-	-	-	-	2.25
Y	5.028	7.015	1.395	-	-	2.18
Ce	5.059	7.063	1.396	-	-	2.18
Gd	5.043	7.049	1.398	-	-	2.18
Tb	5.038	7.030	1.395	-	-	2.18
Dy	5.026	7.014	1.396	-	-	2.18
Ho	5.020	7.003	1.395	-	-	2.18
Er	5.016	6.989	1.393	-	-	2.18
Tm	5.009	6.980	1.393	-	-	2.18
Yb	-	-	-	-	-	2.25
Lu	4.998	6.947	1.390	-	-	2.18

Table 2.2. Structure types occurring in MT_4B_4 systems[a]

M T:	Re	Fe	Ru	Os	Co	Rh[b]	Ir
Sc	-	-	B	-	A	-	-
Y	1	-	B	1	A	A,B	D
La	1	-	D	D	D	-	D
Ce	1	1	B	D	A	-	D
Pr	-	-	B	D	D	B	D
Nd	-	1	B	D	D	A,B	D
Sm	-	1	B	D	D	A,B	D
Eu	-	-	B	D[c]	-	B	D[c]
Gd	1[c]	1	B	1	A	A,B	D
Tb	-	-	B	1	A	A,B	D
Dy	1[c]	-	B	1	A	A,B	D
Ho	1[c]	-	B	1	A	A,B,C	A
Er	-	-	B	1	A	A,B,C	A
Tm	-	-	B	1	A	A,B,C	A
Yb	-	-	B	1	A[c]	B,C	0
Lu	-	-	B	-	A	A,B,C	0
Th	-	-	B	D	-	A	D
U	0	-	B	B	-	-	-

[a]0: Compound formation does not occur at the MT4B4 composition
1: Compound formation occurs; structure not reported
A: Primitive tetragonal $CeCo_4B_4$ structure
B: Body-centered-tetragonal $LuRu_4B_4$ structure
C: Orthorhombic $LuRh_4B_4$ structure
D: Primitive tetragonal $NdCo_4B_4$ structure
-: Synthesis or phase content not reported

[b]The "B" structure is stabilized by substituting several percent of the Rh by Ru

[c]See "Additional References"

Table 2.3. Lattice parameters and superconducting or magnetic critical temperatures for $LuRu_4B_4$-type compounds

M	a_o[Å]	c_o[Å]	c/a	T_c[K]	T_m[K]	Reference
MRu_4B_4						
Sc	7.346	14.895	2.028	7.2	-	2.27,56
Y	7.454	14.994	2.012	1.4	-	2.55,59
Ce	7.470	15.085	2.019	-	-	2.55
Pr	7.505	15.066	2.007	-	-	2.55
Nd	7.502	15.053	2.007	-	1.62	2.55
Sm	7.482	15.049	2.011	-	-	2.55
Eu	7.477	15.035	2.011	-	-	2.55
Gd	7.470	15.009	2.009	-	4.55	2.55
Tb	7.456	14.995	2.011	-	4.30	2.55
Dy	7.453	14.983	2.010	-	2.65	2.55
Ho	7.442	14.975	2.012	-	2.58	2.55
Er	7.438	14.972	2.013	-	2.16	2.55
Tm	7.429	14.965	2.014	-	-	2.55
Yb	7.427	14.962	2.015	-	-	2.55
Lu	7.419	14.955	2.016	2.0	-	2.55,56
Th	7.540	15.143	2.008	-	-	2.55
U	7.459	14.986	2.009	-	-	2.57
$M(Rh_{0.85}Ru_{0.15})_4B_4$						
Y	7.484	14.895	1.990	9.4	-	2.55
Pr	7.543	14.995	1.988	2.3	-	2.55
Nd	7.537	14.969	1.986	-	-	2.55
Sm	7.516	14.945	1.988	-	-	2.55
Eu	7.505	14.932	1.990	1.5	-	2.55
Gd	7.502	14.916	1.988	-	-	2.55
Tb	7.490	14.898	1.989	-	-	2.55
Dy	7.479	14.885	1.990	4.0	1.5	2.55,60
Ho	7.476	14.872	1.989	6.3	-	2.55
Er	7.468	14.862	1.990	7.9	-	2.55
Tm	7.458	14.853	1.992	8.3	-	2.55
Yb	7.449	14.851	1.994	-	0.2	2.55,61
Lu	7.445	14.837	1.993	9.0	-	2.55
MOs_4B_4						
U	7.512	15.053	2.004	-	-	2.57

Table 2.4. Lattice parameters and superconducting transition temperatures for orthorhombic $LuRh_4B_4$-type compounds (after [2.58,67,68])

M	a_o[Å]	b_o[Å]	c_o[Å]	b/a	b/c	T_c[K]
Ho	-	-	-	-	-	1.4
Er	7.444	22.30	7.465	2.996	2.987	4.3
Tm	7.432	22.28	7.455	2.998	2.989	5.4
Yb	7.424	22.26	7.458	2.998	2.985	-
Lu	7.410	22.26	7.440	3.004	2.992	6.2

Table 2.5. Lattice parameters and superconducting or magnetic transition temperatures for orthorhombic $LuRuB_2$-type compounds (after [2.27,74,76])

M	a_o[Å] (±0.006)	b_o[Å] (±0.005)	c_o[Å] (±0.007)	T_c[K]	T_m[K]
$MRuB_2$					
Y	5.918	5.297	6.377	7.80-7.68	-
Tb	5.897	5.311	6.367	-	45.6
Dy	5.886	5.300	6.352	-	21.9
Ho	5.875	5.287	6.332	-	15.5
Er	5.868	5.262	6.323	-	5.21
Tm	5.831	5.254	6.299	-	4.07
Lu	5.809	5.229	6.284	9.99-9.86	-
$MOsB_2$					
Sc	5.647	5.178	6.184	1.34-1.22	-
Y	5.905	5.299	6.391	2.22-1.60	-
Tb	5.885	5.309	6.392	-	39.2
Dy	5.869	5.297	6.369	-	25.3
Ho	5.850	5.286	6.355	-	14.2
Er	5.834	5.274	6.341	-	3.80
Tm	5.817	5.258	6.328	-	2.26
Lu	5.809	5.231	6.318	2.66-2.14	-

Table 2.6. Distribution of structures among MT_3B_2 compounds[a]

M T:	Fe	Ru	Os	Co	Rh	Ir	Ni	Pt
Li	-	-	-	-	-	-	E	-
Mg	-	-	-	-	-	-	E	-
Ca	-	-	-	-	G	G	-	F
Sr	-	-	-	-	G	G	-	F
Ba	-	-	-	-	-	-	-	F
Sc	-	0	-	A	-	B	-	-
Y	-	A	C	A	B	B	1	-
La	-	A,C	-	0	A	A	0	-
Ce	-	A	-	A	A	B	0	-
Pr	-	A	-	0	A	A	-	-
Nd	0	A	-	0	A	B	0[b]	-
Sm	0	A	C	A	A	B	-	-
Eu	-	-	-	-	A	-	-	-
Gd	0	A	C	A	A	B	-	-
Tb	-	A	C	A	B	B	-	-
Dy	-	A	C	A	B	B	-	-
Ho	-	A	C	A	B	B	-	-
Er	-	A	C	A	B	B	-	-
Tm	-	A	C	A	B	B	-	-
Yb	-	A	C	A	B	B	-	-
Lu	-	A	A	A	B	B	-	-
Zr	-	-	-	D	0	1	-	-
Hf	-	-	-	D	0	1	-	-
Th	-	A	C	-	0	A	-	-
U	A	A	A	A	-	A	0	-

[a] 0: Compound formation does not occur at the composition MT_3B_2; 1: Compound formation occurs but the structure is not known; A: Hexagonal $CeCo_3B_2$ structure; B: Monoclinic $ErIr_3B_2$ structure; C: Orthorhombic (?) YOs_3B_2-type; D: Rhombohedral $ZrCo_3B_2$ structure; E: Hexagonal $Mg(Ni_{1-x}Mg_x)_{3-y}B_2$ structure; F: Hexagonal $Ba_{2/3}Pt_3B_2$ structure; G: *Orthorhombic* $Ca(Rh_2\square_1)B_2$ structure; [b] See "Additional References"

Table 2.7. Lattice parameters and superconducting or magnetic critical temperatures of $CeCo_3B_2$-type MT_3B_2 compounds with T = Fe, Ru or Os (after [2.83,85,86])

M	a_o[Å]	a_o[Å]	c_o[Å]	c_o[Å]	T_c[K]	T_m[K]	T_m[K]
MFe_3B_2 [2.83]							
U	5.051		2.997				
MRu_3B_2							
	[2.85]	[2.86]	[2.85]	[2.86]	[2.85]	[2.85]	[2.86]
Y	5.481	5.471	3.028	3.027	*	*	
$La_{0.9}$	5.605		3.006		*	*	
Ce	5.527	5.523	2.991	2.991	*	*	
Pr	5.554	5.532	3.006	3.015	*	6.85	5
Nd	5.544	5.538	3.010	3.010	*	33.2	39
Sm	5.536	5.514	2.997	3.010	*	>45	~80[a]
Gd	5.508	5.493-5.505	3.018	3.024-3.015	*	>45	~100[b]
Tb	5.495	5.485	3.010	3.014	*	>45	83
Dy	5.485	5.474	3.011	3.016	*	40.1	50
Ho	5.474	5.466	3.017	3.017	*	16.1	25
Er	5.467	5.461	3.017	3.016	*	27.3	32
Tm	5.461	5.454	3.012	3.010	*	18.3	27
Yb	5.464	5.454	3.006	3.003	*	*	
Lu	5.454	5.439	3.004	3.016	*	*	
Th	5.526	5.528	3.070	3.070	1.79-1.60	*	
U	5.476	10.950	2.960	5.934	*	*	
$Y_{0.5}Th_{0.5}$	5.499		3.048		1.53-1.40	*	
MOs_3B_2							
Lu	5.457		3.052		4.62-4.45	*	
U	5.523	11.048	2.976	5.948	*	*	
$Lu_{0.5}Th_{0.5}$	5.523		3.062		4.14-3.84	*	

*: No transition observed above 1.2 K
[a]: Estimated from [Ref.2.86, Fig.5]
[b]: In [Ref.2.86, Table I], T_m is listed as 10 K; this is evidently a misprint, according to [Ref.2.86, Fig.3]. We have estimated T_m from this figure.

Table 2.8. Lattice parameters and superconducting or magnetic critical temperatures of $CeCo_3B_2$-type compounds with T = Co, Rh or Ir

M	a_o[Å]	c_o[Å]	a/c	T_c[K]	T_m[K]
MCo_3B_2 [2.79,82,83,95]					
Sc	4.889	2.977	1.642	-	-
Y	5.033	3.038	1.657	-	-
Ce	5.057	3.036	1.666	-	-
Sm	5.101	2.991	1.705	-	40
Gd	5.059	3.019	1.676	-	58
Tb	5.048	3.005	1.680	-	-
Dy	5.028	3.015	1.668	-	-
Ho	5.026	3.029	1.659	-	-
Er	5.003	3.024	1.654	-	20
Tm	4.991	3.019	1.653	-	-
Yb	4.985	3.020	1.651	-	-
Lu	4.959	3.035	1.634	-	-
U	4.956	3.072	1.613	-	-
MRh_3B_2 [2.85,95]					
La	5.480	3.137	1.747	2.82-2.60	*
Ce	5.474	3.085	1.774	*	115
Pr	5.461	3.105	1.759	*	1.68
Nd	5.466	3.109	1.758	*	13.5
Sm	5.437	3.090	1.760	*	>45
Eu	5.601	2.906	1.927	*	41.7
Gd	5.404	3.115	1.735	*	>45
$Y_{0.5}La_{0.5}$	5.434	3.127	1.738	1.88-1.56	*
MIr_3B_2 [2.85,87]					
La	5.543	3.116	1.779	1.65-1.38	*
Th	5.449	3.230	1.687	2.09-1.90	*
U	5.369	3.181	1.688	*	*
Pr	-	-	-	-	-

*: No transition observed above 1.2 K

Table 2.9. Magnetic ordering temperatures and lattice parameters for compounds with $ErIr_3B_2$-type structure (after [2.27,87])

Compound	T_m[K]	a[Å] (±0.006)	b[Å] (±0.009)	c[Å] (±0.004)	β[°] (±0.1)
YRh_3B_2	*	5.377	9.325	3.102	90.9
$TbRh_3B_2$	45	5.390	9.335	3.100	91.0
$DyRh_3B_2$	38.0	5.379	9.331	3.093	90.9
$HoRh_3B_2$	24.2	5.366	9.307	3.100	91.0
$ErRh_3B_2$	20.4	5.362	9.288	3.099	90.9
$TmRh_3B_2$	11.8	5.357	9.285	3.091	90.8
$YbRh_3B_2$	*	5.355	9.280	3.090	90.9
$LuRh_3B_2$	*	5.353	9.276	3.089	90.9
$ScIr_3B_2$	*	5.344	9.307	3.062	92.1
YIr_3B_2	*	5.428	9.406	3.107	92.8
$CeIr_3B_2$	*	5.502	9.526	3.090	90.8
$NdIr_3B_2$	4.72	5.513	9.540	3.084	90.8
$SmIr_3B_2$	12.6	5.492	9.506	3.076	90.8
$GdIr_3B_2$	29.8	5.466	9.473	3.092	91.0
$TbIr_3B_2$	6.1	5.459	9.464	3.084	91.1
$DyIr_3B_2$	17.7	5.437	9.393	3.106	91.1
$HoIr_3B_2$	12.9	5.420	9.393	3.107	91.2
$ErIr_3B_2$	11.9	5.409	9.379	3.101	91.3
$TmIr_3B_2$	5.69	5.404	9.371	3.097	91.3
$YbIr_3B_2$	*	5.401	9.353	3.099	91.3
$LuIr_3B_2$	*	5.394	9.354	3.080	91.4

*No transition observed above 1.2 K.

Table 2.10. Superconducting and magnetic transition temperatures (T_c and T_m) of hexagonal $LaRu_3Si_2$-type compounds (after [2.91,92])

Compound	a[Å]	c[Å]	c/a	T_c[K]	T_m[K]
YRu_3Si_2 [b,c]	5.543(1) [a]	7.152(2)	1.290	3.51-3.48	
$LaRu_3Si_2$ [b]	5.676(1)	7.120(1)	1.254	7.60-7.02	
$ThRu_3Si_2$ [b]	5.608(2)	7.201(3)	1.284	3.98-3.91	
$CeRu_3Si_2$ [b]	*	*		*	*
$PrRu_3Si_2$ [b]	*	*		*	*
$NdRu_3Si_2$ [b]	*	*			17.95
$SmRu_3Si_2$	*	*	polyphase		>25
$EuRu_3Si_2$	*	*	polyphase		>25
$GdRu_3Si_2$	*	*	polyphase		>25
$DyRu_3Si_2$ [b]	*	*			14.95
$HoRu_3Si_2$ [b]	*	*			14.17
$ErRu_3Si_2$ [b]	*	*	polyphase		16.19
$TmRu_3Si_2$ [b]	*	*			7.50
$YbRu_3Si_2$ [b]	*	*		*	*
URu_3Si_2 [b]	*	*		*	*

*No data given.

[a] Error in parentheses in units of least significant digit
[b] Prepared at $MRu_{3.5}Si_2$ composition
[c] Metastable [M. Ishikawa, private communication].

Table 2.11. Lattice parameters and superconducting transition temperatures for hexagonal ternary platinum borides (after [2.84])

Compound	a[Å] (±0.005)	c[Å] (±0.010)	c/a	T_c[K]
$Ba_{0.67}Pt_3B_2$	6.161	5.268	0.855	5.60-5.55
$Sr_{0.67}Pt_3B_2$	6.092	5.184	0.851	2.78-2.69
$Ca_{0.67}Pt_3B_2$	5.989	5.127	0.856	1.57-1.43

Table 2.12. Superconducting transition temperatures and lattice parameters for phosphides with ordered Fe_2P-type structure (after [2.98])

Compound	T_c[K]	a[Å] (±0.006)	c[Å] (±0.004)	c/a	Heat treatment*
TiRuP	1.33 onset	6.325	3.542	0.560	am
ZrRuP	12.34-10.56	6.459	3.778	0.585	am,900°C,a,q
HfRuP	12.70-11.08	6.414	3.753	0.585	s 1000°C,sc
TiOsP	**	6.285	3.625	0.577	s 1000°C,sc
ZrOsP	7.44- 5.70	6.460	3.842	0.595	s 1000°C,sc
HfOsP	6.10- 4.96	6.417	3.792	0.591	s 1000°C,sc

*am, arc-melted; s, sintered; sc, slow-cooled; q, quenched; a, annealed. All heat treatments specified are the last step only. **No transition observed above 1.2 K

Table 2.13. Occurrence of ternary phases among compounds with approximate composition $MT_{1.1}X_4$, where T is a transition metal and X is Ge or Sn. See text for references

	X:	Sn	Ge	Sn	Ge	Sn	Sn	Sn	Sn
M	T:	Ru		Os		Co	Rh	Ir	Pt
Ca		-	-	-	-	I	I	I	I*
Sr		-	-	-	-	-	I	I	-
Sc		-	-	-	-	-	II	II	-
Y		III*	I	III	I	-	II	III	-
La		I	0	-	0	-	I	I	-
Ce		-	I	-	I	-	I	I	-
Pr		-	I	-	I	-	I	I	-
Nd		-	I	-	I	-	I	I	-
Sm		-	I	-	I	-	I	I	-
Eu		-	0	-	I	-	I	-	-
Gd		-	I	-	I	-	I	III	-
Tb		-	I	-	I	-	III	III	-
Dy		-	I	-	I	-	III	III	-
Ho		-	I	-	I	-	II,III	III	-
Er		II,III*	I	III	I	II	II,III	II	-
Tm		-	I	-	I	-	II,III	II	-
Yb		-	I	-	I	I	I,I*	I,II,III	-
Lu		II	I	-	I	-	II	II	-
Th		-	-	-	I	-	I	-	-

-: It is not known whether this phase forms a ternary structure
0: A phase does not form at the composition $MT_{\sim1.1}X_{\sim4}$
I: Primitive cubic $Pr_3Rh_4Sn_{13}$ structure
I*: Phase I, but with Bragg reflection splitting at high Bragg angles
III: A face-centered-cubic structure
III*: A slightly distorted face-centered-cubic structure
II: A superstructure of phase III

Table 2.14. Lattice parameters and superconducting or magnetic critical temperatures for stannides and germanides with the primitive cubic $Pr_3Rh_4Sn_{13}$-type structure

M	T	X	a_o[Å]	T_c[K]	T_m[K]
Stannides (after [2.105,108-110,112,113])					
Ca	$Rh_{1.2}$	$Sn_{4.5}$	9.702	8.7-8.6	-
Sr	Rh_x	Sn_y	9.800	4.3-4.0	-
La	Rh_x	Sn_y	9.745	3.2-3.0	-
Ce	$Rh_{1.2}$	$Sn_{4.0}$	9.710	*	*
Pr	Rh_x	Sn_y	9.693	*	*
Nd	$Rh_{1.2}$	$Sn_{4.1}$	9.676	*	*
Sm	$Rh_{1.2}$	$Sn_{4.3}$	9.657	*	*
Eu	Rh_x	Sn_y	9.749	-	~11
Gd	Rh_x	Sn_y	9.638[a]	-	11.2
Yb	$Rh_{1.4}$	$Sn_{4.6}$	9.675 9.676,9.692	8.6-8.2	-
Th	Rh_x	Sn_y	9.692	1.9-1.7	-
Ca	Ir_x	Sn_y	9.718	7.1-6.8	-
Sr	Ir_x	Sn_y	9.807	5.1-5.0	-
La	Ir_x	Sn_y	9.755	2.6-2.5	-
Ce	Ir_x	Sn_y	9.720	-	-
Pr	Ir_x	Sn_y	9.704	-	-
Nd	Ir_x	Sn_y	9.691	-	-
Sm	Ir_x	Sn_y	9.668	-	-
Yb	Ir_x	Sn_y	9.709	-	-
Ca	Co_x	Sn_y	9.584	5.9	-
Yb	Co_x	Sn_y	9.563	-	-
La	$Ru_{1.5}$	$Sn_{4.5}$	9.772	3.9-3.2	-
Ca	$Pt_{1.4}$	$Sn_{4.3}$	9.76,9.78	*	-

*No transitions observed above 1.1 K.
[a]Electron and X ray diffraction revealed a tetragonal superstructure [2.118].

(Table 2.14, continuation see next page)

Table 2.14 (cont.)

M	a_o[Å] (±0.002)	T_c[K] (±0.1)	T_m[K] (±0.1)
$M_3Ru_4Ge_{13}$ (after [2.111,114])			
Y	8.962	1.7-1.4	-
Ce	9.045	-	6.7
Pr	9.069	-	14.2
Nd	9.056	-	-
Sm	9.018	-	-
Gd	8.990	-	-
Tb	8.966	-	-
Dy	8.951	-	-
Ho	8.948	-	-
Er	8.941	-	1.2
Tm	8.930	-	-
Yb	8.921	-	-
Lu	8.912	2.3-2.2	-
$M_3Os_4Ge_{13}$ (after [2.111,114])			
Y	8.985	3.9-3.7	-
Ce	9.074	-	6.1
Pr	9.097	-	16.0
Nd	9.081	-	1.9
Sm	9.045	-	-
Eu	9.055	-	10.1
Gd	9.024	-	-
Tb	9.002	-	14.1
Dy	8.984	-	2.1
Ho	8.973	-	1.0
Er	8.969	-	1.9
Tm	8.952	-	-
Yb	8.938	-	-
Lu	8.938	3.6-3.1	-
Th	9.095	-	-

Table 2.15. Lattice parameters and superconducting and magnetic transition temperatures for tetragonal phase II and face-centered-cubic phase III M-T-Sn ternary compounds

M	T	X	a_o[Å]	c_o[Å][a]	T_c[K]	T_m[K]
Phase II Compounds (after [2.108,112,113,117,118])						
Er	Ru_x	Sn_y	13.730	9.671	*	-
Lu	Ru_x	Sn_y	13.692	9.639	*	-
Er	Co_x	Sn_y	13.529	9.522	-	-
Sc	Rh_x	Sn_y	13.565	9.558	4.5-4.1	-
Y	Rh_x	Sn_y	13.772	9.707	3.2-3.1	-
Ho	$Rh_{1.2}$	$Sn_{3.9}$	13.747	9.683	-	-
Er	$Rh_{1.1}$	$Sn_{3.6}$	13.733	27.418	1.36	0.46
Tm	$Rh_{1.3}$	$Sn_{4.0}$	13.706	9.659	-	-
Lu	$Rh_{1.2}$	$Sn_{4.0}$	13.693	9.648	4.0-3.9	-
Sc	Ir_x	Sn_y	13.574	9.566	-	-
Er	Ir_x	Sn_y	13.756	9.688	-	-
Tm	Ir_x	Sn_y	13.727	9.677	-	-
Yb	Ir_x	Sn_y	13.757	9.703	-	-
Lu	Ir_x	Sn_y	13.708	9.661	3.2-2.8	-
Phase III Compounds (after [2.105,108,112,113,118,120])						
Y	$Ru_{1.1}$	$Sn_{3.1}$	13.772		1.3 (onset)	-
Er	Ru_x	Sn_y	13.734		*	-
Y	Os_x	Sn_y	13.801		2.5-2.3	-
Er	Os_x	Sn_y	13.760		1.1	0.4
Tb	$Rh_{1.1}$	$Sn_{3.6}$	13.774		-	4.0
Dy	$Rh_{1.1}$	$Sn_{3.6}$	13.750		-	2.08
Ho	$Rh_{1.2}$	$Sn_{3.9}$	13.750		-	-
Er	$Rh_{1.1}$	$Sn_{3.6}$	13.714		-	-
Tm	$Rh_{1.3}$	$Sn_{4.0}$	13.701		-	-
Y	Ir_x	Sn_y	13.773		2.2-2.1	-
Gd	Ir_x	Sn_y	13.811		-	-
Tb	Ir_x	Sn_y	13.781		-	-
Dy	Ir_x	Sn_y	13.766		-	-
Ho	Ir_x	Sn_y	13.750		-	-
Yb	Ir_x	Sn_y	13.751		-	-

*No transitions observed above 1.1 K

[a]The 9.7 Å c_o values should be multiplied by $2\sqrt{2}$ to obtain the true tetragonal c-axis values [2.118].

Table 2.16. Superconducting and magnetic transition temperatures (T_c and T_m) of some compounds with the tetragonal $Sc_2Fe_3Si_5$-type structure (after [2.126,127,129,130])

Compound	a[Å]	c[Å]	c/a	T_c[K]	T_m[K]
$Sc_2Fe_3Si_5$	10.225(5)[a]	5.275(5)	0.516	4.52-4.25	
$Y_2Fe_3Si_5$	10.43(1)	5.47(1)	0.524	2.4 -2.0	
$Sm_2Fe_3Si_5$	10.47(1)	5.55(1)	0.530	*	*
$Gd_2Fe_3Si_5$	10.47(1)	5.513(8)	0.527	*	8.6
$Tb_2Fe_3Si_5$	10.43(1)	5.48(1)	0.525	*	10.3
$Dy_2Fe_3Si_5$	10.423(8)	5.465(8)	0.524	*	4.4
$Ho_2Fe_3Si_5$	10.39(1)	5.44(1)	0.524	*	2.9
$Er_2Fe_3Si_5$	10.385(9)	5.425(8)	0.522	*	2.9
$Tm_2Fe_3Si_5$	10.37(1)	5.40(9)	0.521	~1.3	~1.1
$Yb_2Fe_3Si_5$	10.36(1)	5.385(8)	0.520	*	5.0[b]
$Lu_2Fe_3Si_5$	10.34(1)	5.375(8)	0.520	6.1-5.8	
$Lu_2Ru_3Si_5$	10.62(1)	5.541(8)	0.522	*	*
$Y_2Ru_3Si_5$	10.68(1)	5.623(8)	0.526	*	*
$Lu_2Os_3Si_5$	10.63(1)	5.588(8)	0.526	*	*
$Y_2Os_3Si_5$	10.68(1)	5.658(8)	0.530	*	*
$Y_2Re_3Si_5$	10.88	5.533	0.509	1.76-1.64	

*No transition above 1.0 K

[a]Estimated errors in parentheses in units of least significant digit.
[b]May be due to impurity phase.

Table 2.17. Compounds with the tetragonal $Sc_5Co_4Si_{10}$-type structure (after [2.134-136])

Compound	a[Å]	c[Å]	a/c	T_c[K]	T_m[K]
$Sc_5Co_4Si_{10}$	12.01(1)	3.936(5)	3.051	5.0 -4.8	
$Sc_5Rh_4Si_{10}$	12.325(6)	4.032(3)	3.057	8.54-8.45	
$Sc_5Ir_4Si_{10}$	12.316(5)	4.076(3)	3.022	8.46-8.38	
$Y_5Ir_4Si_{10}$	12.599(8)	4.234(5)	2.976	3.0 -2.3	
$Lu_5Ir_4Si_{10}$	12.475(8)	4.171(4)	2.991	3.76-3.72	
$Tm_5Ir_4Si_{10}$	12.520(8)	4.195(5)	2.985		1.0
$Er_5Ir_4Si_{10}$	12.543(8)	4.203(5)	2.984		2.3
$Ho_5Ir_4Si_{10}$	12.554(8)	4.215(5)	2.978		1.5
$Dy_5Ir_4Si_{10}$	12.569(8)	4.227(5)	2.974		5.0
$Y_5Ir_4Ge_{10}$	12.927(5)	4.308(5)	3.001	2.62-2.58	
$Y_5Os_4Ge_{10}$	13.006(8)	4.297(5)	3.027	8.68-8.41	

References

2.1 B.T. Matthias, M. Marezio, E. Corenzwit, A.S. Cooper, H.E. Barz: Science *175*, 1465 (1972)
2.2 B.T. Matthias, E. Corenzwit, J.M. Vandenberg, H.E. Barz: Proc. Natl. Acad. Sci. USA *74*, 1334-1335 (1977)
2.3 G.K. Shenoy, B.D. Dunlap, F.Y. Fradin (eds.): *Ternary Superconductors* (Elsevier North Holland, New York 1981)
2.4 R. Chevrel, M. Sergent, J. Prigent: J. Solid State Chem. *3*, 515-519 (1971)
2.5 For a review see Ø. Fischer: Appl. Phys. *16*, 1-28 (1978)
2.6 K. Yvon: "Bonding and Relationships Between Structure and Physical Properties in Chevrel-Phase Compounds $M_xMo_6X_8$ (M: Metal, X: S, Se, Te)" in *Current Topics in Materials Science*, Vol.3, ed. by E. Kaldis (North Holland, Amsterdam 1979) Chap.2, pp.53-129
2.7 Ø. Fischer, A. Treyvaud, R. Chevrel, M. Sergent: Solid State Commun. *17*, 721-724 (1975)
2.8 R.N. Shelton, R.W. McCallum, H. Adrian: Phys. Lett. *56*A, 213-214 (1976)
2.9 J.M. Vandenberg, B.T. Matthias: Proc. Natl. Acad. Sci. USA *74*, 1336-1337 (1977)
2.10 R. Odermatt, Ø. Fischer, H. Jones, G. Bongi: J. Phys. C*7*, L13-L15 (1974)
2.11 S. Foner, E.J. McNiff, Jr., E.J. Alexander: Phys. Lett. *49*A, 269-270 (1974)
2.12 K. Okuda, M. Kitawaga, T. Sakakibara, M. Date: J. Phys. Soc. Jpn. *48*, 2157-2158 (1980)
2.13 W.A. Fertig, D.C. Johnston, L.E. DeLong, R.W. McCallum, M.B. Maple, B.T. Matthias: Phys. Rev. Lett. *38*, 987-990 (1977)
2.14 M. Ishikawa, Ø. Fischer: Solid State Commun. *23*, 37-39 (1977)
2.15 D.E. Moncton, D.B. McWhan, J. Eckert, G. Shirane, W. Thomlinson: Phys. Rev. Lett. *39*, 1164-1166 (1977)
2.16 J.W. Lynn, D.E. Moncton, W. Thomlinson, G. Shirane, R.N. Shelton: Solid State Commun. *26*, 493-496 (1978)
2.17 K. Yvon, A. Grüttner:"The Influence of the Formal Electric Charge on the Size of the Transition Metal Atom Cluster in YRh_4B_4, YRu_4B_4 and $PbMo_6S_8$ Related Compounds", in *Superconductivity in d- and f-Band Metals*, ed. by H. Suhl, M.B. Maple (Academic Press, New York 1980) pp.515-519
2.18 Yu.B. Kuz'ma, N.S. Bilonizhko: Soviet Phys.-Cryst. *16*, 897-898 (1972)
2.19 A. Grüttner, K. Yvon: Acta Cryst. B*35*, 451-453 (1979)
2.20 H.C. Hamaker, L.D. Woolf, H.B. MacKay, Z. Fisk, M.B. Maple: Solid State Commun. *31*, 139-144 (1979)
2.21 H.C. Hamaker, L.D. Woolf, H.B. MacKay, Z. Fisk, M.B. Maple: Solid State Commun. *32*, 289-294 (1979)
2.22 L.D. Woolf, D.C. Johnston, H.B. MacKay, R.W. McCallum, M.B. Maple: J. Low Temp. Phys. *35*, 651-669 (1979)
2.23 H.C. Hamaker, H.B. MacKay, L.D. Woolf, M.B. Maple, W. Odoni, H.R. Ott: Phys. Lett. *81*A, 91-94 (1981)
2.24 H.C. Ku, B.T. Matthias, H. Barz: Solid State Commun. *32*, 937-944 (1979)
2.25 Yu.B. Kuz'ma, N.S. Bilonizhko, S.I. Mykhalenko, G.F. Stepanchikova, N.F. Chaban: J. Less-Common Metals *67*, 51-57 (1979) and references therein
2.26 J.M. Vandenberg, B.T. Matthias: Science *198*, 194-195 (1977)
2.27 H.C. Ku: "Superconductivity, Magnetism and Structural Study of Some New Ternary Transition Metal Borides"; Ph.D. Dissertation, University of California, San Diego, CA (1980)
2.28 H.C. Ku, F. Acker, B.T. Matthias: Phys. Lett. *76*A, 399-402 (1980)
2.29 H.C. Ku, F. Acker: Solid State Commun. *35*, 937-943 (1980)
2.30 H.C. Ku, H. Barz: "Superconductivity in Pseudoternary Compounds $RE(Ir_xRh_{1-x})_4B_4$" in [Ref.2.3, pp.209-212]
2.31 J.M. Rowell, R.C. Dynes, P.H. Schmidt: Solid State Commun. *30*, 191-194 (1979)
2.32 J.M. Rowell, R.C. Dynes, P.H. Schmidt: "Ion Damage, Critical Current and Tunneling Studies of $ErRh_4B_4$ Films", in [Ref.2.17, pp.409-418]
2.33 R.C. Dynes, J.M. Rowell, P.H. Schmidt: "The Effect of α-Particle Damage on the Superconducting and Magnetic Transition Temperatures of Ternary Borides", in [Ref.2.3, pp.169-173]

2.34 D.C. Johnston, W.A. Fertig, M.B. Maple, B.T. Matthias: Solid State Commun. *26*, 141-144 (1978)
2.35 M.B. Maple, H.C. Hamaker, D.C. Johnston, H.B. MacKay, L.D. Woolf: J. Less-Common Metals *62*, 251-263 (1978)
2.36 R.H. Wang, R.J. Laskowski, C.Y. Huang, J.L. Smith, C.W. Chu: J. Appl. Phys. *49*, 1392-1394 (1978)
2.37 S. Kohn, R.H. Wang, J.L. Smith, C.Y. Huang: J. Appl. Phys. *50*, 1862-1864 (1979)
2.38 C.Y. Huang, S.E. Kohn, S. Maekawa, J.L. Smith: Solid State Commun. *32*, 929-931 (1979)
2.39 K. Okuda, Y. Nakakura, K. Kadowaki: Solid State Commun. *32*, 185-188 (1979)
2.40 S. Maekawa, J.L. Smith, C.Y. Huang: Phys. Rev. B*22*, 164-167 (1980)
2.41 C.B. Vining, R.N. Shelton: "Pressure Dependence of the Superconducting Transition Temperature of $(Th_{1-x}Y_x)Rh_4B_4$", in [Ref.2.3, pp.189-192]
2.42 L.D. Woolf, M.B. Maple: "Transition from Antiferromagnetism to Ferromagnetism in the Superconducting Pseudoternary System $(Sm_{1-x}Er_x)Rh_4B_4$", in [Ref.2.3, pp.181-184]
2.43 T. Jarlborg, A.J. Freeman, T.J. Watson-Yang: Phys. Rev. Lett. *39*, 1032-1034 (1977)
2.44 A.J. Freeman, T. Jarlborg, T.J. Watson-Yang: J. Magn. Magn. Matls. *7*, 296-298 (1978)
2.45 A.J. Freeman, T. Jarlborg: J. Appl. Phys. *50*, 1876-1879 (1979)
2.46 H.B. MacKay, L.D. Woolf, M.B. Maple, D.C. Johnston: J. Low Temp. Phys. *41*, 639-651 (1980)
2.47 See M.B. Maple: Appl. Phys. *9*, 179-204 (1976) and references therein
2.48 The data points for the $(Er,Gd)Rh_4B_4$ alloys in Fig.2.5 in the dGF range from 5 to 6 lie below the AG curve whereas in the fit of [2.37], these data lie above the AG curve. The origin of this discrepancy can be traced to the assumption made in [2.37] that T_{co} in these alloys is identical to the T_c (8.7 K) of $ErRh_4B_4$
2.49 K. Kumagai, Y. Inoue, Y. Kohori, K. Asayama: "NMR Study of Magnetic Superconductor $(RE)Rh_4B_4$", in [Ref.2.3, pp.185-188]
2.50 R.N. Shelton, D.C. Johnston: "Pressure Dependencies of the Superconducting and Magnetic Critical Temperatures of Ternary Rhodium Borides", in *High-Pressure and Low-Temperature Physics*, ed. by C.W. Chu, J.A. Woollam (Plenum, New York 1978) pp.409-417
2.51 R.N. Shelton, C.U. Segre, D.C. Johnston: Solid State Commun. *33*, 843-846 (1980)
2.52 L.E. DeLong, H.B. MacKay, M.B. Maple: "Pressure Dependence of the Superconducting Transition Temperature of Dilute $(\underline{Lu}RE)Rh_4B_4$ Alloys", in [Ref.2.3, pp.193-196]
2.53 C.W. Chu, C.Y. Huang, S. Kohn, J.L. Smith: J. Less-Common Metals *62*, 245-250 (1978)
2.54 A.C. Lawson: Private communication (1978); quoted in [Ref.2.50]
2.55 D.C. Johnston: Solid State Commun. *24*, 699-702 (1977)
2.56 H.C. Ku, D.C. Johnston, B.T. Matthias, H. Barz, G. Burri, L. Rinderer: Mat. Res. Bull. *14*, 1591-1599 (1979)
2.57 P. Rogl: Monatsh. Chem. *111*, 517-527 (1980)
2.58 K. Yvon, D.C. Johnston: Acta Cryst. B*38*, 247-250 (1982)
2.59 B.T. Matthias, E. Corenzwit, H. Barz: (Unpublished data); referred to in [2.56]
2.60 H.C. Hamaker, M.B. Maple: "Interaction of Superconductivity and Magnetism in $Dy(Ru_xRh_{1-x})_4B_4$", in [Ref.2.3, pp.201-204]
2.61 M.B. Maple: (Private communication); referred to in [Ref.2.65]
2.62 D.C. Johnston: (Unpublished results); Solid State Commun. *42*, 453-456 (1982)
2.63 H.C. Hamaker, M.B. Maple: Physica *108*B, 757-758 (1981)
2.64 H.E. Horng, R.N. Shelton: "Superconductivity and Long Range Magnetic Order in the Body-Centered Tetragonal System $Er(Rh_{1-x}Ru_x)_4B_4$", in [Ref.2.3, pp.213-216]
2.65 D.C. Johnston: Physica *108*B, 755-756 (1981)
2.66 R.N. Shelton, C.U. Segre, D.C. Johnston: "Effect of Pressure on the Superconducting and Magnetic Critical Temperatures of bct Ternary Ruthenium Borides", in [Ref.2.3, pp.205-208]

2.67 D.C. Johnston, H.B. MacKay: Bull. Am. Phys. Soc. *24*, 390 (1979) and (Unpublished results)
2.68 D.C. Johnston, H.B. MacKay, K. Yvon: To be published
2.69 Yu.B. Kuz'ma, N.S. Bilonizhko: Dopov. Akad. Nauk Ukr. RSR, Ser. A (3), 275-277 (1978)
2.70 P. Rogl: Monatsh. Chem. *110*, 235-243 (1979)
2.71 B. Rupp, P. Rogl, R. Sobczak: Mat. Res. Bull. *14*, 1301-1304 (1979)
2.72 S.I. Mykhalenko, Yu.B. Kuz'ma, A.S. Sobolev: Poroshkovaya Met. *169*, 48-50 (1977)
2.73 N.F. Chaban, Yu.B. Kuz'ma, N.S. Bilonizhko, O.O. Kachmar, N.V. Petriv: Dopov. Akad. Nauk Ukr. RSR, Ser. A (10), 873-876 (1979)
2.74 H.C. Ku, R.N. Shelton: Mat. Res. Bull. *15*, 1441-1444 (1980)
2.75 P.C. Donohue, P.E. Bierstedt: Inorg. Chem. *8*, 2690-2694 (1969)
2.76 R.N. Shelton, B.A. Karcher, D.R. Powell, R.A. Jacobson, H.C. Ku: Mat. Res. Bull. *15*, 1445-1452 (1980)
2.77 H.C. Ku, R.N. Shelton: Bull. Am. Phys. Soc. *26*, 342 (1981); Solid State Commun. *40*, 237-240 (1981)
2.78 W.H. Zachariasen: J. Inorg. Nucl. Chem. *35*, 3487-3497 (1973)
2.79 Yu.B. Kuz'ma, P.I. Kripyakevich, N.S. Bilonizhko: Dopov. Akad. Nauk Ukr. RSR, Ser. A (10), 939-941 (1969)
2.80 W. Haucke: Z. anorg. allg. Chem. *244*, 17-22 (1940)
2.81 K. Niihara, S. Yajima: Bull. Chem. Soc. Jpn. *46*, 770-774 (1973)
2.82 P. Rogl: Monatsh. Chem. *104*, 1623-1631 (1973)
2.83 I.P. Val'ovka, Yu.B. Kuz'ma: Inorg. Mat. *14*, 356-359 (1978)
2.84 R.N. Shelton: J. Less-Common Metals *62*, 191-196 (1978)
2.85 H.C. Ku, G.P. Meisner, F. Acker, D.C. Johnston: Solid State Commun. *35*, 91-96 (1980)
2.86 K. Hiebl, P. Rogl, E. Uhl, M.J. Sienko: Inorg. Chem. *19*, 3316-3320 (1980)
2.87 H.C. Ku, G.P. Meisner: J. Less-Common Metals *78*, 99-107 (1981)
2.88 W. Jung: Z. Naturforsch. *32*b, 1371-1374 (1977)
2.89 B. Schmidt, W. Jung: Z. Naturforsch. *33*b, 1430-1433 (1978)
2.90 W. Jung: Z. Naturforsch. *34*b, 1221-1228 (1979)
2.91 H. Barz: Mat. Res. Bull. *15*, 1489-1491 (1980)
2.92 J.M. Vandenberg, H. Barz: Mat. Res. Bull. *15*, 1493-1498 (1980)
2.93 P. Rogl, H. Nowotny: J. Less-Common Metals *67*, 41-50 (1979)
2.94 Yu.V. Voroshilov, P.I. Kripyakevich, Yu.B. Kuz'ma: Soviet Phys. Cryst. *15*, 813-816 (1971)
2.95 S.K. Dhar, S.K. Malik, R. Vijayaraghavan: J. Phys. C: Solid State Phys. *14*, L321-L324 (1981); H. Oesterreicher, F.T. Parker, M. Misroch: Appl. Phys. *12*, 287-292 (1977)
2.96 H.C. Ku: Private communication
2.97 I. Felner, I. Nowik: Phys. Rev. Lett. *45*, 2128-2131 (1980)
2.98 H. Barz, H.C. Ku, G.P. Meisner, Z. Fisk, B.T. Matthias: Proc. Natl. Acad. Sci. USA *77*, 3132-3134 (1980)
2.99 V. Johnson, W. Jeitschko: J. Solid State Chem. *4*, 123-130 (1972)
2.100 S. Rundqvist, F. Jellinek: Acta Chem. Scand. *13*, 425-432 (1959)
2.101 A.E. Dwight, M.H. Mueller, R.A. Conner, Jr., J.W. Downey, H. Knott: Trans. Met. Soc. AIME *242*, 2075-2080 (1968)
2.102 R. Ferro, R. Marazza, G. Rambaldi: Z. anorg. allg. Chem. *410*, 219-224 (1974)
2.103 J. Ball, D.C. Johnston: Unpublished data
2.104 A.E. Dwight, W.C. Harper, C.W. Kimball: J. Less-Common Metals *30*, 1-8 (1973)
2.105 G.P. Espinosa, A.S. Cooper, H. Barz, J.P. Remeika: Mat. Res. Bull. *15*, 1635-1641 (1980)
2.106 G.P. Meisner, H.C. Ku: "The Effect of Oxygen Impurities on the Superconductivity of ZrRuP", in [Ref.2.3, pp.255-258]
2.107 C.U. Segre, R.N. Shelton, R. Klima, R.A. Jacobson: Bull. Am. Phys. Soc. *24*, 504 (1979)
2.108 J.P. Remeika, G.P. Espinosa, A.S. Cooper, H. Barz, J.M. Rowell, D.B. McWhan, J.M. Vandenberg, D.E. Moncton, Z. Fisk, L.D. Woolf, H.C. Hamaker, M.B. Maple, G. Shirane, W. Thomlinson: Solid State Commun. *34*, 923-926 (1980)
2.109 J.M. Vandenberg: Mat. Res. Bull. *15*, 835-847 (1980)

2.110 J.L. Hodeau, J. Chenavas, M. Marezio, J.P. Remeika: Solid State Commun. *36*, 839-845 (1980)
2.111 C.U. Segre, H.F. Braun, K. Yvon: "Properties of $Y_3Ru_4Ge_{13}$ and Isotopic Compounds", in [Ref.2.3, pp.243-246]
2.112 G.P. Espinosa: Mat. Res. Bull. *15*, 791-798 (1980)
2.113 A.S. Cooper: Mat. Res. Bull. *15*, 799-805 (1980)
2.114 C.U. Segre: Private communication
2.115 G.K. Shenoy, F. Pröbst, J.D. Cashion, P.J. Viccaro, D. Niarchos, B.D. Dunlap, J.P. Remeika: Solid State Commun. *37*, 53-55 (1980)
2.116 J. Waser, E.D. McClanahan, Jr.: J. Chem. Phys. *19*, 413-416 (1951)
2.117 K. Andres, J.P. Remeika, G.P. Espinosa, A.S. Cooper: Phys. Rev. B*23*, 1179-1184 (1981)
2.118 J. Chenavas, J.L. Hodeau, A. Collomb, M. Marezio, J.P. Remeika, J.M. Vandenberg: "The Structural Arrangement of Phases I, II, and III in the Superconducting/ Magnetic Rare-Earth Rhodium Stannides", in [Ref.2.3, pp.219-224]
2.119 H.R. Ott, W. Odoni, Z. Fisk, J.P. Remeika: "Low Temperature Properties of $ErRh_xSn_y$ Compounds", in [Ref.2.3, pp.251-254]
2.120 S.E. Lambert, Z. Fisk, H.C. Hamaker, M.B. Maple, L.D. Woolf, J.P. Remeika, G.P. Espinosa: "Re-entrant Superconductivity in the Rhodium and Osmium Ternary Stannide Compounds", in [Ref.2.3, pp.247-250]
2.121 G.P. Meisner: Physica *108*B, 765-766 (1981)
2.122 W. Jeitschko, D. Braun: Acta Cryst. B*33*, 3401-3406 (1977)
2.123 B.T. Matthias, V.B. Compton, E. Corenzwit: J. Phys. Chem. Solids *19*, 130-133 (1961)
2.124 B.S. Chandrasekhar, J.K. Hulm: J. Phys. Chem. Solids *7*, 259-267 (1958)
2.125 G. Bongi, R. Flükiger, A. Treyvaud, Ø. Fischer, H. Jones, D. Schneider: J. Low Temp. Phys. *17*, 223-246 (1974)
2.126 H.F. Braun: Phys. Lett. *75*A, 386-388 (1980)
2.127 H.F. Braun: "Superconducting Ternary Silicides and Germanides", in [Ref.2.3, pp.225-231]
2.128 O.I. Bodak, B.Ya. Kotur, V.I. Yarovets, E.I. Gladyshevskii: Soviet Phys. Cryst. *22*, 217-219 (1977)
2.129 H.F. Braun, C.U. Segre, F. Acker, M. Rosenberg, S. Dey, P. Deppe: J. Magn. Magn. Mater. *25*, 117-123 (1981)
2.130 H.F. Braun, C.U. Segre: Bull. Am. Phys. Soc. *26*, 343 (1981); C.U. Segre, H.F. Braun: Phys. Lett. *85*A, 372-374 (1981)
2.131 J.D. Cashion, G.K. Shenoy, D. Niarchos, P.J. Viccaro, A.T. Aldred, C.M. Falco: J. Appl. Phys. *52*, 2180-2182 (1981)
2.132 J.D. Cashion, G.K. Shenoy, D. Niarchos, P.J. Viccaro, C.M. Falco: Phys. Lett. *79*A, 454-456 (1980)
2.133 C.U. Segre, H.F. Braun: "Nonlinear Pressure Effects in Superconducting Rare Earth-Iron-Silicides", in *Physics of Solids Under High Pressure*, ed. by J.S. Schilling, R.N. Shelton (North-Holland, Amsterdam 1981) pp.381-384
2.134 H.F. Braun, K. Yvon, R.M. Braun: Acta Cryst. B*36*, 2397-2399 (1980)
2.135 H.F. Braun, C.U. Segre: "Ternary Superconductors of the $Sc_5Co_4Si_{10}$ Type", in [Ref.2.3, pp.239-242]
2.136 H.F. Braun, C.U. Segre: Solid State Commun. *35*, 735-738 (1980)
2.137 H.F. Braun, G. Burri, L. Rinderer: J. Less-Common Metals *68*, P1-P8 (1979)
2.138 B.T. Matthias, J.K. Hulm: Phys. Rev. *89*, 439-441 (1953)
2.139 J.C. Armici, M. Decroux, Ø. Fischer, M. Potel, R. Chevrel, M. Sergent: Solid State Commun. *33*, 607-611 (1980)
2.140 F. Acker, H.C. Ku: J. Low Temp. Phys. *42*, 449-458 (1981)
2.141 K. Yvon, A. Paoli: Solid State Commun. *24*, 41-45 (1977)
2.142 A. Perrin, R. Chevrel, M. Sergent, Ø. Fischer: J. Solid State Chem. *33*, 43-47 (1980)
2.143 T. Jarlborg, A.J. Freeman: Phys. Rev. Lett. *44*, 178-182 (1980)
2.144 W. Hönle, H.G. von Schnering, A. Lipka, K. Yvon: J. Less-Common Metals *71*, 135-145 (1980)
2.145 S. Yashonath, M.S. Hegde, P.R. Sarode, C.N.R. Rao, A.M. Umarji, G.V. Subba Rao: Solid State Commun. *37*, 325-327 (1981)
2.146 S. Ramasamy, T. Nagarajan, A.M. Umarji, G.V. Subba Rao: Phys. Lett. *81*A, 71-74 (1981)

2.147 B.T. Matthias: "Ternary Superconductors", in [Ref.2.3, pp.3-4]
2.148 K. Yvon: "Structure and Superconductivity of Ternary Compounds", in [Ref.2.3, pp.15-20]
2.149 B.T. Matthias: "Superconductivity", in *Critical Materials Problems in Energy Production* (Academic, New York 1976) Chap.21, pp.663-681
2.150 M. Sergent, Ø. Fischer, M. Decroux, C. Perrin, R. Chevrel: J. Solid State Chem. *22*, 87-92 (1977)
2.151 C. Perrin, R. Chevrel, M. Sergent, Ø. Fischer: Mat. Res. Bull. *14*, 1505-1515 (1979)

3. Critical Fields of Ternary Molybdenum Chalcogenides

M. Decroux and Ø. Fischer

With 26 Figures

One of the most striking properties of the ternary molybdenum chalcogenides (TMC) is the very high upper critical field found in several of them [3.1-5].

For example, $PbMo_6S_8$ ($T_c \simeq 15$ K) which has $H_{c2}(0) > 60$ Tesla, $SnMo_6S_8$($T_c \simeq 14$ K) with $H_{c2}(0) \simeq 45$ Tesla and $LaMo_6Se_8$ ($T_c \simeq 11$ K) with $H_{c2}(0) \simeq 45$ Tesla, all have upper critical fields higher than observed in any other material. This immediately raises two questions: (i) why do these materials have such high critical fields?, and (ii) can they be used for the production of ultrahigh magnetic fields? With respect to the first question it is presently well understood that the particular cluster structure [Ref.3.6, Chaps.1 and 2] favours narrow conduction bands [Ref. 3.6, Chap.6] which in turn give small Fermi velocities and thus high critical fields [3.7-10]. Nevertheless, several questions remain unanswered such as the role of spin paramagnetism and that of impurities in the determination of H_{c2}. A large part of this article is, therefore, devoted to these problems.

The crucial question concerning a possible application is whether it will be possible to draw wires of such seemingly complicated materials and whether these wires can be made to have high current carrying capacities. Recently there has been progress in this direction [3.10-13] and, in the last section, we shall give a short summary of the present status of this field. Of course, closely related to this discussion is the critical current studies realized in thin films [3.14] which is dealt with in [Ref.3.6, Chap.5].

3.1 Overview of Experimental Results

The critical field H_{c2} observed in superconductors is determined by the combined effect of the external magnetic field on the orbits and on the spins of the conduction electrons. In the discussion of H_{c2}, it is useful to introduce the orbital critical field H^*_{c2} which is the critical field the material would have if there were no interactions between the external field and the electron spins. Likewise, we introduce the paramagnetic critical field H_p which is the critical field the sample would have had in the absence of orbital interactions. For a pure BCS superconductor, H_p is given by the Chandrasekhar-Clogston limit H_{po} [3.15,16]

$$H_{po} = 1.84\ T_c[\text{Tesla}] \quad . \tag{3.1}$$

MAKI [3.17] has shown that for a BCS superconductor, H_{c2} is given by

$$H_{c2} = \frac{H^*_{c2}H_{po}}{(2H^{*\,2}_{c2} + H^2_{po})^{1/2}} \quad . \tag{3.2}$$

This means that H_{c2} is always lower than the smaller of H^*_{c2} and H_p. For most superconductors $H^*_{c2} \ll H_{po}$ and we have $H_{c2} \simeq H^*_{c2}$, i.e., the spin paramagnetic effects are unimportant. However, for materials where the ratio H_{c2}/T_c is high, the latter effect becomes important.

In Table 3.1, we present critical field data for a selection of ternary molybdenum chalcogenides.

Table 3.1. A selection of published critical field data on some TMC's. The differences between the results of several groups are probably due to differences in sample preparation. For the very high field measurements the precision is less good and in some measurements, dynamical effects may have influenced the data

Compounds		T_c [K]	$(dH_{c2}/dT)^{T_c}$ [T/K]	$H_{c2}(0)$ [T]	$H^*_{c2}(0)$ [T]	H_{po} [T]
Mo_6Se_8	[3.21]	6.3	1.7	8.6	7.5	11.6
Mo_6Se_8	[3.21]	6.2	3.6	16.5	15.5	11.4
$LaMo_6Se_8$	[3.5]	11.3	7.0	-	54.8	20.8
$LaMo_6Se_8$	[3.29]	10.8	7.5	44.5	56.1	19.9
$LaMo_6S_8$	[3.10]	5.8	1.5	5.4	6.0	10.8
$Cu_{1.8}Mo_6S_8$	[3.31]	10.0	2.5	19.2	17.3	18.4
$Cu_{1.8}Mo_6S_8$	[3.31]	10.5	1.9	15.3	13.8	19.3
$SnMo_6S_8$	[3.2]	13.4	3.7	34.4	34.4	24.7
$SnMo_6S_8$	[3.29]	11.8	4.7	34.0	38.4	21.7
$SnMo_6S_8$	[3.27]	11.7	3.7	36.0	30.0	21.5
$PbMo_6S_8$	[3.2]	14.4	6.0	60.0	59.9	26.5
$PbMo_6S_8$	[3.30]	12.6	5.8	50.0	50.5	23.2
$PbMo_6S_8$	[3.28]	13.3	6.0	60.0	55.0	24.5

H^*_{c2} has been estimated here using the weak coupling formula [3.18]

$$H^*_{c2} = 0.693(dH_{c2}/dT|_{T_c})T_c \quad . \tag{3.3}$$

From Table 3.1 we first note, as was already mentioned in the introduction, that several of these materials have very high upper critical fields, some being much higher than observed in any other superconductor to date. A comparison between $PbMo_6S_8$ and several other high field superconductors is shown in Fig.3.1 (note especially the differences in the initial slopes). From Table 3.1 it is also clear

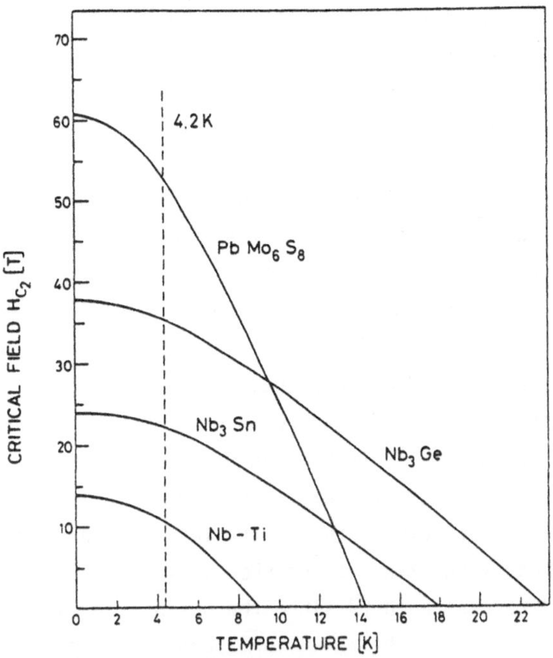

Fig.3.1. Comparison of the critical fields of $PbMo_6S_8$, Nb_3Ge, Nb_3Sn and Nb-Ti [3.10]

that not all TMC's have high critical fields. For instance, the $REMo_6S_8$ compounds for RE being trivalent rare earth all have rather low critical fields and in those materials, spin-paramagnetic effects are only important if the rare-earth ions are magnetic and thus produce large effective fields when they align in the external field [3.19].

For some of the high field TMC's, one finds that H_{po} is considerably lower than the observed H_{c2}. Thus, (3.2) is obviously not applicable in these materials. Qualitatively this can be understood as being a result of impurities producing spin-orbit scattering which increases H_p above H_{po} [3.10,20]. In the case of strong spin-orbit scattering, one finds

$$H_p \simeq 1.3\sqrt{\lambda_{so}}H_{po} \quad , \tag{3.4}$$

where $\lambda_{so} = 2h/3\tau_{so}k_BT_c$ is the spin-orbit scattering parameter. However, the analysis of the temperature dependence of H_{c2} apparently gives unreasonably high λ_{so} values [3.10]. The reason for this is not yet fully understood, but is related to the fact that the orbital critical field H^*_{c2}, as calculated by (3.3), is too low [3.21]. This can be seen from Table 3.1 where for some of the compounds like Mo_6Se_8, H_{c2} is higher than H^*_{c2}.

If one compares H_{c2} values obtained by different authors one finds an important scatter of the data. This indicates that H_{c2} may be very sensitive to details in the preparation procedure. A careful study of Mo_6Se_8 doped with small quantities of impurities has shown that less than 1 at % impurities makes H_{c2} increase by more than a factor of 2 [3.21]. For $PbMo_6S_8$ it has proven even harder to control H_{c2}, but we believe that a similar situation exists.

The study of H_{c2} in compounds containing magnetic ions has revealed many interesting new phenomena. In the antiferromagnetic compounds to be discussed in Chap.5, an anomalous decrease of H_{c2} has been observed as the superconductor enters the antiferromagnetic state [3.22]. A completely different situation is found in the high field superconductor series $Sn_{1-x}Eu_xMo_6S_8$ where an important increase of H_{c2} is observed as the Eu concentration increases [3.23]. This effect, which is accompained by an anomalous temperature dependence of H_{c2}, has been explained as a result of the Jaccarino-Peter compensation effect [3.24].

Finally, we should mention that the high field measurements have been made in pulsed magnetic fields [3.7] and in most cases the resistive transition is observed. Characteristic of these compounds is a rather large transition width, which in polycrystalline materials is typically 10% of H_{c2}. We believe that this is due to the large sensitivity of H_{c2} to impurities or defects and to the large anisotropy found in these materials [3.25,26]. These large transitions show that one has to be careful when comparing results by different authors since the definitions used for H_{c2} may not be equivalent. The same care must be taken in discussing temperature dependences of H_{c2}. In most results reported here, the midpoint of the resistive transition has been used to define H_{c2}. In the nonmagnetic systems, the temperature dependence is independent of the definition used for H_{c2}, but in some magnetic compounds this may not always be true.

This chapter is organized in the following way: in the next section we include the theory of H_{c2} for the reader who is not familiar with this subject and discuss several aspects which are important for the detailed understanding of the experimental situation in the TMC's. Section 3.3 discusses the role of impurities in Mo_6Se_8 and $PbMo_6S_8$. The anisotropy of H_{c2} has been observed in several TMC's and the results are summarized in Sect.3.4. The temperature dependence of H_{c2} and the question of the paramagnetic limitation is presented in Sect.3.5. The following section discusses compounds with paramagnetic ions and in particular, the Jaccarino-Peter effect in $Sn_{1-x}Eu_xMo_6S_8$. Finally, in Sect.3.7 the recent progress in wire-production and studies of the critical current are presented.

3.2 Summary of Theory

3.2.1 Ginzburg-Landau Theory

The Ginzburg-Landau (G-L) theory is extremely useful in many different situations, for example, when a qualitative description is needed or when spatial variations of the superconducting order parameter are considered. In this section, we only want to recall its simplest aspects which are related to the upper critical field H_{c2}. Let us just add here that the Ginzburg-Landau theory is very useful in discussing the problem of coexistence of magnetism and superconductivity, and that a complete description of this aspect is developed in Chap.9.

The simplest form of the free energy density f_s of a superconductor in the presence of a magnetic field is [3.32]

$$f_s = \frac{1}{2}\alpha_s|\Delta(r)|^2 + \frac{1}{4}\beta_s|\Delta(r)|^4 + \frac{1}{2m}|(\frac{\hbar}{i}\nabla - 2eA)\Delta(r)|^2 + \frac{1}{2}\mu_0 H^2 \quad . \tag{3.5}$$

Minimization with respect to $\Delta(r)$ gives the well-known Ginzburg-Landau equation

$$\alpha_s\Delta(r) + \beta_s|\Delta(r)|^2\Delta(r) + \frac{1}{2m}(\frac{\hbar}{i}\nabla - 2eA)\Delta(r) = 0 \quad . \tag{3.6}$$

Here, $\alpha_s = \alpha_s'(T - T_c)$ $(\alpha_s' > 0)$ and $\beta_s \simeq$ constant.

The critical field H_{c2} of a type II superconductor is obtained from the linear part of this equation $[\Delta(r) \to 0]$ [3.32]

$$\frac{1}{2m}(\frac{\hbar}{i}\nabla - 2eA)\Delta(r) = -\alpha_s\Delta(r) \quad . \tag{3.7}$$

This equation is equivalent to the Schrödinger equation for a free electron in a magnetic field. The critical field is obtained as the highest field for which (3.7) has a solution. One easily sees that this corresponds to the lowest eigenfunction of the corresponding harmonic oscillator problem. In this way one finds

$$H_{c2}^* = \frac{m|\alpha_s|}{e\hbar} \quad . \tag{3.8}$$

We write H_{c2}^* here to specify that it is the critical field obtained from orbital effects only. In the following, we will always give H in units of Tesla and reserve the letter B for the magnetic induction in the presence of magnetic ions.

Using

$$\xi^2 = \frac{\hbar^2}{2m|\alpha_s|} \quad , \tag{3.9}$$

one obtains the well-known formula [3.32]

$$H_{c2}^* = \frac{\phi_0}{2\pi\xi^2} \quad , \tag{3.10}$$

where ξ is the Ginzburg-Landau coherence length and ϕ_0 the flux quantum ($\phi_0 = 2.07 \cdot 10^{-15}$ Tm^2). To get an idea of the order of magnitude of ξ, let us give two examples:

Nb : $H_{c2} \simeq 0.26$ T $\quad \xi \simeq 360$ Å

$PbMo_6S_8$: $H_{c2} \simeq 60$ T $\quad \xi \simeq 23$ Å .

In both high field superconductors and magnetic superconductors, it is important to consider the spin polarization effects. The easiest way to do this is to consider a coupling of $\Delta(r)$ to a uniform magnetization M and add a term $\frac{1}{2}\gamma M^2\Delta^2(r)$ to the free energy density f_s (see [3.33] and Chap.9). The linearized Ginzburg-Landau equation then reads

$$\frac{1}{2m}\left(\frac{\hbar}{i}\nabla - 2eA\right)\Delta(r) + \gamma_s M^2\Delta(r) = -\alpha_s\Delta(r) \quad . \tag{3.11}$$

By defining $\tilde{\alpha}_s = \alpha_s + \gamma_s M^2$ and using (3.8), we easily obtain the solution to (3.11):

$$H_{c2} = \frac{m|\tilde{\alpha}_s|}{e\hbar} = \frac{m|\alpha_s|}{e\hbar} - \frac{m\gamma_s}{e\hbar} M^2$$

$$H_{c2} = H^*_{c2} - \frac{m\gamma_s}{e\hbar} M^2 \quad . \tag{3.12}$$

Thus the real critical field H_{c2} is reduced with respect to the orbital critical field H^*_{c2} by an amount proportional to M^2. This fact is, of course, very important in a magnetically-ordered system and is one primary reason for the destruction of superconductivity by ferromagnetic order. This point will be discussed further in Chaps.5,9. Here we shall now concentrate on the paramagnetic case where

$$M = \chi(T)H \quad . \tag{3.13}$$

Equation (3.12) then becomes

$$H_{c2}(T) = H^*_{c2}(T) - a(T)H^2_{c2}(T) \tag{3.14}$$

with $a(T) = m\gamma_s\chi^2(T)/e\hbar$. This is now an equation for $H^*_{c2}(T)$ in terms of $H^*_{c2}(T)$ given by (3.10) and a(T). In this chapter we shall consider (3.14) for two cases: (i) nonmagnetic high field superconductors (like $PbMo_6S_8$) where the term $\frac{1}{2}\gamma_s M^2\Delta^2(r)$ in the free energy density represents the coupling of the external field to the conduction electron spins, and $\chi(T)$ is the Pauli susceptibility; (ii) superconductors with a regular lattice of magnetic ions, like $ErMo_6S_8$, where $\chi(T)$ has a Curie-Weiss like behaviour. In the special case of $Sn_{1-x}Eu_xMo_6S_8$, which is treated in Sect.3.6, both these aspects become important.

An important consequence of (3.14) is that spin effects change the temperature dependence of $H_{c2}(T)$ [even if a(T) = const], and thus determining $H_{c2}(T)$ experimentally is a way of getting information about the influence of these effects on the superconducting state.

As a special case, let us consider the situation where the orbital effects can be neglected compared to the spin effects. The critical field is, in this case, denoted H_p and can be obtained from (3.11,13) by neglecting the first term in (3.11):

$$H_p = \frac{1}{\chi}\left(\frac{|\alpha_s|}{\gamma_s}\right)^{1/2} \sim (T_c - T)^{1/2} \quad . \tag{3.15}$$

Note here that the temperature dependence of H_p close to T_c is $(T_c - T)^{1/2}$, thus $H_p(T)$ has an infinite slope at T_c contrary to $H_{c2}(T)[\sim(T_c - T)]$.

Let us now turn to anisotropic superconductors. The simplest generalization of (3.7) is to replace (1/m) by an effective mass tensor $(\Gamma)_{ij}$ and write [3.34]

$$\frac{1}{2}\sum_{ij}\left(\frac{\hbar}{i}\nabla_i - 2eA_i\right)\Gamma_{ij}\left(\frac{\hbar}{i}\nabla_j - 2eA_j\right)\Delta(r) = -\alpha_s\Delta(r) \quad . \tag{3.16}$$

For a cubic system, the effective mass tensor must be proportional to the unity tensor and (3.16) reduces to (3.7), i.e., the critical field is isotropic. In the ternary molybdenum chalcogenides, we have hexagonal symmetry and Γ takes the form

$$\Gamma = \frac{1}{M}\begin{pmatrix}1 & 0 & 0\\ 0 & 1 & 0\\ 0 & 0 & \varepsilon^2\end{pmatrix} , \quad \varepsilon^2 = \frac{M}{m} \tag{3.17}$$

where m and M are masses parallel and perpendicular to the hexagonal c-axis, respectively.

The solution to (3.16) can in this case be written [3.34]

$$H^*_{c2}(\theta) = \frac{M|\alpha_s|}{e\hbar}\,\frac{1}{\delta(\theta)} \tag{3.18}$$

$$\delta(\theta) = (\cos^2\theta + \varepsilon^2\sin^2\theta)^{1/2} \quad . \tag{3.19}$$

Here θ is the angle between the c-axis and the field. Defining the parallel and perpendicular coherence distance, $\xi_{\parallel}$ and $\xi_{\perp}$ as

$$\xi^2_{\parallel} = \frac{\hbar^2}{2m|\alpha_s|} \qquad \xi^2_{\perp} = \frac{\hbar^2}{2M|\alpha_s|} \quad , \tag{3.20}$$

we get

$$H^*_{c2\parallel} = \frac{\phi_0}{2\pi\xi^2_{\perp}} \qquad H^*_{c2\perp} = \frac{\phi_0}{2\pi\xi_{\perp}\xi_{\parallel}} \tag{3.21}$$

and

$$H^*_{c2}(\theta) = \frac{H^*_{c2\parallel}}{\delta(\theta)} \quad . \tag{3.22}$$

In compounds where the paramagnetic limitation is important we have to replace α_s by $\tilde{\alpha}_s$ as in (3.12). Equation (3.18) for $H_{c2}(\theta)$ then becomes

$$H_{c2}(\theta) = \frac{1}{\delta(\theta)}[H^*_{c2\parallel} - aH^2_{c2}(\theta)] \quad . \tag{3.23}$$

Therefore, we find that the angular dependence of H_{c2} is modified by the paramagnetic limitation.

3.2.2 Microscopic Theory

a) *Isotropic, Weak Coupling Superconductors*

Subsequent to his derivation of the Ginzburg-Landau equations from the microscopic theory, GOR'KOV calculated the orbital critical field H^*_{c2} for a pure BCS superconductor [3.36]. Somewhat later, de GENNES [3.37] and MAKI [3.38] calculated H^*_{c2} in the dirty limit, i.e., when $\ell \ll \xi_0$. ℓ is the mean free path in the normal state and

$$\xi_0 = 0.18 \frac{\hbar v_F}{k_B T_c} \tag{3.24}$$

is the BCS coherence length. HELFAND and WERTHAMER finally worked out a complete theory for H^*_{c2} valid for all values of ℓ [3.39]. Their result is that $H^*_{c2}(T)$ is given by

$$\ln \frac{1}{t} = \sum_{\nu=-\infty}^{\infty} \left[\frac{1}{|2\nu + 1|} - \frac{(t/\hat{h}^{1/2}) J(\alpha_\nu)}{1 - (\lambda_{tr}/h^{1/2}) J(\alpha_\nu)} \right] , \tag{3.25}$$

where

$$J(\alpha_\nu) = 2 \int_0^\infty d\omega \, \exp(-\omega^2) \, \tan^{-1}(\alpha_\nu \omega) \tag{3.26}$$

and

$$\alpha_\nu = \frac{\hat{h}^{1/2}}{|2\nu + 1| t + \lambda_{tr}} \tag{3.27}$$

$$t = T/T_c \tag{3.28}$$

$$\hat{h} = \frac{2eH^*_{c2}}{\hbar} (\hbar v_F / 2\pi k_B T_c)^2 \tag{3.29}$$

$$\lambda_{tr} = \frac{\hbar}{2\pi k_B T_c \tau} \simeq 0.88 \xi_0 / \ell \quad . \tag{3.30}$$

Here τ is the relaxation time.

This result simplifies considerably in the dirty limit ($\lambda_{tr} \gg 1$) where $J(\alpha_\nu)$ can be developed and (3.25) can be put in the form

$$\ln(\tfrac{1}{t}) = \psi(\tfrac{1}{2} + \tfrac{\rho}{2t}) - \psi(\tfrac{1}{2}) \tag{3.31}$$

$$\rho = \frac{\hat{h}}{3\lambda_{tr}} = \frac{e v_F^2 \tau H^*_{c2}}{3\pi k_B T_c} \quad . \tag{3.32}$$

$\psi(x)$ is the digamma function and ρ the pairbreaking parameter. Equation (3.31) appears very often in superconductivity when pairbreaking effects are present. The temperature dependence predicted by this equation is very close to the one found for the pure limit from (3.25). Note that in the dirty limit, λ_{tr} drops out of the temperature dependence.

The expressions (3.29,32) can be used to relate H^*_{c2} and (dH^*_{c2}/dT) to normal state parameters. Using [3.38]

$$-\frac{d\hat{h}}{dt} = \begin{cases} 1.426 \; ; & \lambda_{tr} = 0 \quad (3.33) \\ 1.216\lambda_{tr} ; & \lambda_{tr} \gg 1 \quad (3.34) \end{cases}$$

$$\hat{h}(0) = \begin{cases} 1 \; ; & \lambda_{tr} = 0 \quad (3.35) \\ 0.84\lambda_{tr} \; ; & \lambda_{tr} \gg 1 \; , \quad (3.36) \end{cases}$$

we find

$$H^*_{c2}(0) \simeq \begin{cases} \dfrac{2\pi^2 k_B^2 T_c^2}{e\hbar v_F^2} \; ; & \lambda_{tr} = 0 \quad (3.37) \\ 2.64 \, \dfrac{k_B T_c}{e\tau v_F^2} \; ; & \lambda_{tr} \gg 1 \quad (3.38) \end{cases}$$

$$-\left.\frac{dH^*_{c2}}{dT}\right|_{T_c} \simeq \begin{cases} 28 \, \dfrac{k_B^2 T_c}{e\hbar v_F^2} \; ; & \lambda_{tr} = 0 \quad (3.39) \\ 3.81 \, \dfrac{k_B}{e\tau v_F^2} \; ; & \lambda_{tr} \gg 1 \; . \quad (3.40) \end{cases}$$

Expressed in terms of the electronic specific heat coefficient γ and the normal state resistivity ρ, one finds

$$H^*_{c2}(0) = \begin{cases} 0.97 \cdot 10^{35} \dfrac{\gamma^2}{S^2} T_c^2 \; ; & \lambda_{tr} = 0 \quad (3.41) \\ 3.04 \cdot 10^3 \, \rho_o \gamma T_c \; ; & \lambda_{tr} \gg 1 \quad (3.42) \end{cases}$$

$$-\left.\frac{dH^*_{c2}}{dT}\right|_{T_c} = \begin{cases} 1.38 \cdot 10^{35} \dfrac{\gamma^2}{S^2} T_c \; ; & \lambda_{tr} = 0 \quad (3.43) \\ 4.40 \cdot 10^3 \, \rho_o \gamma \; ; & \lambda_{tr} \gg 1 \; , \quad (3.44) \end{cases}$$

where S is the Fermi surface area. In the intermediate case $\lambda_{tr} \simeq 1$, H^*_{c2} is essentially given by the sum of the two expressions. One finds [3.40]

$$H^*_{c2}(0) = \frac{R(\infty)}{R(\lambda_{tr})} \left[0.83 \cdot 10^{35} \left(\frac{\gamma T_c}{S}\right)^2 + 3.04 \cdot 10^3 \rho_o \gamma T_c \right] \; . \quad (3.45)$$

All quantities are given in MKSA units (ρ_o in [Ωm] and γ in [$\mathrm{Jm^{-3}\,K^{-2}}$]).

Note that (3.43,44) are also valid if spin paramagnetism is present. So far we have only discussed orbital effects, coming about through the kinetic energy

$(\underline{p} - e\underline{A})^2/2m$. The coupling of the external field to the electron spins through the Zeeman energy, $-g\mu_B(\underline{H}\underline{\sigma})$, also leads to pairbreaking effects and thus to a critical field H_{po}. At T = 0, one finds in a clean BCS superconductor that H_{po} is given by (3.1). The temperature dependence of H_{po} was discussed by SARMA [3.41] who showed that when one approaches T_c the transition becomes of second order. In a real superconductor this result will be changed by spin-orbit scattering effects. The latter makes the spin susceptibility in the superconducting state nonzero [3.42] and enhances the paramagnetic limiting field H_p above H_{po}.

WERTHAMER, HELFAND and HOHENBERG (WHH) [3.20] worked out the theory for H_{c2} including both orbital and spin paramagnetic effects as well as nonmagnetic and spin-orbit scattering. The result of their calculation can be expressed as follows in the limit $\lambda_{tr} \gg 1$, $\lambda_{so} < \lambda_{tr}$:

$$\ln\left(\frac{1}{t}\right) = \left(\frac{1}{2} + \frac{i\lambda_{so}}{4\gamma}\right)\psi\left(\frac{1}{2} + \frac{h + \frac{1}{2}\lambda_{so} + i\gamma}{2t}\right) + \left(\frac{1}{2} - \frac{i\lambda_{so}}{4\gamma}\right)\psi\left(\frac{1}{2} + \frac{h + \frac{1}{2}\lambda_{so} - i\gamma}{2t}\right) \tag{3.46}$$

$$\gamma = \left[(\alpha h)^2 - \frac{1}{4}\lambda_{so}\right] \tag{3.47}$$

$$\lambda_{so} = \frac{2\hbar}{3\pi k_B T_c \tau_{so}} \simeq 1.17\xi_0/\ell_{so} \tag{3.48}$$

$$\alpha = \frac{2\hbar}{2mv_F^2\tau} \tag{3.49}$$

$$h = \frac{2eHv_F^2\tau}{6\pi k_B T_c} = \frac{\hat{h}}{3\lambda_{tr}} \quad . \tag{3.50}$$

Here τ_{so} is the relaxation time for spin-orbit scattering, α is the Maki parameter describing the relative importance of spin paramagnetic and orbital effects in the absence of spin-orbit scattering, and m is the free electron mass (coming from the Bohr magneton). In terms of the dirty limit orbital critical field $H^*_{c2}(T = 0)$, and the Chandrasekhar-Clogston limit H_{po}, one finds

$$h \simeq 0.281 \frac{H_{c2}(T)}{H^*_{c2}(T=0)} \tag{3.51}$$

$$\alpha = \sqrt{2}\,\frac{H^*_{c2}(T=0)}{H_{po}} \quad . \tag{3.52}$$

The Maki parameter can also be obtained from

$$\alpha = -0.528 \left.\frac{dH_{c2}}{dT}\right|_{T_c} \tag{3.53}$$

or, using (3.44),

$$\alpha = 2.32 \cdot 10^3 \rho_0 \gamma \quad . \tag{3.54}$$

The influence of the paramagnetic pair breaking on H_{c2} as compared with the orbital effects is seen in the temperature dependence. In Fig.3.2, $\tilde{h} = \pi^2 h/4$ is shown (h normalized to unity slope at T_c) for $\lambda_{so} = 1$ and for different values of α. For $\alpha = 0$, we have the orbital critical field, and when α increases, the temperature-dependence changes as the influence of the paramagnetic pair breaking increases. The role of λ_{so} is to reduce the effect of the paramagnetic pair breaking. In Fig. 3.3, we show $\tilde{h}$ for $\alpha = 3$ and for different values of λ_{so}. As $\lambda_{so} \to \infty$, we recover the orbital critical field. Since α can be determined by $(dH_{c2}/dT)_{T_c}$, the temperature dependence of H_{c2} can, in principle, be used to determine λ_{so}. However, the values thus obtained are usually very high, i.e., most superconductors have a critical field which seems to be too close to the orbital limit [3.9]. The reasons for the high values of λ_{so} are not always clear. We shall come back to this point later.

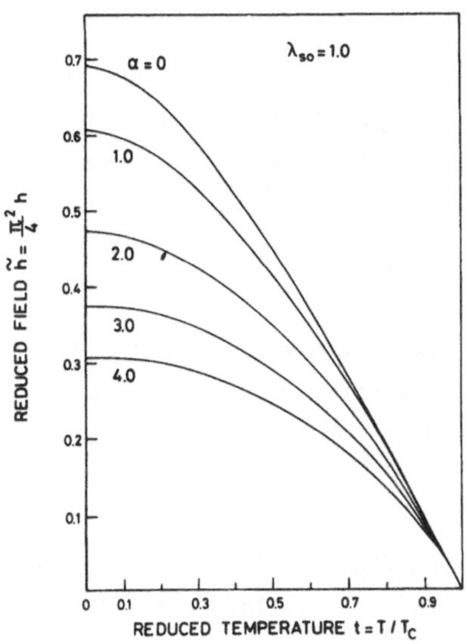

Fig.3.2. Reduced critical field $h = \pi^2 h/4$ for $\lambda_{so} = 1$ and different values of α

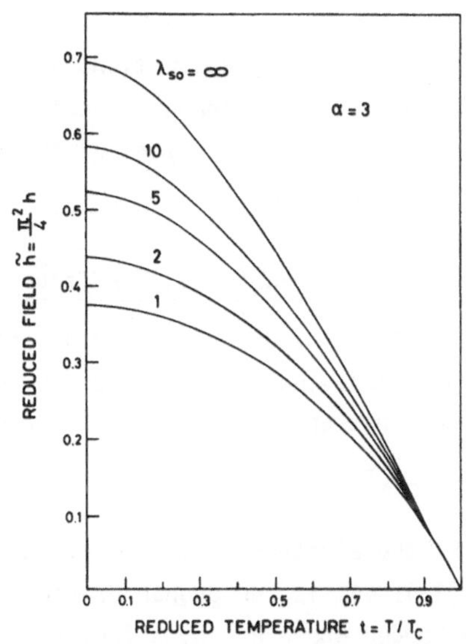

Fig.3.3. Reduced critical field $h = \pi^2 h/4$ for $\alpha = 3$ and different values of λ_{so}

In the limit $\lambda_{so} \gg 1$ (which often seems to be the case), FULDE and MAKI [3.43] have shown that (3.46) reduces to the simpler form (3.31) with a pairbreaking parameter

$$\rho = h + \frac{\alpha^2}{\lambda_{so}} h^2 \quad . \tag{3.55}$$

Using (3.31) and its solution $\rho(t)$ to define the orbital critical field $H^*_{c2}(T)$ and substituting for $\rho(t)$ and h in (3.55), we get

$$H_{c2}(T) = H^*_{c2}(T) - 0.22 \frac{\alpha}{\lambda_{so}T_c} H^2_{c2}(T) \quad . \tag{3.56}$$

We thus find the result (3.14) obtained from the G-L theory. The second term on the right-hand side describes the reduction of the (orbital) critical field due to spin paramagnetic pairbreaking. The qualitative features of h displayed in Figs.3.2,3 can be easily understood from this equation.

When the orbital effects can be neglected, the critical field determined by spin effects alone, $H_p(T)$, becomes from (3.56) [set the left-hand side equal to zero]

$$H_p(T) = 1.33\sqrt{\lambda_{so}}H_{po}\left(\frac{\rho(T/T_c)}{\rho(0)}\right)^{1/2} \quad . \tag{3.57}$$

This equation is valid in the limit $\lambda_{so} \gg 1$. $H_p(T)$ has an infinite slope at $T = T_c[\rho(t) \sim (1 - t) \Rightarrow H_p(T) \sim (T_c - T)^{1/2}]$ as we found from G-L theory.

Most high field superconductors can be qualitatively described by the theory as outlined above. However, closer analysis of data often show disagreements in the temperature dependence, and as mentioned previously the values of λ_{so} come out too high.

The main simplifying assumptions in the above theory are: weak-coupling, spherical Fermi surface, no exchange effects in the electron gas and one conduction band.

In the following two sections we shall discuss what happens when these restrictions are removed.

b) *Strong-Coupling and Exchange Effects*

The strong-coupling theory for the orbital critical field was worked out by RAINER et al. [3.44,45]. Their results are as follows: (i) The Fermi velocity v_F appearing in the equation for H^*_{c2} has to be renormalized by the phonon dressing

$$v_F \longrightarrow v_F/(1 + \lambda) \quad , \tag{3.58}$$

where λ is the electron-phonon coupling parameter. The density of states appearing in γ in (3.41-44) is the phonon dressed density of states. (ii) The critical field H^{sc}_{c2} is enhanced above H^*_{c2} by a factor $\eta(T)$:

$$H^{sc}_{c2}(T) = \eta(T)H^*_{c2}(T) \quad . \tag{3.59}$$

The first correction is a trivial renormalization; however, it will be important to also include it in weak coupling superconductors where $\lambda < 1$. The second correction becomes particularly important in materials with $\lambda \gg 1$. $\eta(T)$ is a function of the order 1-1.3. Its temperature dependence is generally weak and for most superconductors $\eta(0) < \eta(T_c)$ such that in a plot where (dH_{c2}/dT) is normalized to one, H^{sc}_{c2} will fall below H^*_{c2}. In materials where the Eliashberg function $\alpha^2F(\omega)$ has a dominant contribution for very low frequencies, as in certain amorphous alloys, RAINER et al. found $\eta(0) > \eta(T_c)$.

Perhaps the most important effect of the strong-coupling corrections is that they modify the size of the spin paramagnetic pairbreaking effects. This is because the spin susceptibility in the normal states is not renormalized by $(1 + \lambda)$ [3.39]. Thus, the density of states that appears in the calculation of α is the "bare" density of states. Since α is normally obtained from (dH_{c2}/dT), one can take this correction into account by replacing (3.53) by

$$\alpha = - \frac{0.528}{1 + \lambda} \left. \frac{dH_{c2}}{dT} \right|_{T_c} \quad . \tag{3.60}$$

Exchange enhancement effects in the conduction electron gas will also affect superconductivity. Their influence on T_c has been investigated by several authors, but to our knowledge a detailed calculation of H_{c2} has not been carried out. However, we expect that the most important effect on H_{c2} will be an enhanced paramagnetic pairbreaking. In fact, the spin susceptibility in the normal state will be enhanced by

$$S = 1/[1 - IN(0)] \quad ,$$

where I is the exchange constant and N(0) the density of states. This can be taken into account by replacing α by $\alpha \cdot S$ in (3.43). Combining this result with (3.60) we finally get

$$\alpha = - 0.528 \frac{S}{1 + \lambda} \left. \frac{dH_{c2}}{dT} \right|_{T_c} \quad . \tag{3.61}$$

In addition to these corrections, we expect that an exact calculation will add factors of order unity (like η).

To summarize this section we find that strong coupling effects only weakly influence the temperature dependence of H_{c2}, except in anomalously strong-coupling materials. For a comparison with experiment, the weak coupling formula (3.46) can be used provided (i) the critical field is multiplied by a factor η, (ii) v_F is replaced by $v_F^* = v_F/(1 + \lambda)$ and (iii) (3.61) is used to calculate the Maki parameter α.

In the latter correction, there will be a partial cancellation of the correction due to phonons by the exchange enhancement effects. (For the materials considered here we expect both S and $1 + \lambda$ to take a value between one and three.) The result of neglecting the correction to α is that the value of λ_{so}, determined by a fit of the experimental data to the theory, is wrong by a factor $[(1 + \lambda)/S]^2$. [This is correct for the limit $\lambda_{so} \gg 1$ as can be seen from the pairbreaking parameter ρ (3.55)].

c) *Anisotropy and Multiband Effects*

The basis for treating anisotropy effects on H_{c2} was made by HOHENBERG and WERTHAMER [3.46] who treated the nonlocality by expanding the kernel of the linearized gap equation in powers of the momentum operator $\underline{\pi} = \frac{h}{i} \underline{\nabla}_r - 2e\underline{A}$. The anisotropy of the system influences H_{c2} in two ways. First, it introduces an anisotropy of H_{c2} and, secondly, it modifies the temperature dependence. The Hohenberg-Werthamer treatment of the anisotropy leads to the following equation for H_{c2}:

$$\ln(\tfrac{1}{t}) = \sum_{\nu=-\infty}^{\infty} \left[\frac{1}{|2\nu + 1|} - \left(\frac{s_\omega}{1 - (s_\omega \lambda_{tr}/t)}\right)\right] , \tag{3.62}$$

where s_ω is the lowest eigenvalue of

$$\frac{\pi T}{|\tilde{\omega}| N(0)} \int dqN(q)\left[1 - \left(\frac{\underline{v}(q)\underline{\pi}}{2\tilde{\omega}}\right)^2 + \left(\frac{\underline{v}(q)\underline{\pi}}{2\tilde{\omega}}\right)^4 - ..\right]\Delta(r) = s_\omega \Delta(r) \tag{3.63}$$

$$\omega = (2\nu + 1)\pi T \qquad \tilde{\omega} = \omega + \frac{\mathrm{sgn}}{2\tau} \quad .$$

N(q) is the density of states at point $\underline{q}$ at the Fermi surface, v(q) is the velocity at that point and the integral $\int dq$ is to be taken over the Fermi surface.

The first two terms in the bracket on the left-hand side of (3.63) correspond to the dirty limit and also to the effective mass approximation. To see this, we introduce the effective mass tensor

$$\Gamma_{ij} = \int dqN(q)v_i(q)v_j(q) \tag{3.64}$$

and (3.63) becomes

$$\frac{1}{2}\sum_{ij} \Gamma_{ij}\pi_i\pi_j\Delta(r) = -\alpha_s\Delta(r) \quad , \tag{3.65}$$

where

$$s_\omega = \frac{\pi T}{|\tilde{\omega}|}\left(1 + \frac{\alpha_s}{2|\tilde{\omega}|^2 N(0)}\right) . \tag{3.66}$$

Equation (3.65) is identical to (3.16) and gives the angular dependence of H_{c2}. To get the temperature dependence of H_{c2}, we substitute $\alpha(H_{c2})$ obtained from (3.65) into (3.66) and the latter expression into the gap equation (3.62). In the case of hexagonal symmetry, we get from (3.18)

$$\alpha = \frac{e\hbar H^*_{c2}}{M}\,\delta(\theta) \quad . \tag{3.67}$$

Assuming the dirty limit $\lambda_{tr} \gg 1$, we get from (3.62)

$$\ln(\tfrac{1}{t}) = \sum_{\nu=-\infty}^{\infty}\left(\frac{1}{|2\nu + 1|} - \frac{1}{|2\nu + 1| + \frac{h}{t\delta(\theta)}}\right) \tag{3.68}$$

which is equivalent to (3.31) with the pair breaking parameter

$$\rho = \frac{h}{\delta(\theta)} = \frac{ev^2_{F\perp}\tau H^*_{c2}}{3\pi k_B T_c \delta(\theta)} , \tag{3.69}$$

where v_F is the Fermi velocity perpendicular to the c-axis. From this equation we obtain

$$H^*_{c2}(T,\Theta) = H^*_{c2}(T,0)/\delta(\theta) \quad ,$$

where $H^*_{c2}(T,0)$ is given by the dirty limit expression [(3.38) or (3.42)] using the average Fermi velocity perpendicular to the c-axis $\langle v_F^2 \rangle$. Note that the effective mass anisotropy is independent of the mean free path, which determines the absolute value of $H^*_{c2}(T,\theta)$.

The remaining terms in (3.63) are the nonlocal corrections. These corrections disappear in the dirty limit but may affect H^*_{c2} if $\lambda_{tr} \lessapprox 1$. In the isotropic case, these corrections are small; and are responsible for the small difference in the temperature dependence between the pure and the dirty limits. The reason for the smallness of this correction is due to the alternating signs in the series producing a partial cancellation of these terms. In an anisotropic material, this is no longer true and more important corrections to the temperature dependence may occur. At the same time, these terms produce an anisotropy different from the effective mass anisotropy. This is most easily observed in cubic materials where the effective mass anisotropy dissappears [3.47]. The corrections to the temperature dependence of H^*_{c2} generally result in an enhancement of H^*_{c2} at lower temperatures above the dirty limit (or the pure limit) curve. In pure and anisotropic materials this enhancement may be quite substantial (20-30%) but it is always strongly reduced as $\lambda_{tr} \gg 1$. As an example, for $\lambda_{tr} = 5$, a few percent enhancement may still remain [3.48,49].

Equations (3.62,63) do not contain gap anisotropy due to an anisotropy of the electron-phonon interaction [3.50]. These effects may also become important in pure materials, but impurity scattering will tend to smear out the gap anisotropy and so in dirty materials this contribution will not be important.

Closely related to the anisotropy effects are the multiband effects. Real superconductors have several bands at the Fermi surface and these bands may have different Fermi velocities v_F and different scattering times τ. ENTEL and PETER [3.51] did a model calculation where they divided the Fermi surface into many different sections having different v_F's and τ's. They showed that substantial deviations from the isotropic one band H^*_{c2} may occur. It is important to notice that in this picture, the deviations may also remain in the dirty limit, as long as the interband scattering is weak. DECROUX has shown that in a two-band superconductor, important deviations from the one band H^*_{c2} may occur, but where the interband scattering parameter λ_{inter} becomes much larger than one, one recovers the one-band picture [3.52].

3.3 The Role of Impurities

It is well-known from theory and experiments on other systems that impurities may have an important influence on H_{c2}, both through changes in the electronic structure and a reduction of the mean free path. In the TMC's it is useful to classify the effects according to where the impurities are located in the structure:

(i) If we substitute for M in $M_xMo_6X_8$, it appears that the influence on H_{c2} is only through the change in charge transfer from the M-site to the Mo_6 cluster. Thus, if the impurity M' and M have the same valency, both T_c and H_{c2} vary only very little and the dependence on x in $M_{1-x}M'_xMo_6X_8$ is nearly linear. As an example, we show in Fig.3.4 T_c versus x in $Pb_{1-x}Yb_xMo_6S_8$. Exceptions to this rule are found when structural phase transitions are involved like in $Sn_{1-x}Eu_xMo_6S_8$ (Sect.3.6).

(ii) The effect of impurities is much more dramatic when we substitute for Mo or X. Here, both scattering effects and important changes in the electronic structure occur. As an example, we show in Fig.3.5 T_c as a function of x in $PbMo_6S_{8-x}Se_x$ [3.53]. The large minimum centered around x = 4 is characteristic for all such systems and we believe it is related to the destruction of the 3-fold symmetry around the ternary axis [3.54].

In the following, we shall mainly discuss this second type of impurities.

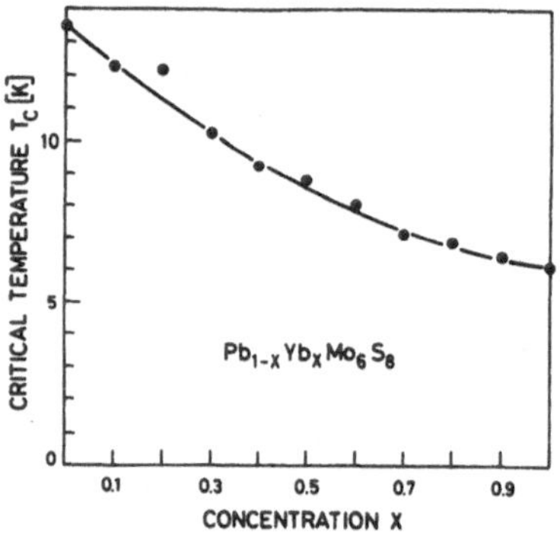

Fig.3.4. Critical temperature as a function of x in the series $Pb_{1-x}Yb_xMo_6S_8$

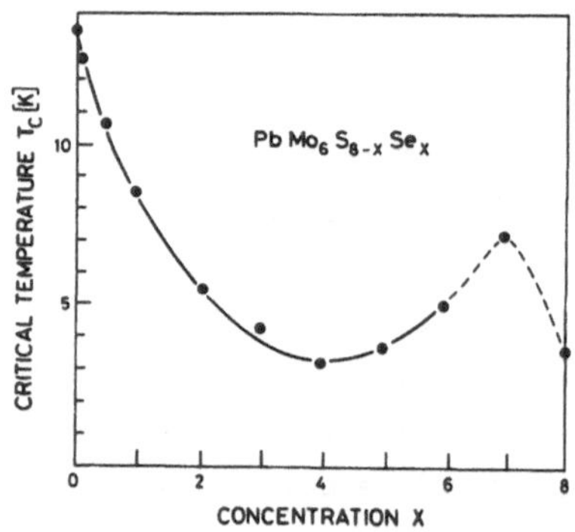

Fig.3.5. Critical temperature as a function of x in the series $PbMo_6S_{8-x}Se_x$ [3.53]

3.3.1 Mo_6Se_8

In order to study the influence of small amounts of impurities on H_{c2}, several series of pseudobinary compounds were made: $Mo_{6-y}A_ySe_{8-x}X_x$ with A = Rh, Ru and X = S, Te, Cl, Br, I [3.52]. The variation of T_c and the initial slope $(dH_{c2}/dT)_{T_c}$ are shown in Figs.3.6-9. At low concentrations all systems show the same general trend, a fast increase in $(dH_{c2}/dT)_{T_c}$ and a relatively slow variation of T_c. At

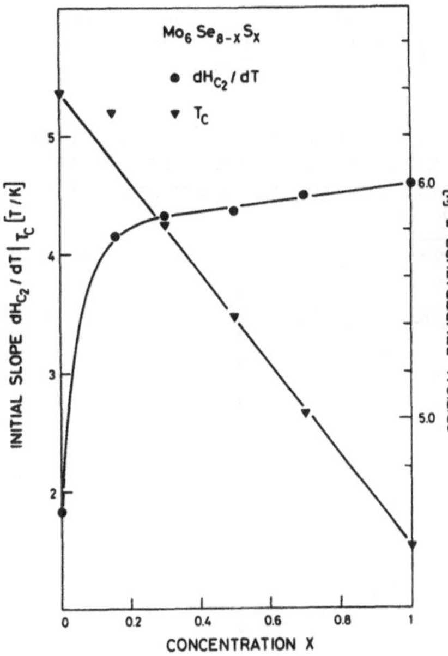

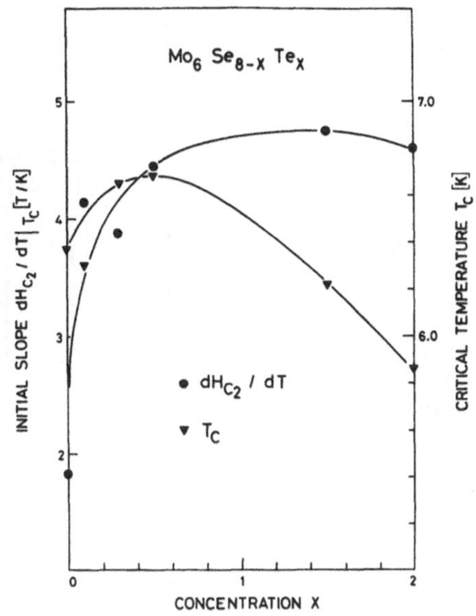

◄◄Fig.3.6
◄Fig.3.7

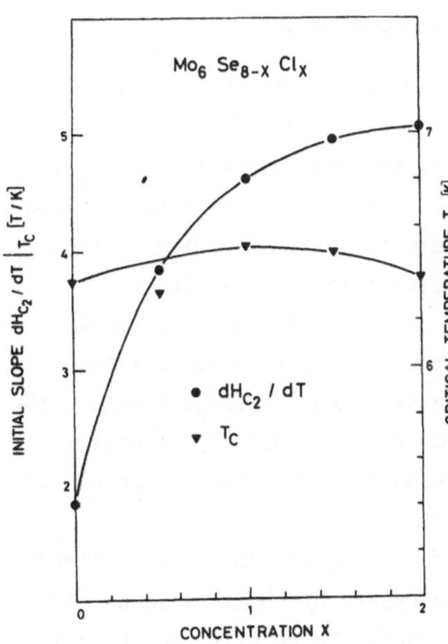

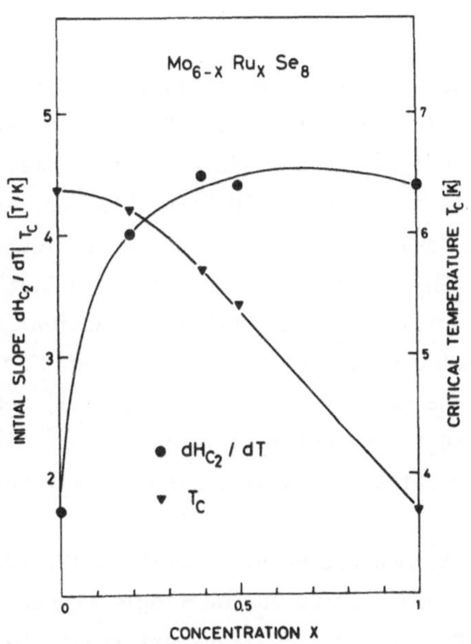

◄◄Fig.3.8
◄Fig.3.9

Fig.3.6. Critical temperature and initial slope of the critical field $(dH_{c2}/dT)_{T_c}$ as a function of x in the series $Mo_6Se_{8-x}S_x$ [3.52]

Fig.3.7. Critical temperature and initial slope of the critical field $(dH_{c2}/dT)_{T_c}$ as a function of x in the series $Mo_6Se_{8-x}Te_x$ [3.52]

Fig.3.8. Critical temperature and initial slope of the critical field $(dH_{c2}/dT)_{T_c}$ as a function of x in the series $Mo_6Se_{8-x}C\ell_x$ [3.52]

Fig.3.9. Critical temperature and initial slope of the critical field $(dH_{c2}/dT)_{T_c}$ as a function of x in the series $Mo_{6-x}Ru_xSe_8$ [3.52]

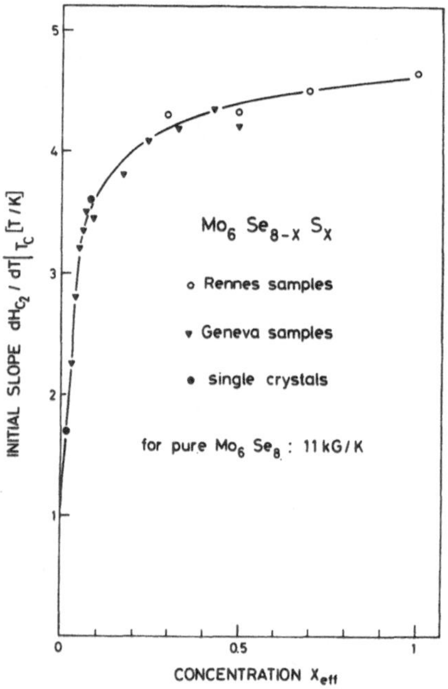

◄Fig.3.10. Initial slope of the critical field $(dH_{c2}/dT)_{T_c}$ as a function of x_{eff} for the series $Mo_6Se_{8-x}S_x$ [3.21,52]

Fig.3.11. Density of states at the Fermi level as determined from specific heat measurements for the series $Mo_6Se_{8-x}S_x$ [3.21,55]

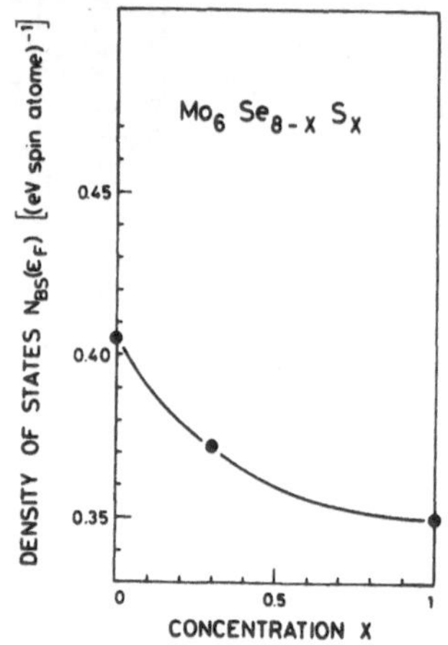

higher values of x or y, the initial slope saturates at a value between 4.5 and 5.0 Tesla/K. The case of $Mo_6Se_{8-x}S_x$ was studied in more detail for $0 < x < 0.2$ [3.20]. Using $T_c(x)$ as a measure for the effective impurity concentration x_{eff}, the results shown in Fig.3.10 were obtained. As expected, x_{eff} is slightly larger than x as a result of uncontrolled impurities, defects, etc. According to these results, Mo_6Se_8 should have $(dH_{c2}/dT)_{T_c} \simeq 1.1$[Tesla/K] if it could be produced without impurities. For a further discussion see Sect.3.5. In this system the specific heat coefficient γ was determined for several sulfur concentrations [3.21,55]. The density of states (Fig.3.11) decreases roughly as T_c, however, the fast increase in the critical field is not reflected in the density of states and we believe that $(dH_{c2}/dT)_{T_c}$ essentially reflects the behaviour of the residual resistivity.

Resistivity measurements, unfortunately, could not be made on these sintered samples but were carried out on single crystals prepared by chemical vapour transport using Cℓ, Br, or I as transport agents [3.21,52]. The results are shown in Fig.3.12. The resistivities at room temperature for these two samples were the same to within the uncertainty of 30% and equal to 600 μΩcm. The temperature dependence of the two samples were different and the residual resistivity of the Cl-transported sample was twice that of the Br-transported sample. This correlates very well with the critical field measurements which showed that the initial slope of $Mo_6Se_8(C\ell)[(dH_{c2}/dT)_{T_c} = 3.6$[Tesla/K]]was about twice that of Mo_6Se_8(Br) [$(dH_{c2}/dT)_{T_c} = 1.7$[Tesla/K]]. We interpret these results in terms of a doping of

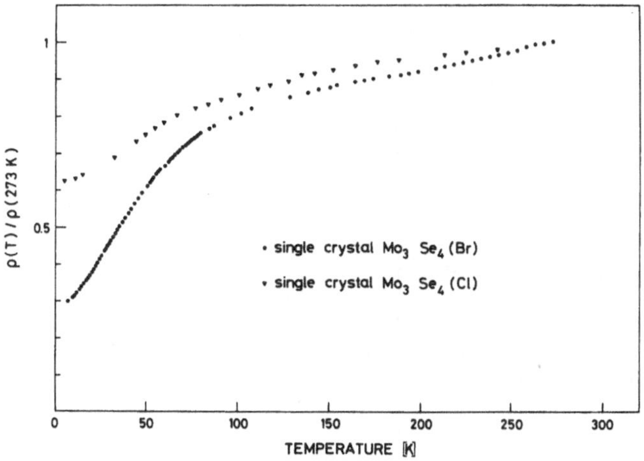

Fig.3.12. Resistance versus temperature for two single crystals of Mo_6Se_8 [3.21,52]

the single crystals by the transport agent. X ray fluorescence measurements on Mo_6Se_8(Cl) did, in fact, show that chlorine was present in this crystal.

We are now in a position to check (3.44). Using the observed initial slope and specific heat coefficient, we can calculate ρ_0. For Mo_6Se_8(Cl) this gives $\rho_0 = 350\ \mu\Omega$cm which agrees reasonably well with the direct measurement which gave $\rho_0 = 280 \pm 80\ \mu\Omega$cm.

An important conclusion of this work is that the impurities have the effect of taking away the low temperature anomaly. If we assume that at "high" impurity concentration the high temperature linear behaviour continues down to low temperature, we find $\rho_0 \simeq 400\ \mu\Omega$cm which correlates well with the saturation of $(dH_{c2}/dT)_{T_c}$ at 4.5-5.0[Tesla/K].

From these results it is possible to estimate normal state parameters using (3.44) to determine ρ_0 and the following relations to get v_F, ℓ and ξ_0:

$$v_F = \frac{k_B^2}{\hbar}\left(\frac{\pi^4}{3}\right)^{1/3} n^{2/3}\,\frac{S}{S_F}\,\frac{1}{\gamma} \tag{3.70}$$

$$\ell = \frac{(3\pi^2)^{1/3}\,\hbar}{e^2}\,\frac{S_F}{n^{2/3}S\rho_0} \tag{3.71}$$

$$\xi_0 \simeq 1.38\cdot 10^{-12}\,\frac{v_F}{T_c}\quad . \tag{3.72}$$

In order to make these estimates, we take for n the value of $5\cdot 10^{28}\ m^{-3}$ which is the average between the values for $PbMo_6S_8$ and $CuMo_6S_8$ determined by WOOLLAM et al. [3.56]. For S/S_F we take arbitrarily 0.4 (for similar estimations for $PbMo_6S_8$ and $CuMo_6S_8$ [Ref.3.6, Chap.5]). With these assumptions we find the results given in Table 3.2. Because of the uncertainty of n, we believe that these results are only correct to within a factor of two.

Table 3.2. Superconducting and normal state parameters for $Mo_6Se_8(Br)$ and $Mo_6Se_8(C\ell)$ single crystals as well as for several polycrystalline samples in the series $Mo_6Se_{8-x}S_x$ [3.21,52]

Compounds	T_c [K]	$(dH_{c2}/dT)_{T_c}$ [T/K]	γ [$Jm^{-3}K^{-2}$]	ρ [Ωm]	v_F [m/s]	ℓ_{tr} [Å]	ξ_0 [Å]	ξ_0/ℓ
$Mo_6Se_8(Br)$	6.30	1.70	232	$1.80 \cdot 10^{-6}$	$1.34 \cdot 10^5$	13.1	294	22
$Mo_6Se_8(Cl)$	6.18	3.60	230	$3.53 \cdot 10^{-6}$	$1.35 \cdot 10^5$	6.7	301	45
Mo_6Se_8	6.25	2.25	232	$2.18 \cdot 10^{-6}$	$1.34 \cdot 10^5$	10.8	296	27
$Mo_6Se_{7.99}S_{0.01}$	6.24	2.65	231	$2.58 \cdot 10^{-6}$	$1.35 \cdot 10^5$	9.1	299	33
$Mo_6Se_{7.97}S_{0.03}$	6.23	3.11	231	$3.03 \cdot 10^{-6}$	$1.35 \cdot 10^5$	7.8	299	38
$Mo_6Se_{7.95}S_{0.05}$	6.04	-	230	-	$1.35 \cdot 10^5$	-	308	-
$Mo_6Se_{7.97}S_{0.07}$	6.08	3.45	230	$3.38 \cdot 10^{-6}$	$1.35 \cdot 10^5$	7.0	306	44
$Mo_6Se_{7.9}S_{0.1}$	6.15	3.50	230	$3.43 \cdot 10^{-6}$	$1.35 \cdot 10^5$	6.9	303	44
$Mo_6Se_{7.85}S_{0.15}$	5.75	3.81	227	$3.78 \cdot 10^{-6}$	$1.37 \cdot 10^5$	6.2	329	53
$Mo_6Se_{7.8}S_{0.2}$	5.65	4.07	226	$4.06 \cdot 10^{-6}$	$1.38 \cdot 10^5$	5.8	337	58
$Mo_6Se_{7.6}S_{0.4}$	5.25	4.35	221	$4.43 \cdot 10^{-6}$	$1.41 \cdot 10^5$	5.3	371	70

The most important result here is that the mean free path is very low, being of the order of 10 Å even in the samples which we believe have few impurities. Since ξ_o is on the order 300 Å, we are always in the dirty limit, i.e., $\lambda_{tr} \sim \xi_o/\ell \gg 1$.

The explanations for the anomalous concentration dependence are still highly speculative. Saturation due to ℓ becoming equal to interatomic distances [3.57] is possible but is not clearly reflected in the temperature dependence (see below). It is interesting to note, however, that saturation occurs approximately at a concentration where there is, on the average, one impurity per Mo_6Se_8 unit. Thus, one is led to surmise that scatterings on impurities situated on the same Mo_6Se_8 unit are correlated such that two impurities do not contribute much more to the scattering than one impurity. In view of the cluster nature of these compounds, this is not unreasonable.

In addition to this saturation, the order of magnitude of the changes in ℓ per impurity atom appears to be very high. Usually, a few percent of impurities in a compound having $\ell \sim 10$ Å would not have much effect on ℓ. The reason why this is not the case here is also unclear; however, we speculate that correlation effects in these narrow band materials may be important and may lead to an enhancement of the scattering cross section of an impurity.

The temperature dependence of the resistivity shown in Fig.3.11 is similar to those observed for other high temperature high field superconductors such as V_3Si and Nb_3Sn with the only difference that the anomaly appears at a somewhat lower temperature [3.58,59]. We can fit the temperature dependence reasonably well with the empirical expression [3.60]

$$\rho(T) = \rho_o + BT + C\, e^{-T_o/T} \quad , \tag{3.73}$$

where B ranges from 0.2 to 0.6 [$\mu\Omega$cmK^{-1}] and C is strongly concentration dependent. T_o is about 45 K. Different explanations for this anomalous temperature dependence have been given in the literature. One explanation is that ℓ saturates when it becomes of the order of the interatomic distances [3.57]. This does not seem to apply directly to the present situation since there remains a linear term in $\rho(T)$, even after the anomalous contribution has saturated. Another possible explanation would be that the electron phonon interaction is frequency dependent, and that the electrons couple more to the high frequency phonons than to the low frequency phonons This picture is not inconsistent with the tunneling results which suggest precisely such a situation [3.61]. In this picture, the impurities would have the effect of mixing the phonon-modes and thus smear out the frequency dependence of $\alpha^2(\omega)$. This, in turn, would eliminate the low temperature anomaly.

3.3.2 $PbMo_6S_8$

The large scatter of H_{c2} values reported for $PbMo_6S_8$ suggests that a similar concentration dependence is present in this system. However, the control of H_{c2} is more difficult in this case. It is especially difficult to reproduce the "low" H_{c2} values which are sometimes found in an uncontrolled manner. In Table 3.3 we present the results on a series of samples, all of which were produced under exactly the same conditions. This study gave T_c = 13.6 K and $(dH_{c2}/dT)_{T_c}$ = 5.65 [Tesla/K] for $PbMo_6S_8$. Impurities on the chalcogen site always gave a reduction of T_c (see Fig. 3.5) and an increase of the initial slope, so that for $PbMo_6S_7Se$, we have $(dH_{c2}/dT)_{T_c}$ = 7.85 [Tesla/K]. However, since T_c = 8.3 K, the extrapolated H_{c2} value is only 45 Tesla. The H^*_{c2} values given in Table 3.3 are calculated from (3.3) but are believed to be close to the actual critical field [3.2,9,10].

Table 3.3. Critical temperatures and critical fields of $PbMo_6S_8$, "pure" and containing different impurities [3.52]

Compounds	T_{cex} [K]	$(dH_{c2}/dT)_{T_c}$ [T/K]	$H^*_{c2}(0)$ [T]
$PbMo_{5.4}Nb_{0.6}S_8$	14.2	4.75	46.6
$PbMo_{5.8}Nb_{0.2}S_8$	14.0	4.90	47.5
$PbMo_6S_8$	13.6	5.65	53.0
$PbMo_{5.9}Nb_{0.1}S_8$	13.6	6.30	59.0
$PbMo_6S_{7.8}C\ell_{0.2}$	12.2	6.80	57.2
$PbMo_6S_{7.9}Se_{0.1}$	12.5	5.95	51.5
$PbMo_6S_{7.8}Se_{0.2}$	12.1	6.65	55.5
$PbMo_6S_{7.5}Se_{0.5}$	10.5	6.90	50.0
$PbMo_6S_7Se$	8.3	7.85	45.0

In the samples doped with Nb, an increase of T_c was observed and simultaneously a decrease of $(dH_{c2}/dT)_{T_c}$. We believe that Nb occupies neither the Mo sites nor the S sites, but has a similar effect to the impurities substituted for Pb [3.62]. If a different series is made, the values for T_c and $(dH_{c2}/dT)_{T_c}$ may change. However, we found that an increase in T_c always leads to a decrease in the initial slope. For a sample with T_c > 14 K, the highest slope that we have found was 5.8 [Tesla/K]. This was in a sample of nominal composition $Gd_{0.2}PbMo_6S_8$ [3.3,10].

We believe that the observed increase in $(dH_{c2}/dT)_{T_c}$ with impurity concentration on the chalcogen site has a similar origin as the one observed in Mo_6Se_8, i.e., results from an increase in ρ. Unfortunately, since the Fermi level is probably located close to a peak in the density of states, the impurities have the effect of reducing T_c so that only a very modest gain in H_{c2} is obtained.

3.4 Anisotropy

3.4.1 $M_xMo_6X_8$ Compounds

Some of these compounds have been obtained in the form of single crystals and we have measured the anisotropy on four materials: $PbMo_6S_8$, $PbMo_6Se_8$, $SnMo_6Se_8$ and Mo_6Se_8 [3.25,26]. The three former crystals were made using the Stockbarger-Bridgman technique ([3.63] and [Ref.3.6, Chap.4]). Single crystals were cut from melted samples containing a few large crystals. The Mo_6Se_8 crystals were made by chemical vapour transport as described in Sect.3.3. The measurements were made using the resistive transition and the ternary axis was identified by Laue patterns.

These measurements showed that, contrary to our initial expectations, there is a 10-20% anisotropy of the critical field between the orientations parallel and perpendicular to the ternary axis. The angular dependence follows the predictions of the effective mass model (3.18,19). In Figs.3.13,14, we show the critical field versus the angle θ between the magnetic field and the ternary axis for $SnMo_6Se_8$ and Mo_6Se_8. The dashed curves were calculated from (3.18), fitting the mass ratio ε^2. This mass ratio is given for each of the 4 investigated compounds in Table 3.4. We note that the order of magnitude of the anisotropy is about the same in the 4 compounds. However, the three ternary compounds have $\varepsilon < 1$, whereas Mo_6Se_8 was found to have $\varepsilon > 1$. The reason for this is probably found in the difference in electronic structure between Mo_6Se_8 and the ternary compounds, which is reflected in the intra-cluster distances given in Table 3.4 ([Ref.3.6, Chap.3]). In fact, the three ternary compounds are isoelectronic and it is not unreasonable that the anisotropy turns out to be similar in the three. It is not known how the anisotropy is influenced by the impurity effects described in the previous section.

One possible source of error in these measurements is the measuring current. It was not always possible to keep this current perpendicular to the magnetic field. In order to check this point, we varied the value of the current as well as its direction by changing the arrangement of contacts. No influence on the anisotropy was detected. In $PbMo_6S_8$ the current was always kept perpendicular to the magnetic field.

It should be noted that the transition width in the single crystals were considerably smaller than what is usually observed in polycrystalline samples. In the beginning, the transitions in the Mo_6Se_8 crystal were even much narrower, but after a few thermal cycles, they broadened and stabilized at about 1.1 Tesla at 4.2 K. We conclude that the transition widths in the polycrystals are partly due to the anisotropy.

No temperature dependence of the anisotropy was observed and no significant anisotropy was detected in the plane perpendicular to the ternary axis. All of these results point to the validity of the effective mass approximation in these compounds. This is not too surprising, since we saw in the last section that these materials are

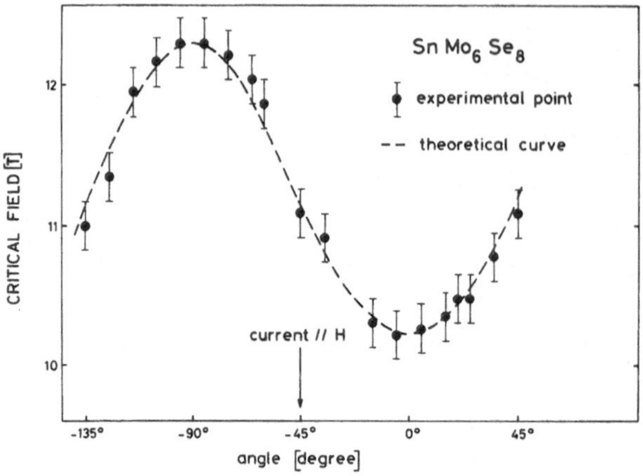

Fig.3.13. Anisotropy of the critical field of $SnMo_6S_8$ at 4.2 K. The angle is measured between the ternary axis and the magnetic field [3.25]

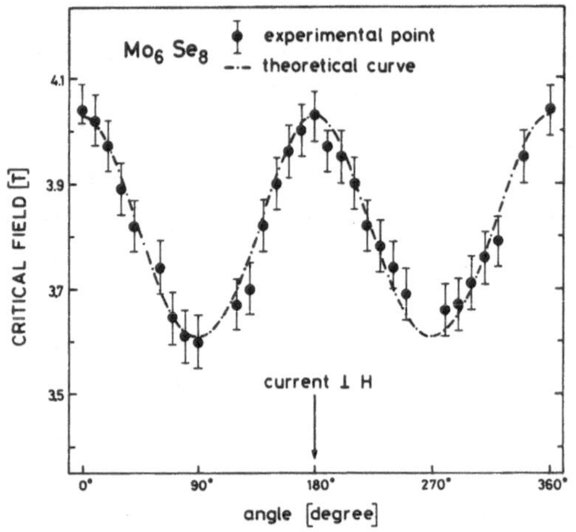

Fig.3.14. Anisotropy of the critical field of Mo_6Se_8 at 4:2 K. The angle is measured between the c-axis and the magnetic field [3.26]

Table 3.4. Superconducting and lattice-parameters for the $PbMo_6S_8$, $PbMo_6Se_8$, $SnMo_6Se_8$ and Mo_6Se_8 single crystals used in the anisotropy measurements [3.25,26,57]

Samples		$PbMo_6S_8$	$PbMo_6Se_8$	$SnMo_6Se_8$	Mo_6Se_8
T_c	[K]	12.00	6.75	6.82	6.30
$(dH_{c2\perp}/dT)_{T_c}$	[T/K]	7.0	-	5.2	-
$(dH_{c2\parallel}/dT)_{T_c}$	[T/K]	5.9	-	4.4	-
$H_{c2\perp}$(4.2 K)	[T]	-	9.8	12.3	3.6
$H_{c2\parallel}$(4.2 K)	[T]	-	8.0	10.2	4.0
Anisotropy ratio ε^2		0.67	0.67	0.69	1.25
Rhombohedral lattice parameters		a_R=6.542 Å α_R=89°14'	a_R=6.719 Å α_R=89°9'	a_R=6.768 Å α_R=89°18'	a_R=6.660 Å α_R=91°43'
$d_{Mo-Mo}^{intra\ (1)}$		2.732 Å	2.734 Å	2.775 Å	2.836 Å
$d_{Mo-Mo}^{intra\ (2)}$		2.679 Å	2.697 Å	2.680 Å	2.684 Å

rather "dirty" ($\ell < \xi_0$) in which case corrections to the effective mass model tend to disappear, as discussed in Sect.3.2.2. What *was* surprising in the beginning was the large size of the anisotropy. Early band structure calculations indicated a nearly cubic symmetry of the Fermi surface (only a small distortion due to the rhombohedral angle being different from 90°). The most likely interpretation of this anisotropy is that hybridization of antisymmetric bands produce noncubic distortions of the Fermi surface. The order of magnitude of these effects is large enough to explain the observed behavior [3.65]. However, a quantitative comparison with theory [Ref.3.6, Chap.6] is not yet possible. Note finally that the paramagnetic limitation should not have a significant influence on the anisotropy. From (3.23) we find, assuming a paramagnetic reduction of H_{c2} of about 10-20% (Sect.3.4.2), that the correction to the anisotropy is only a few percent. This justifies the use of the orbital critical field formula (3.18,19).

Table 3.4 contains a summary of the data obtained on the four compounds. Note the high values of $(dH_{c2}/dT)_{T_c}$ and the rather modest value of T_c for $PbMo_6S_8$. We take this to be an indication that the crystal contained a certain amount of defects of unknown nature.

3.4.2 $T\ell_2Mo_6Se_6$

The structures with condensed clusters [3.66,67, and 3.6, Chaps.2,3] are expected to show more pronounced anisotropy effects. An extreme case is found in the linear chain compound $T\ell_2Mo_6Se_6$ [3.68]. In this material the Mo-atoms are arranged in infinite chains of triangles in staggered positions [3.69,70]. The intrachain distances are about 2.7 Å, whereas the shortest Mo-Mo interchain distance is as large as 6.34 Å. The electronic interchain coupling therefore results from hybridization with the Se p-states. However, the coupling has to pass by two intermediate Se-atoms so it is very weak. The bandstructure of this material is discussed in [Ref.3.6, Chap.6].

Crystals with the approximate dimensions $0.1 \times 0.1 \times 5$ mm^3 can be prepared relatively easily and anisotropy measurements were carried out on such crystals. These crystals were most often not real single crystals, but rather should be considered as bundles of single crystals with approximately the same orientation. The critical temperatures varied between about 3 K and 6 K. The critical field is, as expected, very anisotropic. In Fig.3.15 we show H_{c2} versus θ, the angle between the chain-axis and the magnetic field [3.68]. The angular dependence fits an effective mass picture reasonably well with the mass ratio $\varepsilon^2 \simeq 676$. There was no temperature dependence of the anisotropy and we did not observe any significant anisotropy in the basal plane. This latter result may, however, be due to the sample being composed of fibers having different orientations in the basal plane. Note that in a material like this there would be a significant change in the angular dependence of H_{c2} if the critical field was paramagnetically limited (3.23). Thus we conclude that there is not an important reduction of H_{c2} due to spin paramagnetic effects. The

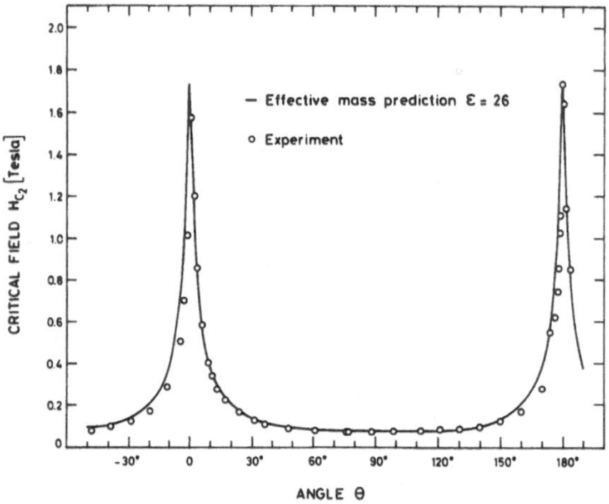

Fig.3.15. Anisotropy of the critical field of $TlMo_6Se_6$ at 2.2 K. The angle is measured between the c-axis and the magnetic field [3.68]

observed anisotropy of the low temperature resistivity $\rho_\perp/\rho_{||} \leq 10^3$ is in good agreement with the mass ratio ε^2. The extrapolated values for H_{c2} at T = 0 for the sample displayed in Fig.3.15 where $H_{c2\perp} = 0.2 \pm 0.02$ Tesla and $H_{c2||} = 5.2 \pm 0.2$ Tesla. From these values we estimate $\xi_\perp$ and $\xi_{||}$ from (3.21) and find $\xi_\perp = 78$ Å and $\xi_{||} = 2030$ Å. Of particular interest here is the large value of $\xi_{||}$. This shows that both the mean free path ℓ and the BCS coherence length are very large, indicating a large mobility and a rather small effective mass in the direction of the chain. Furthermore, using the relation 3.44, where ρ is the perpendicular resistivity and H_{c2} the parallel field, we can estimate γ and find $\gamma \simeq 3.5 \cdot 10^{-4}$ mJ/K^{2}g. This gives a very low value for the density of states and is in agreement with more recent specific heat measurements [3.71]. These results, together with the observation of a relatively large magnetoresistance, suggest that the effective number of carriers in this material is low, and that these carriers have a low mass and a high mobility in the direction of the chains.

In our investigation of this system we found two different kinds of behavior represented by sample A and sample B [3.68]. Sample A had a T_c of 2.8 K and its properties are the ones discussed above. Sample B had a T_c of 5.8 K and it showed, in many respects, different behavior. There was no difference in the X ray patterns between the two at room temperature, and so far, no phase transition has been detected [3.72]. We conclude that further investigations are necessary in order to clarify these differences.

3.5 The Temperature Dependence of H_{c2} and the Question of the Paramagnetic Limitation

In Fig.3.16 we show the temperature dependence of H_{c2} for a certain number of TMC's [3.10]. The initial slopes were obtained in static fields and the high field measurements were made with pulsed magnetic fields. The theoretical curves were

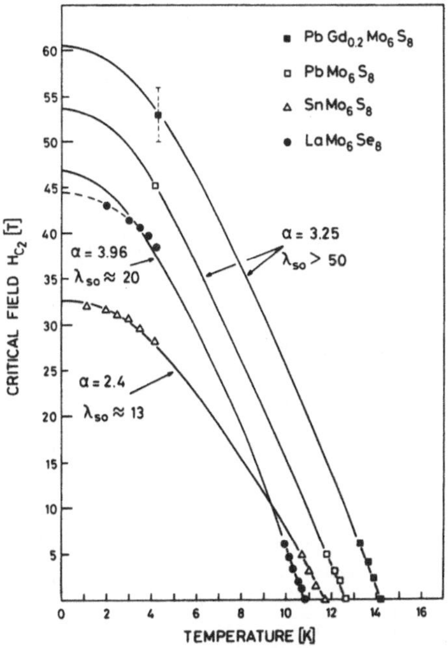

Fig.3.16. Upper critical field versus temperature for several $M_xMo_6X_8$ compounds [3.10]

calculated from (3.46). The Maki parameter α was determined from the initial slope using (3.53) and λ_{so} was deduced from a fit to the data. We have shown here our own results. The data obtained by FONER et al. [3.2,5] are in reasonable agreement with ours.

From these data it is clear that the critical field H_{c2} is very close to the orbital critical field H_{c2}, although H_{c2} is more than a factor of two higher than the Chandrasekhar-Clogston limit H_{po}. The explanation for this is that spin-orbit scattering makes the paramagnetic pairbreaking less effective. The λ_{so} values given in Fig.3.16 give the strength of the spin orbit scattering necessary to explain the experimental data. Note that these values are much higher than the ones needed to make $H_p > H_{c2}$ using (3.57) . The only problem with these data is that λ_{so} is much higher than one would anticipate from simple considerations. λ_{so} can be related to the mean free path due to spin-orbit scattering alone (ℓ_{so}) through (3.48). Since λ_{tr} includes all scattering mechanisms, we should always have

$$\lambda_{so} < \frac{4}{3}\lambda_{tr}$$

and we would expect λ_{so} to be of the order 0.1 λ_{tr}. However, a conservative estimate for $PbMo_6S_8$ gives $1 < \lambda_{tr} < 10$ (see, for instance, [Ref.3.6, Chap.5]); thus, using λ_{so} from Fig.3.16 we find $\lambda_{so} \gg \lambda_{tr}$. Another way to say this is that the spin-orbit mean free path ℓ_{so} would have to be less than the interatomic distances and much less than the transport mean free path ℓ. This, of course, is not possible.

For the reasons given above one is forced to look for alternative explanations. As discussed in Sect.3.2.2, strong coupling and electron-electron interaction cor-

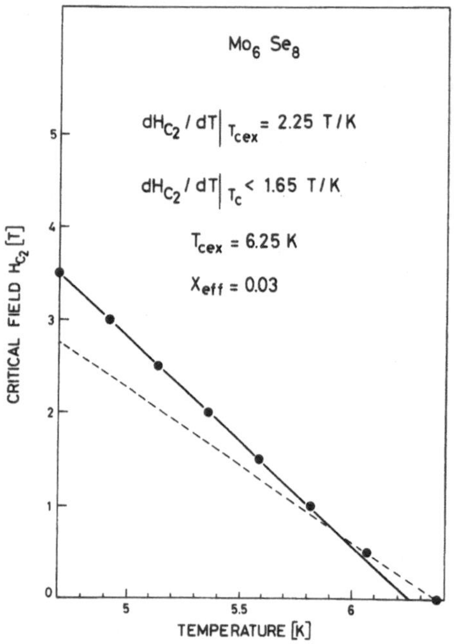

Fig.3.17

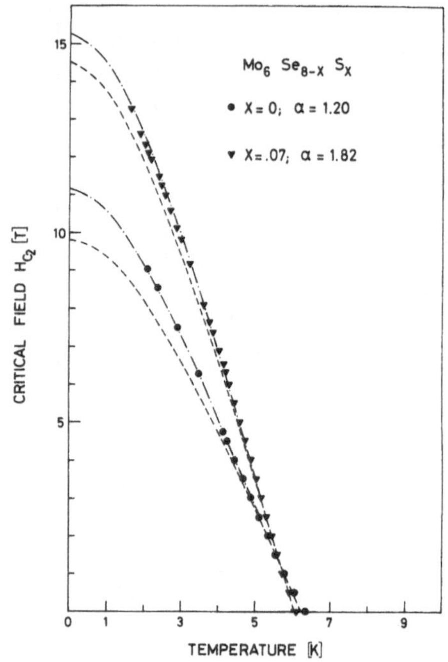

Fig.3.18

Fig.3.17. Critical field versus temperature for Mo_6Se_8 showing the curvature near T_c and illustrating the definition of T_{cex} and x_{eff} [3.52]

Fig.3.18. Critical field versus temperature for x = 0 and x = 0.07 in the series $Mo_6Se_{8-x}S_x$. The dashed line was calculated from (3.31) and the dash-dotted line was obtained using the modified temperature dependence of the orbital critical field [3.21,52]

rections leads to a correction factor $[S/(1 + \lambda)]^2$ for λ_{so}. In the TMC's, λ is of the order unity [Ref.3.6, Chap.8] and we believe that $S \simeq 1.5$ is a reasonable value for the Stoner factor [3.8]. Thus, we are led to believe that the λ_{so} values given in Fig.3.16 are about a factor of two too large. But we see that this is still much too large to solve the problem, and we note that in the A-15 materials a similar problem occurs. However, in these materials the numbers are less extreme and ORLANDO et al. [3.40] have demonstrated that, contrary to the case discussed here, for the A-15 materials, strong coupling effects may explain the discrepancies.

In order to investigate this further, we decided to study the temperature dependence in detail. Since the pulsed field measurements are always less precise than static field investigations, we decided to carry out this study on Mo_6Se_8 where the measurements can be made in static fields. To this end, we studied the series $Mo_6Se_{8-x}S_x$ for which the initial slopes of H_{c2} are given in Fig.3.10. The effective impurity concentration x_{eff} was determined in the following way. Close to T_c, H_{c2} versus T shows a positive curvature (Fig.3.17). We interpreted this as being a result of a slight inhomogeneity of the real impurity/defect distribution. Very close to T_c we observed the cleaner part of the sample, but at temperatures below 0.9 T_c, the critical field corresponded to the dirtier part. Since this is the part of the sample which is studied here, we redefined T_c by an extrapolation of H_{c2} from below 0.9 T_c as indicated by the solid line in Fig.3.17. With this T_c, x_{eff} was determined from the $T_c(x)$ relation. In Fig.3.18 we show $H_{c2}(T)$ for x = 0 and x = 0.07. The dashed curves are the theoretical curves for $\lambda_{so} = \infty$ (i.e., the orbital critical field) fitted to the initial slope of H_{c2}. We see that the x = 0 sample clearly exceeds the orbital critical field calculated from (3.46). For the x = 0.07 sample this deviation is less important and as we continue to increase x [thus $(dH_{c2}/dT)_{T_c}$], the measured curve finally falls below the calculated curve. We interpreted these results in the following manner. The orbital critical field H^*_{c2} was, for reasons to be discussed below, enhanced above the weak coupling values, denoted henceforth as H^{*o}_{c2} (3.31). In the low field material (x = 0), the paramagnetic reduction of H_{c2} was small and H_{c2} reflected essentially the orbital critical field. As the initial slope increased upon doping, the paramagnetic pairbreaking increased and the low temperature enhancement appeared to be reduced and finally disappeared.

To check this assumption, we start from the general equation (3.14) which is valid for weak paramagnetic limitation. We take for a(T) the result of the microscopic theory (3.56) and write [3.21]

$$H^*_{c2}(T) = \beta(T)H^{*o}_{c2}(T) \quad , \tag{3.74}$$

where $H^{*o}_{c2}(T)$ is given by (3.31,32) and $\beta(T)$ is an enhancement factor to be determined. We can now rewrite (3.14) in the following form:

$$\frac{H_{c2}}{H^{*o}_{c2}} = \beta(T) - B\left(\frac{\alpha H_{c2}}{H^{*o}_{c2}}\right)^2 T_c \quad . \tag{3.75}$$

B is a constant which in the weak coupling limit takes the value 1.32 $k_B\tau_{so}/\hbar$. Thus, a plot of (H_{c2}/H^{*o}_{c2}) versus $(\alpha H_{c2}/H^{*o}_{c2})^2 T_c$ should give a straight line. Each point in this plot corresponds to a different concentration and the H_{c2} values have to be taken at the same reduced temperature $t = T/T_c$. In Fig.3.19 we show the results for t = 0.32. In spite of the difficulty in comparing different samples, the points fall very well on a straight line from which we determine $\beta(0.32) = 1.2$ and $B = 7 \cdot 10^{-3}$ K^{-1}. From this we find that there is a 12-20% paramagnetic reduction of H_{c2} in these materials. Similar plots can be obtained at other temperatures and allows us to determine $\beta(t)$. A fit to these data gives [3.21,52]

$$\beta(t) = \frac{\beta_o}{1 + (\beta_o - 1)t^2} \quad ; \quad \beta_o \simeq 1.23 \quad . \tag{3.76}$$

As an additional check, we can now go back and calculate H_{c2} using (3.46) modified by multiplying the orbital terms in the argument of the ψ function by $[\beta(t)]^{-1}$. We

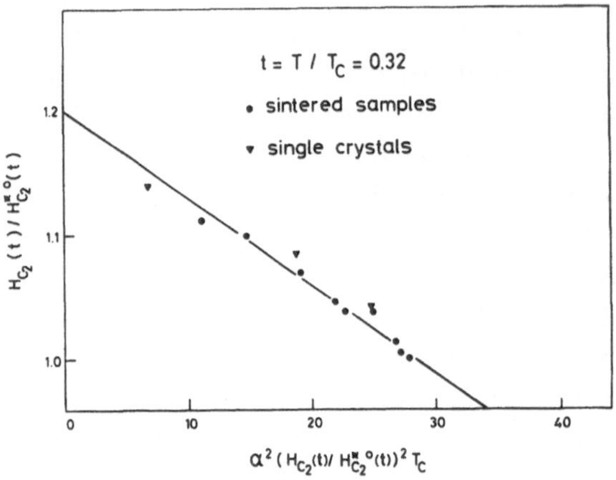

Fig.3.19. (H_{c2}/H^{*o}_{c2}) versus $\alpha^2(H_{c2}/H^{*o}_{c2})^2T_c$ for 3 single crystals and 9 polycrystals in the series $Mo_6Se_{8-x}S_x$ [3.21,52]

can now fit H_{c2} for all samples in this series, as well as for other samples such as the single crystals $Mo_6Se_8(Br)$ and $Mo_6Se_8(C\ell)$ using the same function $\beta(t)$. As an example we show the results for x = 0 and x = 0.07 in Fig.3.18 (dash-dotted curve). If we assume the weak coupling formula for B, we can calculate λ_{so} and find $\lambda_{so} \simeq 6.5$. If we compare with λ_{tr}, which from Sect.3.3 can be determined to be larger than 17, we find that with this modification of the analysis the apparent conflict mentioned in the beginning of this section disappears.

We conclude that a paramagnetic reduction *is* present in this series, but that it is partially masked by an enhancement of H^*_{c2}. We believe that this is rather general behavior in the TMC's. For instance an enhancement of H_{c2} can also be seen in $Cu_xMo_6S_8$, a relatively low field superconductor [Ref.3.6, Chap.5]. The case of $PbMo_6S_8$ is much more difficult to check since the necessity of measuring H_{c2} in pulsed magnetic fields makes a detailed investigation virtually impossible. However, we feel it is plausible to assume that a similar situation prevails, but we are, of course, unable to judge whether λ_{so} will come out low enough that it can be reasonably interpreted as an ordinary spin-orbit scattering parameter. For $SnMo_6S_8$, λ_{so} has been determined from the study of magnetic impurities (Sect.3.6).

In the above we have not assumed anything about the reason for the H^*_{c2} enhancement. In principle, there are several possible causes as discussed in Sect.3.2.2. An enhancement due to strong coupling will only take place in very strong coupling materials. This does not seem to be the case for $PbMo_6S_8$. If we take the electron-phonon interaction to be frequency independent, one finds even $\eta(0) < \eta(T_c)$, [see (3.59)]; thus we do not expect that strong-coupling effects can explain the observed behavior [3.73]. Another possible cause, which we discard, is the enhancement due to nonlocal corrections [higher-order terms in (3.63)]. Since Mo_6Se_8 is well within the dirty limit, we expect these effects to be negligible. The fact that the anisotropy follows the effective mass model supports this assumption. The most plausible cause for this behavior is in our opinion that we have a two or multiband situation where the intra-band scattering is strong ($\lambda_{tr} \gg 1$), but where the inter-band scattering is weak. Such a situation will usually lead to an enhancement of H^*_{c2} [3.52], as demonstrated by ENTEL and PETER [3.51]. In view of these results we suggest that other properties of the TMC's be discussed within a multiband model, as well.

3.6 Compounds with Magnetic Ions. The Jaccarino-Peter Effect

One of the intriguing and exciting properties of the TMC's is the possibility of replacing the third element M by a magnetic rare-earth ion without destroying the superconductivity [3.74]. The study of the critical fields in such compounds has revealed several interesting new phenomena [3.22,23]. In this chapter we shall discuss the behavior of H_{c2} in the paramagnetic state. The aspects related to magnetic ordering [3.22,75,76] will be treated in Chaps.5,9.

We shall assume here that the exchange interaction between the magnetic ions and the conduction electrons can be described by

$$\mathcal{H}_{ex} = \frac{1}{N} \sum_i \Gamma(g - 1)(\underline{J}_i \underline{\sigma}) \quad . \tag{3.77}$$

Γ is the exchange constant, $\underline{J}_i$ the total angular momentum of the magnetic ions and $\underline{\sigma}$ the conduction electron spin. For our discussion, we shall retain only the first and second-order effects in Γ. The first-order effect can be described as the result of an effective field (exchange field) H_J acting on the conduction electron spins [3.19]

$$H_J = x \frac{\Gamma(g - 1)\langle J_Z \rangle}{g_e \mu_B} = \frac{(g - 1)\Gamma}{g g_e N \mu_B^2} M(H,T) \quad . \tag{3.78}$$

x is the concentration of magnetic ions and M(H,T) is their magnetization.

The second order-effects are described by the well-known Abrikosov-Gor'kov pair-breaking parameter [3.77]

$$\rho_{AG} = \frac{(g - 1)^2}{8 k_B T_{co}} N(0) J(J + 1) \overline{\Gamma^2} \quad . \tag{3.79}$$

When introduced for ρ into (3.31) we obtain the critical temperature $T_c(x)$ [$T_{co} = T_c(x = 0)$]. In addition to these two effects, one has to take into account that the field acting on the orbits of the conduction electrons is $B/\mu_0 = H + M(H,T)$.

All these effects can easily be incorporated into the microscopic theory and leads to a generalized equation (3.46) [3.19,43]. However, since we are dealing with materials with strong spin-orbit scattering, we will use the simplified expressions valid for $\lambda_{so} \gg 1$ due to FULDE and MAKI [3.43].

This leads to the multiple pairbreaking picture [3.78] and a generalization of (3.56) [3.19,76]:

$$H_{c2}(T) = H^*_{c2}(T) - M(H_{c2};T) - 3.56\rho_{AG}H^*_{c2}(0)$$

$$- 0.22 \frac{\alpha}{\lambda_{so}T_{co}}[H_{c2}(T) + H_J(H_{c2};T)]^2 \quad . \tag{3.80}$$

This is an implicit equation for $H_{c2}(T)$ in which the terms depending on M(H,T) may lead to a temperature dependence of H_{c2} that is radically different from that of $H^*_{c2}(T)$ [3.19,78].

One such example of an anomalous temperature dependence is found in $ErMo_6S_8$ [3.22], as discussed in Chap.5. In that compound, $H_{c2} \ll H_J$ so that the last term in (3.80) is $\sim H_J^2 \sim M^2(H,T)$. As the temperature is lowered, H_J grows quickly and the paramagnetic reduction of H_{c2} becomes so important that the critical field passes through a maximum and finally decreases towards lower temperatures [3.22,76]. Such effects can only be seen at temperatures and fields such that M(H,T) is not saturated.

A completely different situation arises in compounds where $|H_{c2}| \simeq |H_J|$ and where $H_J < 0$ (i.e., $\Gamma < 0$). In this case a compensation of the internal field H_J by the external field will occur in the last term in (3.80) and thus lead to a higher critical field [3.19]. This is basically the same effect as originally discussed by JACCARINO and PETER [3.24]. They suggested that a ferromagnet, where superconductivity is suppressed by the spin polarization due to H_J, can be made superconducting in a high field if $H_J < 0$. From (3.80) it is clear that one condition necessary to observe this behavior is $H^*_{c2} \geq |H_J|$. Since $|H_J|$ is of the order 30-50 Tesla in the compounds studied here, this effect can only be observed in high field superconductors. In particular, the two ferromagnetic compounds showing re-entrant behavior, $HoMo_6S_8$ and $ErRh_4B_4$ (see the two following chapters), have low critical fields, and it would be hopeless to look for the Jaccarino-Peter effect in these materials. However, this effect can also be seen in paramagnetic compounds, and it has been observed in the series $Sn_{1-x}Eu_xMo_6S_8$ [3.23].

In order to get a feeling for how this compensation effect will influence H_{c2}, we shall now present a qualitative discussion of (3.80). To this end we will neglect the second and third term since they are not important for a qualitative understanding. Thus we have

$$H_{c2}(T) = H^*_{c2}(T) - a[H_{c2}(T) - |H_J(H_{c2};T)|]^2 \quad . \tag{3.81}$$

Let us first consider $H_{c2}(T = 0)$ as a function of the concentration x of magnetic ions. In this case, the magnetization will be saturated and H_J proportional to x. For simplicity we shall also assume that H^*_{c2} is independent of x. Equation (3.81) then tells us that as x increases from 0, the paramagnetic reduction of H_{c2} will decrease, i.e., H_{c2} will increase. This increase will continue until $|H_J| = H^*_{c2}$.

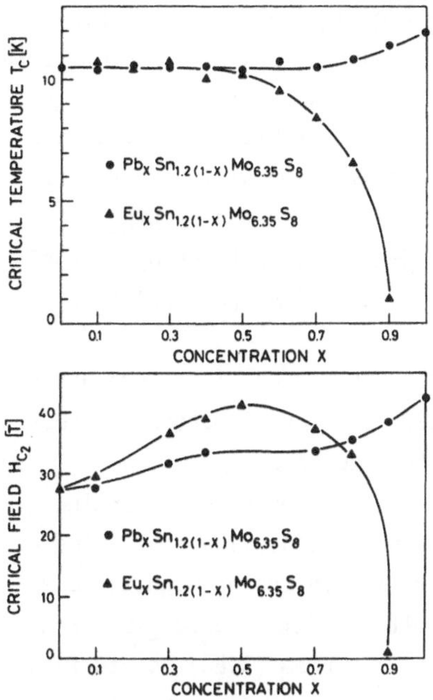

Fig.3.20

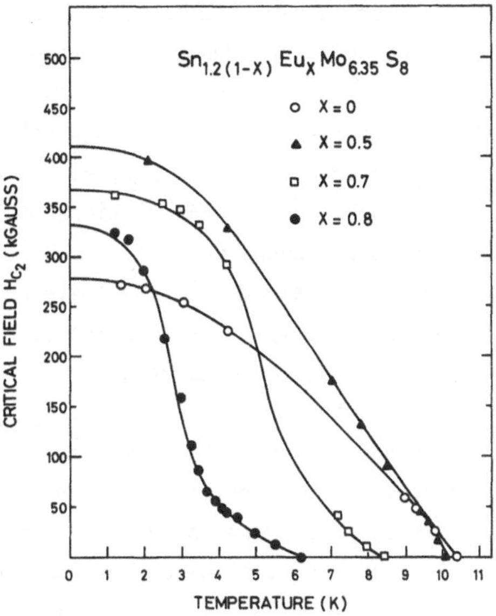

Fig.3.21

Fig.3.20. Critical temperature and critical field (T → 0) in the series $Sn_{1.2(1-x)}Eu_xMo_{6.35}S_8$ versus x. Critical field (T → 0) in the series $Sn_{1.2(1-x)}Pb_xMo_{6.35}S_8$ versus x [3.23]

Fig.3.21. Critical field versus temperature for several alloys in the series $Sn_{1.2(1-x)}Eu_xMo_{6.35}S_8$ [3.23]

At this point H_{c2} is equal to H^*_{c2} and passes through a maximum. A determination of this maximum allows us to directly determine H_J and thus the exchange constant Γ. Note that the total increase in H_{c2} is equal to the paramagnetic reduction of H_{c2} in the compound without magnetic ions. Thus in a material such as, for instance, a low field superconductor where there is no significant paramagnetic pairbreaking, there can be no increase in the critical field due to this effect. We now turn to the temperature dependence and consider the case $|H_J(H_{c2}, T = 0)| \geq H_{c2}(0)$. When we start from T_c and lower the temperature, H_{c2} will first be given by H^*_{c2}, then H_J will rapidly grow and the paramagnetic pairbreaking term in (3.81) becomes very large. In this region H_{c2} versus T flattens out. As T is lowered further, H_J saturates and as H_{c2} increases the paramagnetic pairbreaking is reduced. This leads to an increase in H_{c2} which leads to a further decrease in the paramagnetic pairbreaking, etc. The result of this is that H_{c2} increases very quickly as T decreases, therefore leading to a characteristically positive curvature in $H_{c2}(T)$.

In Fig.3.20 we show H_{c2} (x, T = 0) as well as $T_c(x)$ for $Sn_{1-x}Eu_xMo_6S_8$ [3.79]. H_{c2} increases as expected and passes through a maximum at $x \simeq 0.5$. This gives H_J (x = 0.5) = 41 Tesla and Γ = 40 meV. This value is in reasonable agreement with the values derived for the other $REMo_6S_8$ compounds from T_m and T_c (see Chap.5). Note that the value found above is $\Gamma(q = 0)$, whereas the one determined from T_c is $\overline{\Gamma(q)}$. The latter is expected to be typically a factor of two smaller than $\Gamma(q = 0)$. In Fig.3.21 we show the temperature dependence of $H_{c2}(T)$. At high concentrations one can see the characteristic temperature dependence with a positive curvature. This explanation for the anomalous H_{c2} behavior has been confirmed by NMR and Mossbauer measurements by FRADIN et al. [3.80], by EPR investigations by ODERMATT [3.81], and band structure calculations by JARLBORG and FREEMAN [3.9,82] (for further details see Chaps.6,7).

To analyze the data in a quantitative manner, one has to take into account the corrections to the simplified equation (3.81) [3.19]. The most important correction here is that H_{c2} depends on x. To illustrate this we have plotted in Fig.3.20 $H_{c2}(x;T = 0)$ for the series $Sn_{1.2(1-x)}Pb_xMo_6S_8$. Using this series to determine $H^*_{c2}(x)$, we find that the enhancement of H_{c2} due to the compensation is 7.5 Tesla. The paramagnetic reduction of H_{c2} is therefore about 18% in the nonmagnetic compounds. This compares reasonably well with the result found for $Mo_6Se_{8-x}S_x$ from our detailed study of the temperature dependence of H_{c2}. From (3.80) we can then calculate λ_{so}. This gives $\lambda_{so} \simeq 8$. Taking into account a probable correction $[S/(1 + \lambda)]^2 \simeq 0.5$, we finally estimate $\lambda_{so} \simeq 4$. We therefore approach reasonable values for λ_{so} although it is probably still somewhat large to be interpreted as an ordinary spin-orbit scattering parameter. The temperature dependence of H_{c2} in these materials have not been investigated with the same precision as in the series $Mo_6Se_{8-x}S_x$. However, from the above results we deduce that the orbital critical field of $SnMo_6S_8$ is enhanced above the WHH curve by about 20% at T = 0, similar to what was found for the pseudobinary series. From these data we can now calculate the saturation values of $H_J(x)$ for $x \leq 0.5$ using (3.80). This gives the result shown in Fig.3.22. As expected, H_J varies approximately linearly with x confirming the interpretation of a compensation effect [3.83]. For $x > 0.5$, the sudden decrease of T_c makes the interpretation more difficult. If we assume that the behavior of H_{c2} close to T = 0 corresponds to a $T_c \simeq 10$ K, we get for H_J the values given in Fig.3.22 for $x > 0.5$. The good correspondence with the data for $x \leq 0.5$ suggests that this interpretation is correct. However, the reasons for such behavior is not yet clear. We will come back later to the anomalous concentration dependence of T_c shown in Fig.3.20.

Similar behavior to that discussed above was also found in $Pb_{1-x}Eu_xMo_6S_8$ [3.22]. In the latter system we found a critical field above 60 Tesla at 2K for a compound with the nominal composition of $Pb_{0.7}Eu_{0.3}Gd_{0.2}Mo_6S_8$. Taking into account the large widths of the resistive transition, this means that one would need almost 70 Tesla to drive the material completely normal at T = 0.K. These latter measurements were made with a very fast coil having a rise time of 3 ms. It has turned out later on that in our sintered samples, dynamical effects may have influenced the results.

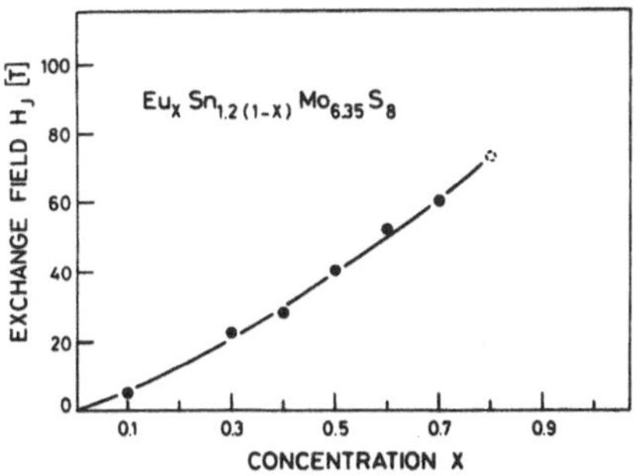

Fig.3.22. Exchange field H_J versus x in the series $Sn_{1.2(1-x)}Eu_xMo_{6.35}S_8$ as calculated from the critical field [3.23,83]

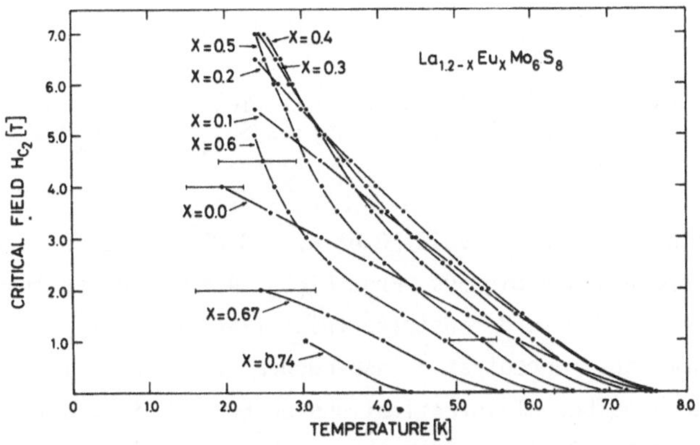

Fig.3.23. Critical field versus temperature for different Eu concentrations in the series $La_{1.2-x}Eu_xMo_6S_8$ [3.84]

These effects, which are sample dependent, are probably related to the sintered nature of our samples and these measurements need to be confirmed as soon as better samples become available.

More recently, TORIKACHVILI and MAPLE [3.84] reported a compensation effect in $La_{1.2-x}Eu_xMo_6S_8$. The temperature dependence of H_{c2} is shown in Fig.3.23. In this system, we expect a large increase of H^*_{c2} as x increases since we replace a trivalent atom by a divalent one. However, the compensation effect shows itself in the positive curvature which happens in the region where M begins to saturate and where the last term in (3.80) is still large.

An interesting possibility is that the paramagnetic pairbreaking may become so large that it already destroys superconductivity at low fields. But at higher fields the compensation effect will reduce the paramagnetic pairbreaking again and superconductivity will reappear. This means that one would have fieldinduced superconductivity [3.19] as originally proposed by JACCARINO and PETER [3.24]. Such an effect may appear in a narrow temperature interval in the series $Sn_{1-x}Eu_xMo_6S_8$ for $x > 0.8$. In Fig.3.24, a $H_{c2}(T)$ curve calculated from the theory developed in [3.19] using the parameters derived from the analysis on $Sn_{1-x}Eu_xMo_6S_8$ and assuming $x = 0.9$ [3.83] is shown. We see that there is a narrow temperature interval around $T = 1.3$ K where field-induced superconductivity is expected. An

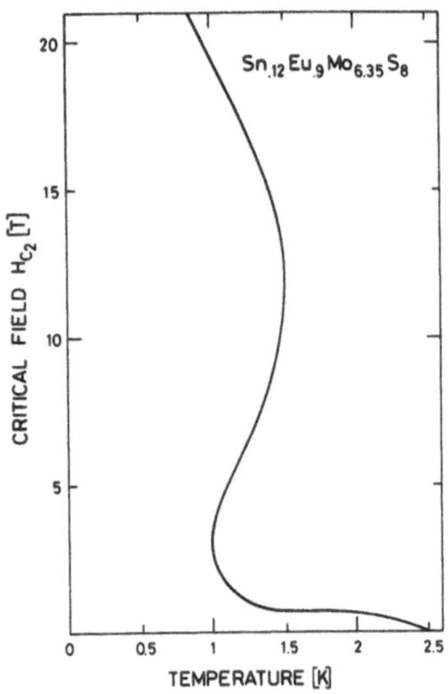

Fig.3.24. Calculated critical field versus temperature for x = 0.9 in the series $Sn_{1.2(1-x)}Eu_xMo_6S_8$ [3.83]

indication for such behavior has been reported by ISINO et al. [3.85]. However, broad transitions make the interpretation ambiguous. From Fig.3.24 it is clear that one needs very sharp transitions in order to demonstrate this effect properly.

Let us finally discuss the anomalous concentration dependence of T_c shown in Fig.3.20. There has recently been a number of investigations on $EuMo_6S_8$ in order to understand the absence of superconductivity. If one bases the discussion on the behavior known from other $REMo_6S_8$ compounds, one would expect a T_c of the order 10 K for $EuMo_6S_8$ [3.10] and an essentially linear dependence of T_c upon x in the series $Sn_{1-x}Eu_xMo_6S_8$. HARRISON et al. [3.86] and CHU et al. [3.87] reported that $EuMo_6S_8$ becomes superconducting under pressure with a maximum T_c of about 11 K. BAILLIF et al. [3.88] found that there is a structural phase transition in this compound at about 110 K and suggest that the low temperature phase is semiconducting. This constitutes an alternative explanation of the anomalous temperature dependence of the resistance reported by MAPLE et al. [3.89]. In the latter paper, the low temperature increase was interpreted as a sign of either a Kondo effect or valence instabilities. Hall effect measurements [3.86] suggest that the number of carriers is much lower in the low temperature phase than in the high temperature one. Thus, there must at least be gaps opening up at certain parts of the Fermi surface. It is reasonable to assume that this phase transition is closely related to the anomalous behavior of the transition temperature in $Sn_{1-x}Eu_xMo_6S_8$. The exact role of the phase transition, as well as the question of eventual magnetic or valence fluctuation effects, remains to be clarified. For a further discussion of these points see Chap.5 also.

3.7 Critical Currents and the Problem of Wire Production

The high critical field of some of the TMC's make them obvious candidates for applications. The comparison between $PbMo_6S_8$ and other high field superconductors shown in Fig.3.1 clearly demonstrates that if one could produce wires or tapes of $PbMo_6S_8$ with high current-carrying capacities, this material would be of great interest for high magnetic field production.

The problems of wire production are many: the material in the final wire must have a composition so that H_{c2} and T_c are optimal (see Sect.3.3); the microstructure must be such that the critical current is high; since the compounds are very brittle, a support material must be used; the compounds are also rather reactive when in contact with other metals so that not all metals can be used as support materials, etc.

Let us first discuss the problems directly related to the wire production. The first attempt to produce $PbMo_6S_8$ wires was carried out by DECROUX et al. [3.11]. In a two-step diffusion method, they fabricated a $PbMo_6S_8$ surface layer on a molybdenum wire. The critical current was only modestly high: 3.10^7 A/m^2 at 4 Tesla and 4.2 K. LUHMAN and DEW-HUGHES reported results on $PbMo_6S_8$ powder drawn to wires in a silver matrix [3.90]. Their T_c and H_{c2} values were high, but the critical current was very low: $\sim 8 \cdot 10^6$ A/m^2 at 4 Tesla and 4.2 K.

In spite of the low J_c values reported in [3.90], we believe that powder metallurgical methods may finally lead to good wires. As mentioned above, the problem is to find a support material which has good mechanical properties and which does not react with the superconducting phase. This latter condition was the reason for choosing silver as a support material in [3.90]. Unfortunately, copper cannot be used in direct contact with $PbMo_6S_8$ since small quantities of Cu impurities destroys the superconducting properties [3.62]. The same is, for instance, true for Fe. ROSSEL et al. investigated the possibility of using tantalum as a support material [3.12]. Ta is a very ductile material and can easily be drawn into wires, and they started from 2 mm Ta-tubes filled with a mixture of MoS_2, PbS and Mo powders (grain size < 50 μm). These tubes were drawn to a final outer diameter of between 0.25 and 0.5 mm. The $PbMo_6S_8$-phase was then reacted at 700°-900°C for 10 to 120 minutes. These wires had T_c = 12.5 K and $(dB_{c2}/dT)_{T_c} \simeq 3$ Tesla/K. It turned out that higher temperatures and longer reaction times led to an undesired reaction with tantalum resulting in a decrease in T_c. In spite of these shortcomings, these wires showed an order of magnitude increase in J_c with respect to the $PbMo_6S_8$-silver wires.

From the point of view of reaction with the superconductor, molybdenum is probably the best matrix material. SEEBER et al. [3.13] studied this possibility. The main problem with Mo is that it is not ductile enough to be drawn at room temperature, and needs, in fact, a rather high deformation temperature. Furthermore, the ductility of molybdenum is very sensitive to both oxygen and nitrogen and the surface has to be protected during the deformation process. The Mo-based wires were produced in the following way: a 10 mm o.d. Mo-tube was filled with a mixture of MoS_2, PbS, and Mo-powders. This tube was sealed and then inserted into a stainless steel tube of 14 mm o.d. This tube was welded at both ends in order to have protection against oxidation during the hot drawing process. The diameter was first reduced to 3 mm by forging at 750°C and then to 0.7 mm by drawing at 600°C. After every 30% reduction of the cross section, a recovery anneal at 850°C for 30 min was necessary. After a final heat treatment at 900°C, the samples had $T_c \simeq$ 12-13 K and $(dH_{c2}/dT)_{T_c} \simeq$ 3 Tesla/K. The critical currents were generally somewhat better than in the Ta-wires with a best result of 2.10^8 A/m^2 at 4 Tesla and 4.2 K.

Of course, these wires are still not as good as one would like them to be. However, in view of the rather modest effort that has been spent on these problems so far, we feel that these results demonstrate that wires of $PbMo_6S_8$ can be produced and that there is good hope that the superconducting properties will be good enough to compete with other technological superconductors.

Critical current studies on thin films of $CuMo_6S_8$ and $PbMo_6S_8$ have been carried out in great detail by ALTEROVITZ and WOOLLAM [3.14]. Their results are described in [Ref.3.6, Chap.5] and we refer the reader to that chapter for a detailed discussion.

A convenient way of investigating the critical current in these materials is by inductive methods. In this case the measurement can be carried out on bulk samples and thus one avoids the complications due to wire drawing and the corresponding difficulty in interpreting the results. ALEKSEEVSKII et al. [3.91] report relatively high current densities in such materials. They found $J_0 = 10^8$ A/m^2 at 10 Tesla for a Ga-doped $PbMo_6S_8$ sample. ROSSEL [3.92] has recently completed a thorough study of the critical current of bulk samples using the AC-method of CAMBELL [3.93]. He achieved $J_c = 2.10^8$ A/m^2 at 10 Tesla in hot-pressed samples. These samples have the advantage that they are very dense; however, the grains are relatively large and one would expect that J_c could be increased considerably if the grain size could be reduced. In Fig.3.25 we show the critical current density as a function of field for the different investigations discussed here.

This brings us to the obvious question of how high the critical current density can be made in $PbMo_6S_8$. Pinning theories are still in a rather incomplete state. However, as BRANDT has pointed out recently [3.94], for high field materials the maximum critical current density can be estimated by a direct summation of the individual pinning forces. Using this fact, BRANDT calculated the maximum critical current assuming an optimal arrangement of pinning centers and found

$$J_c^{Max} = 3.65 \cdot 10^{12} \mu_0^{3/2} H_c^2 H_{c2}^{-1/2} (1 - h) \quad , \tag{3.82}$$

where H_c is the thermodynamical critical field and $h = H/H_{c2}$. Using this formula,

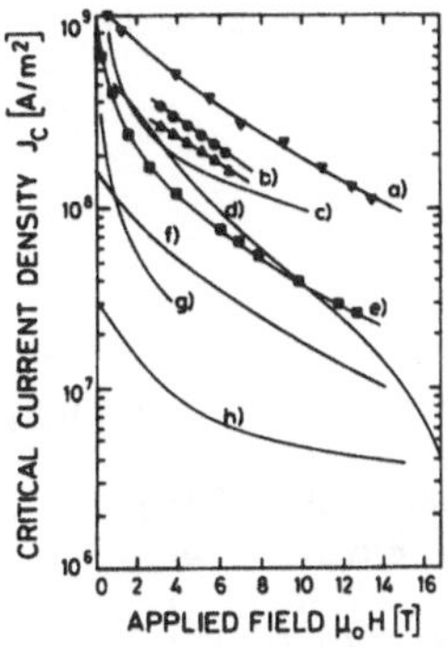

Fig.3.25. Critical current density versus magnetic field for $PbMo_6S_8$ reported in the different investigations disccussed in the text [3.13,92]: (a) inductive measurements on a hot pressed bulk sample [3.92]; (b) Mo-based wires [3.13]; (c) inductive measurement on a sintered $PbGa_2Mo_6S_8$ sample [3.91]; (d) sputtered thin film [3.95]; (e) Ta-based wires [3.12]; (f) sputtered thin film [3.97]; (g) Mo wires with surface layer of $PbMo_6S_8$ [3.11]; (h) Ag-based wires [3.90]

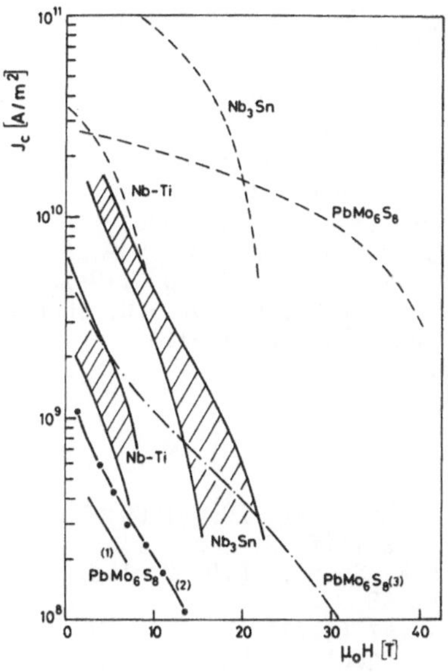

Fig.3.26. Comparison of critical current densities in $PbMo_6S_8$, Nb_3Sn and Nb-Ti. The dashed lines are upper limits estimated from (3.82). The hatched areas represent the present values for Nb-Ti and Nb_3Sn. The dot-dashed line represents the critical current density extrapolated from measurements on thin films [3.96]. From [3.13,92]

J_c^{max} has been estimated for NbTi, Nb_3S_n and $PbMo_6S_8$ [3.14,92]. The results are displayed in Fig.3.26 together with typical values for NbTi and Nb_3Sn as well as the best results for $PbMo_6S_8$. Considering first the J_c^{max} values, we see that $PbMo_6S_8$ is potentially better than NbTi above about 3 Tesla and better than Nb_3Sn above 18-20 Tesla. It would, of course, be unreasonable to expect that one could achieve J_c values as high as J_c^{max}. The ratio J_c^{max}/J_c^{obs} for NbTi and Nb_3Sn give a reasonable estimate of what can be obtained in practical wires. Thus, one is led to the conclusion that critical currents of the order 10^9 A/m^2 at 20 T should be possible. Similar values, although slightly lower, were estimated by ALTEROVITZ and WOOLLAM on the basis of their investigations on thin films. Their estimate is also shown in Fig.3.26 and more details can be found in [Ref.3.6, Chap.5].

In conclusion, we note that $PbMo_6S_8$ has the potential of becoming a practical superconductor. Wires can be produced and from the data obtained, as well as estimates based on simple relationships, we expect the critical currents to be high enough for practical applications.

References

3.1 R. Odermatt, Ø. Fischer, H. Jones, G. Bongi: J. Phys. C (Sol. Stat. Phys.) *7*, L13 (1974)
3.2 S. Foner, E.J. McNiff, Jr., E.J. Alexander: Phys. Lett. *49*A, 269 (1974)
3.3 Ø. Fischer, H. Jones, G. Bongi, M. Sergent, R. Chevrel: J. Phys. C. Sol. State Phys. *7*, L450 (1974)
3.4 Ø. Fischer: Proc. Lt 14, Vol.5, ed. by M. Krusius and M. Vuorio (North Holland, Amsterdam 1975) p.172
3.5 S. Foner, E.J. McNiff, Jr., R.N. Shelton, R.W. McCallum, M.B. Maple: Phys. Lett. *57*A, 345 (1976)
3.6 Ø. Fischer, M.B. Maple (eds.): *Superconductivity in Ternary Compounds I*, Topics in Current Phys., Vol.32 (Springer, Berlin, Heidelberg, New York 1982)
3.7 Ø. Fischer: Coll. Int. CENRS *242*, 79 (1974)
3.8 O.K. Anderson, W. Klose, H. Nohl: Phys. Rev. B*17*, 1209 (1978)
3.9 T. Jarlborg, A.J. Freeman: Phys. Rev. Lett. *44*, 178 (1980)
3.10 Ø. Fischer: Appl. Phys. *16*, 1 (1978)
3.11 M. Decroux, Ø. Fischer, R. Chevrel: Cryogenics *17*, 291 (1977)
3.12 C. Rossel, B. Seeber, Ø. Fischer: Phys. Stat. Sol. (a) *59*, K43 (1980)
3.13 B. Seeber, C. Rossel, Ø. Fischer: In *Ternary Superconductors*, ed. by G.K. Shenoy, B.D. Dunlap, F.Y. Fradin (North Holland, Amsterdam 1981) pp.119-124
3.14 S.A. Alterovitz, J.A. Woollam: In *Ternary Superconductors*, ed. by G.K. Shenoy, B.D. Dunlap, F.Y. Fradin (North Holland, Amsterdam 1971) pp.113-118
3.15 B.S. Chandrasekhar: Appl. Phys. Lett. *1*, 7 (1962)
3.16 A.M. Clogston: Phys. Rev. Lett. *9*, 266 (1962)
3.17 K. Maki: Physics *1*, 127 (1964)
3.18 R.R. Hake: Appl. Phys. Lett. *10*, 186 (1967)
3.19 Ø. Fischer: Helv. Phys. Acta *45*, 229 (1972)
3.20 N.R. Werthamer, E. Helfand, P.C. Hohenberg: Phys. Rev. *147*, 295 (1966)
3.21 M. Decroux, Ø. Fischer, C. Rossel, B. Lachal, R. Baillif, R. Chevrel, M. Sergent: In *Ternary Superconductors*, ed. by G.K. Shenoy, B.D. Dunlap, F.Y. Fradin (North Holland, Amsterdam 1981) pp.65-68
3.22 M. Ishikawa, Ø. Fischer: Solid State Commun. *24*, 747 (1977)
3.23 Ø. Fischer, M. Decroux, S. Roth, R. Chevrel, M. Sergent: J. Phys. C*8*, L474 (1975)
3.24 V. Jaccarino, M. Peter: Phys. Rev. Lett. *9*, 290 (1962)
3.25 M. Decroux, Ø. Fischer, R. Flükiger, B. Seeber, R. Delesclefs, M. Sergent: Solid State Commun. *25*, 393 (1978)
3.26 M. Decroux, B. Seeber, Ø. Fischer, R. Delesclefs, R. Flükiger: J. Physique *39*, C6-363 (1978)
3.27 N.E. Alekseevskii: Cryogenics *257* (1980)
3.28 K. Okuda, M. Kitagawa, T. Sakakibara, M. Date: J. Phys. Soc. Jpn. *48*, 2157 (1980)
3.29 Ø. Fischer, M. Decroux, M. Sergent, R. Chevrel: J. Physique *39*, C2-257 (1978)
3.30 Ø. Fischer, M. Decroux, R. Chevrel, M. Sergent: In *Superconductivity in d- and f-band Metals"*, ed. by D.H. Douglass (Plenum Press, New York 1976) p.175
3.31 S.A. Alterovitz, J.A. Woollam: Solid State Commun. *25*, 141 (1978)
3.32 See, for instance, M. Tinkham: *Introduction to Superconductivity* (McGraw Hill, New York 1975)
3.33 For a detailed discussion of spin paramagnetism in the Ginzburg-Landau theory, see N.R. Werthamer: In *Superconductivity*, ed. by R.D. Parks (Marcel Dekker, New York 1969)
3.34 L.P. Gor'kov, T.K. Melik-Barkhudarov: Sov. Phys.-JETP *18*, 1031 (1964)
3.35 R.C. Morris, R.V. Coleman, R. Bhandari: Phys. Rev. B*5*, 895 (1972)

3.36 L.P. Gor'kov: Sov. Phys.-JETP *10*, 998 (1960)
3.37 P.G. de Gennes: Phys. Cond. Mat. *3*, 79 (1964)
3.38 K. Maki: Physics *1*, 21 (1964)
3.39 E. Helfand, N.R. Werthamer: Phys. Rev. *147*, 288 (1966)
3.40 T.P. Orlando, E.J. McNiff, Jr., S. Foner, M.R. Beasley: Phys. Rev. B*19*, 4545 (1979)
3.41 G. Sarma: J. Phys. Chem. Solids *24*, 1029 (1963)
3.42 A.A. Abrikosov, L.P. Gor'kov: Sov. Phys.-JETP *12*, 1243 (1961)
3.43 P. Fulde, K. Maki: Phys. Rev. *141*, 275 (1966)
3.44 D. Rainer, G. Bergmann: J. Low Temp. Phys. *14*, 501 (1974)
3.45 D. Rainer, G. Bergmann, U. Eckhardt: Phys. Rev. B*8*, 5324 (1973)
3.46 P.C. Hohenberg, N.R. Werthamer: Phys. Rev. *153*, 493 (1967)
3.47 For recent references on anisotropy see *Anisotropy Effects in Superconductors*, ed. by H. Weber (Plenum, New York 1977)
3.48 D.W. Youngner, R.A. Klemm: Phys. Rev. B*21*, 3890 (1980)
3.49 H.W. Pohl, H. Teichler: Phys. Stat. Sol. (b) *75*, 205 (1976)
3.50 M. Peter, J. Ashkenazi, M. Dacorogna: Helv. Phys. Acta *50*, 267 (1977)
3.51 P. Entel, M. Peter: J. Low. Temp. Phys. *22*, 613 (1976)
3.52 M. Decroux: Thesis, University of Geneva (1980)
3.53 R. Chevrel, M. Sergent, Ø. Fischer: Mat. Res. Bull. *10*, 1169 (1975)
3.54 Ø. Fischer, B. Seeber, M. Decroux, R. Chevrel, M. Potel, M. Sergent: In *Superconductivity in d- and f-band Metals*, ed. by H. Suhl, M.B. Maple (Academic, New York 1980) p.485
3.55 B. Lachal: Private communication
3.56 J.A. Woollam, S.A. Alterovitz, E.J. Hangland: Phys. Lett. *68*A, 122 (1978)
3.57 P.B. Allen, W.E. Pickett, K.M. Ho, M.L. Cohen: Phys. Rev. Lett. *40*, 1532 (1978)
3.58 M. Milewitz, S.J. Williamson: Phys. Rev. B*13*, 5199 (1976)
3.59 Z. Fisk, G.W. Webb: Phys. Rev. Lett. *36*, 1084 (1976)
3.60 D.W. Woodard, G.D. Cody: Phys. Rev. *136*, A166 (1964)
3.61 U. Poppe, H. Wuhl: J. Low. Temp. Phys. *43*, 371 (1981); see also Ref.3.6, Chap.8
3.62 M. Sergent, R. Chevrel, C. Rossel, Ø. Fischer: J. Less Comm. Metals *58*, 179 (1978)
3.63 R. Flükiger, R. Baillif, E. Walker: Mat. Res. Bull. *13*, 743 (1978)
3.64 S.V. Vonsovsky, Y.A. Izyumov, E.Z. Kurmaev: *Superconductivity of Transition Metals*, Springer Ser. in Solid-State Sci., Vol.27 (Springer, Berlin, Heidelberg, New York 1982)
3.65 O.K. Andersen: Private communication
3.66 B. Seeber, M. Decroux, Ø. Fischer, R. Chevrel, M. Sergent, A. Gruttner: Sol. State Commun. *29*, 419 (1979)
3.67 M. Potel, R. Chevrel, M. Sergent, M. Decroux, Ø. Fischer: C.R. Acad. Sc. Paris *288*, C-429 (1979)
3.68 J.C. Armici, M. Decroux, Ø. Fischer, M. Potel, R. Chevrel, M. Sergent: Sol. State Commun. *33*, 607 (1980)
3.69 M. Potel, R. Chevrel, M. Sergent: Acta Cryst. B*36*, 1545 (1980)
3.70 W. Höhnle, H.G. von Schnering, A. Lipka, K. Yvon: J. Less Common Metals *71*, 135 (1980)
3.71 B. Lachal: Private communication
3.72 H.W. Meul: Private communication
3.73 D. Rainer: Private communication
3.74 Ø. Fischer, A. Treyvaud, R. Chevrel, M. Sergent: Sol. State Commun. *17*, 721 (1975)
3.75 M. Ishikawa, Ø. Fischer, J. Muller: J. Physique *39*, C6-1379 (1978)
3.76 Ø. Fischer, M. Ishikawa, M. Pelizzone, A. Treyvaud: J. Physique *40*, C5-89 (1979)
3.77 A.A. Abrikosov, L.P. Gor'kov: Sov. Phys.-JETP *12*, 1243 (1961)
3.78 R.D. Parks: In *Superconductivity*, Vol.2, ed. by P.R. Wallace (Gordon and Breach, New York 1969) p.625
3.79 We write here the stoichiometric formula. The nominal formula for this series was $Sn_{1.2(1-x)}Eu_xMo_{6.25}S_8$ as specified in the figures
3.80 F.Y. Fradin, G.K. Shenoy, B.D. Dunlap, A.T. Aldred, C.W. Kimball: Phys. Rev. Lett. *38*, 719 (1977)
3.81 R. Odermatt: Helv. Phys. Acta *54*, 1 (1981)

3.82 T. Jarlborg, A.J. Freeman: J. of Magn. and Magn. Mat. *15-18*, 1579 (1980)
3.83 M. Decroux: Diploma Thesis, University of Geneva (1975)
3.84 M.S. Torikachvili, M.B. Maple: Solid State Commun. *40*, 1 (1981)
3.85 M. Isino, N. Kobayashi, M. Muto: In *Ternary Superconductors*, ed. by G.K. Shenoy, B.D. Dunlap, F.Y. Fradin (North Holland, Amsterdam 1981) p.95
3.86 D.W. Harrison, K.C. Lim, H.D. Thompson, C.Y. Huang, P.D. Hambourger, H.L. Luo: Phys. Rev. Lett. *46*, 280 (1981)
3.87 C.W. Chu, S.Z. Huang, C.H. Lin, R.L. Meng, W.K. Wu, P.H. Schmidt: Phys. Rev. Lett. *46*, 276 (1981)
3.88 R. Bailiff, A. Dunand, J. Muller, K. Yvon: Phys. Rev. Lett. *47*, 672 (1981)
3.89 M.B. Maple, L.E. De Long, W.A. Fertig, D.C. Johnston, R.W. McCallum, R.N. Shelton: In *Valence Instabilities and Related Narrow-Band Phenomena*, ed. by R.D. Parks (Plenum, New York 1977) p.17
3.90 T. Luhman, D. Dew-Hughes: J. Appl. Phys. *49*, 936 (1978)
3.91 N.R. Alekseevskii, N.M. Dobrovol'skii, D. Eckert, V.I. Tsebro: Sov. Phys.-JETP *45*, 599 (1977); J. Low Temp. Phys. *29*, 565 (1977)
3.92 C. Rossel: Thesis, University of Geneva (1981) to be published
3.93 A.M. Cambell: J. Phys. C*2*, 1492 (1969); C*4*, 3186 (1971)
3.94 E.H. Brandt: Phys. Lett. *77*A, 484 (1980)
3.95 S.A. Alterovitz, J.A. Woollam, L. Kammerdiner, H.L. Luo: J. Low Temp. Phys. *30*, 797 (1978)
3.96 S.A. Alterovitz, J.A. Woollam: Cryogenics *19*, 167 (1979)
3.97 P. Przyslupski, R. Horyn, B. Gren: J. Low Temp. Phys. *38*, 83 (1980)

4. Superconductivity, Magnetism and Their Mutual Interaction in Ternary Rare Earth Rhodium Borides and Some Ternary Rare Earth Transition Metal Stannides

M. B. Maple*, H. C. Hamaker, and L. D. Woolf*

With 27 Figures

Along with the rare-earth molybdenum chalcogenides $RE_xMo_6X_8$ (X = S or Se), the rare-earth rhodium borides $RERh_4B_4$ have been studied extensively in recent years in connection with the interplay between superconductivity and magnetism. The purpose of this chapter is to review the experiments that have been carried out on the $RERh_4B_4$ compounds since their discovery some five years ago. The experimental situation for the $RE_xMo_6X_8$ compounds is discussed in Chap.5 by Ishikawa, Fischer and Müller, while the theoretical aspects of magnetic superconductors are considered in Chap.9 by Fulde and Keller.

4.1 Background

In 1977, MATTHIAS and coworkers discovered a set of compounds with the formula MRh_4B_4, where M is Y, Th or a RE element [4.1]. They reported that some of the compounds (M = Y, Th, Nd, Sm, Er, Tm and Lu) were superconducting, while others (M = Gd, Tb, Dy and Ho) were ferromagnetic. The compounds with M = La, Ce, Pr, Eu and Yb could not be formed. Shortly thereafter, the crystal structure of the MRh_4B_4 compounds was determined by VANDENBERG and MATTHIAS [4.2] to be the same as the primitive tetragonal structure of the $CeCo_4B_4$-type compounds that were investigated earlier by KUZ'MA and BILONIZHKO [4.3].

Two special features make the $RERh_4B_4$ compounds (and the $RE_xMo_6X_8$ compounds, as well) uniquely suited for investigations of the mutual interaction between superconductivity and magnetism: (i) the exchange interaction between the conduction electron spins and the RE magnetic moments is relatively weak, and (ii) the RE ions form an ordered sublattice. The absolute value of the exchange interaction parameter $\mathcal{J}$ for $RERh_4B_4$ compounds is only about 0.01 eV-atom [4.4,5], almost an order of magnitude smaller than typical values encountered in binary RE metallic systems such as the rare-earth dialuminides $REAl_2$ [4.6]. The exchange interaction parameter $\mathcal{J}$ is defined in terms of the usual exchange interaction Hamiltonian which, for RE systems, is given by

$$\mathcal{H}_{int} = -2\mathcal{J}(g_J - 1)\underline{J}\cdot\underline{s} \quad . \qquad (4.1)$$

*Research supported by the U. S. Department of Energy under Contract No. DE-AT03-76ER70227

Here, g_J and $\underline{J}$ are, respectively, the Landé g-factor and total angular momentum operator for the Hund's rule ground state of the RE ion under consideration, and $\underline{s}$ is the conduction electron spin density at the site of the RE ion. The small magnitude of $\mathcal{J}$ enables the $RERh_4B_4$ compounds to retain their superconductivity, even in the presence of a relatively large concentration of RE magnetic moments ($\sim$11 at.%), and results in ordering of the RE magnetic moments via the indirect RKKY interaction [4.7] at relatively low temperatures that are comparable to superconducting transition temperatures. The ordered sublattice of the RE ions leads to long-range magnetic ordering as evidenced by sharp magnetic ordering temperatures T_M and well-defined features in the physical properties at T_M. This greatly facilitates the analysis of effects that arise from the competition between the superconducting and magnetic order parameters in these materials.

The superconducting and magnetic critical temperatures of the $RERh_4B_4$ compounds are plotted vs RE in Fig.4.1 [4.1,8-14] and tabulated in Table 4.1. With increasing RE atomic number, the low temperature behavior switches from superconductivity for Nd and Sm, to ferromagnetism for Gd, Tb, Dy and Ho, and back to superconductivity for Er, Tm and Lu. Moreover, all of the superconducting $RERh_4B_4$ compounds in which the RE 4f electron shell is partially filled undergo some type of magnetic ordering below their superconducting transition temperatures T_c at temperatures T_M in the vicinity of 1 K. Whereas $ErRh_4B_4$ becomes ferromagnetic [4.8,9], $NdRh_4B_4$ [4.10,11], $SmRh_4B_4$ [4.12] and $TmRh_4B_4$ [4.13,14] exhibit antiferromagnetic transitions.

If one disregards the ferromagnetic $RERh_4B_4$ compounds near the middle of the RE series, T_c would appear to display a minimum near Gd at the half-filled 4f electron shell. This general behavior is in accordance with the theory of ABRIKOSOV and GOR'KOV (hereafter AG) [4.15] which predicts that the depression of the superconducting transition temperature by paramagnetic impurities scales with the deGennes factor $\mathcal{G} = (g_J - 1)^2 J(J + 1)$ which attains a maximum value at Gd. As will be discussed later, the magnitude of $\mathcal{J}$ has been estimated for the $RERh_4B_4$ compounds from the depression of T_c of $LuRh_4B_4$ by RE solutes with partially filled 4f electron shells.

Besides the depression of T_c, other properties (e.g., the magnetic ordering temperature and spin disorder resistivity) of an isostructural series of metallic RE compounds in which the RE magnetic moments interact with one another via the indirect RKKY mechanism should scale with the deGennes factor [4.16], as observed, for example, in $REAl_2$ compounds [4.6]. However, the peak in the magnetic ordering temperatures of the $RERh_4B_4$ compounds occurs at RE = Dy rather than Gd, as would be expected from the variation of $\mathcal{G}$ with RE. This suggests that other factors including crystalline electric field (CEF), magnetoelastic and possibly even dipolar interactions affect the superconducting and magnetic properties of the $RERh_4B_4$ compounds. Normal state properties such as magnetic susceptibility, heat capacity and thermal expansion are, in fact, replete with features that can be attributed to

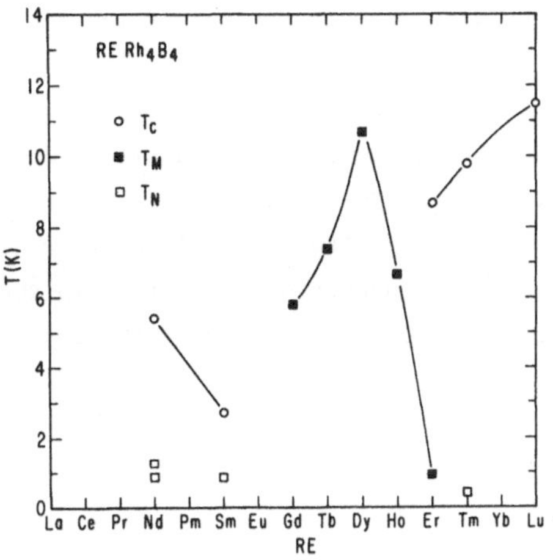

Fig.4.1. Superconducting T_C, ferromagnetic T_M, and antiferromagnetic T_N transition temperatures vs RE for $RERh_4B_4$ compounds. From [4.1,8-14]

Table 4.1. Superconducting and magnetic transition temperatures for $RERh_4B_4$

RE	T_c[K]	T_M[K]	T_N[K]
Y	10.8		
Nd	5.3		1.31,0.89
Sm	2.7		0.87
Gd		5.8	
Tb		7.4	
Dy		10.7	
Ho		6.7	
Er	8.7	0.93	
Tm	9.8		0.4
Lu	11.5		

CEF effects where the splittings of the energy levels of the RE ions are also of the order of T_c and T_M. These features include pronounced Schottky anomalies in the specific heat, reduced entropies of ordering and magnetic moments (compared to Hund's rule values) and magnetic anisotropy. The role of dipolar contributions has not yet been assessed with certainty.

A possible source of the comparatively small conduction electron spin-RE magnetic moment exchange interaction in the $RERh_4B_4$ compounds is the Rh_4B_4 molecular units or "clusters" that, along with the RE ions, are the building blocks of the primitive tetragonal $CeCo_4B_4$-type crystal structure of these materials (Fig.2.1). It has even been conjectured by VANDENBERG and MATTHIAS [4.17] that clusters may be responsible for the relatively high superconducting critical temperatures of the $RERh_4B_4$ compounds and other high temperature superconductors as well.

In the $RERh_4B_4$ structure, the Rh-Rh intracluster distances are nearly the same as in the element and are smaller than the Rh-Rh intercluster distances. The relative confinement of the Rh 4d electrons within the clusters presumably results in a low 4d electron density at the site of the RE ion which accounts for the small value of $|\mathcal{J}|$. It also produces narrow peaks in the density of states at the Fermi level [4.18] which could contribute to the high values of T_c. Finally, the external and internal modes of the clusters themselves may modify the electron-phonon interaction in such a way as to enhance superconductivity, as discussed, for example, in [Ref.4.19, Chaps.1,7,8].

4.2 Ternary $RERh_4B_4$ Compounds

4.2.1 Nonmagnetic Superconducting $RERh_4B_4$ Compounds: RE = Lu, Y

The highest superconducting critical temperatures among the MRh_4B_4 compounds are displayed by the nonmagnetic compounds $LuRh_4B_4$ (T_c = 11.5 K), which has a completely filled 4f electron shell, and YRh_4B_4 (T_c = 10.8 K) which has no 4f electrons. The low temperature heat capacities of these two compounds in zero applied magnetic field are shown in Fig.4.2 [4.20,21]. The normal state heat capacity data for both compounds could not be fit by a modified Debye model, but could be described well by a molecular crystal model similar to the one employed by BADER et al. [4.22] for the compounds $PbMo_{5.1}S_8$ and $SnMo_5S_6$. In this model, the primitive tetragonal structure of the $RERh_4B_4$ compounds is considered to be composed of RE atoms coupled to Rh_4B_4 clusters. WOOLF et al. [4.21] have also measured the specific heat of $LuRh_4B_4$ in applied fields of 15 and 30 kOe, and KUMAGAI et al. [4.23] have reported specific heat measurements on YRh_4B_4. Various superconducting and normal state parameters for these compounds are tabulated in Tables 4.1,2. The magnetic susceptibilities of $LuRh_4B_4$ [4.20] and YRh_4B_4 [4.31] each exhibit a strong temperature dependence, presumably indicative of a rapid variation of the density of states near the Fermi level with temperature.

The shape of magnetization vs applied magnetic field isotherms at temperatures below T_c indicates that $LuRh_4B_4$ is a type II superconductor. From these data various superconducting and normal state properties have been estimated [4.30]. Thermal conductivity λ data on $LuRh_4B_4$ show that a considerable contribution to λ at T_c arises from phonon conduction, so that below T_c this contribution increases due to the reduced scattering by normal electrons [4.32].

The Ginsburg-Landau parameter κ defined by $H_{c2}(0) = \sqrt{2}\kappa H_c(0)$ has a value of ~ 9 for both $LuRh_4B_4$ and YRh_4B_4.

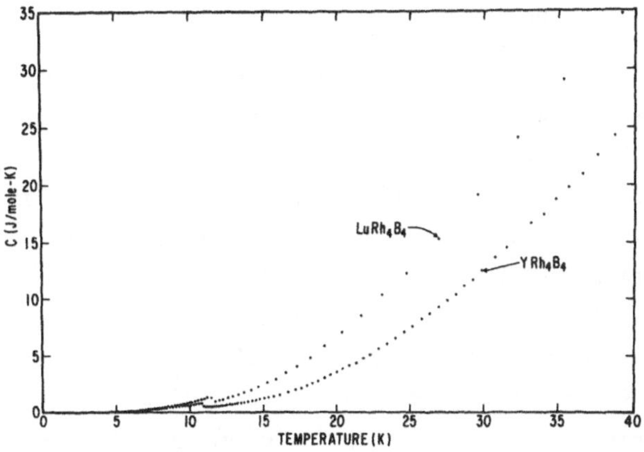

Fig.4.2. Heat capacity C vs temperature for $LuRh_4B_4$ [4.20] and YRh_4B_4 [4.21]

Table 4.2. Superconducting and normal state parameters for $LuRh_4B_4$ and YRh_4B_4

Parameter	Units	$LuRh_4B_4$	YRh_4B_4
T_c	K	11.5[a]	10.8[a]
ΔC	mJ/mole-K	474[a]	412[a]
γ	$mJ/mole\text{-}K^2$	26,[a] 31[b]	22,[a] 28[c]
$H_c(0)$	Tesla	0.187,[a] 0.179[b]	0.160[a]
$N(E_F)$	states/eV-atom-spin direction	0.61[a]	0.51,[a] 0.66[c]
$N_b(E_F)$	states/eV-atom-spin direction	0.32[d]	0.32[d]
λ		0.91[a,d]	0.59[a,d], 1.06[c,d]
$\Delta C/\gamma T_c$		1.60[a]	1.76,[a] 1.12[c]
$2\Delta(0)/k_BT_c$		4.1,[a] 3.8,[e] 4.2,[f] 3.9[g]	3.9[a]
$H_{c2}(0)$	Tesla	2.8,[a] 2.2[b]	2.1,[a] 2.2[h]
$\kappa(0)$		9.8,[a] 8.7[b]	9.7[a]

ΔC: jump in specific heat at T_c
$H_c(0)$: thermodynamic critical field at T = 0
λ: electron-phonon coupling parameter determined from $N(E_F)=N_b(E_F)(1+\lambda)$
$\Delta(0)$: superconducting energy gap at T = 0
[a] [4.20,21,24]
[b] [4.25]
[c] [4.23]
[d] [4.26]: calculation performed for YRh_4B_4
[e] [4.27]: data taken on $ErRh_4B_4$
[f] [4.28]: data taken on $ErRh_4B_4$
[g] [4.29]: data taken on $(Er_{0.58}Ho_{0.42})Rh_4B_4$
[h] [4.30]

4.2.2 Rate of Depression of T_c in $(Lu_{1-x}RE_x)Rh_4B_4$

The magnetic ordering temperatures of the $RERh_4B_4$ compounds shown in Fig.4.1 peak at $DyRh_4B_4$ [4.1] rather than $GdRh_4B_4$, as would be expected if the RE magnetic moments were to couple primarily via the RKKY interaction [4.7] in the absence of CEF effects. MACKAY et al. [4.5] noted that both the magnetic ordering temperature T_M in the RKKY model and the initial depression rate of T_c by magnetic impurities in superconductors in the AG theory are proportional to $N(E_F)\mathcal{J}^2$ [4.15,16] where $N(E_F)$ is the bare density of states at the Fermi level per spin direction. Thus, if the anomalous behavior of T_M with RE were due to variations of $N(E_F)\mathcal{J}^2$ with RE, this could be discerned by measuring the initial depression rate of T_c with x in the $(Lu_{1-x}RE_x)Rh_4B_4$ system for all RE elements with partially filled 4f electron shells. The depression rates of T_c, shown in Fig.4.3, were found to scale with the deGennes factor $\mathcal{G}$ for RE ions from Gd to Tm, implying that the quantity $N(E_F)\mathcal{J}^2$ is roughly constant for the heavy RE compounds. These authors then invoked CEF effects as a possible cause of the anomalous magnetic ordering temperatures. DELONG et al. [4.33] have also measured the pressure dependence of T_c in the $(Lu_{1-x}RE_x)Rh_4B_4$ system. Using the AG theory for the depression rate of T_c caused by Gd, for which CEF effects are absent in this system, a value for $|\mathcal{J}| \sim 2.3 \times 10^{-2}$ eV-atom was calculated [4.5].

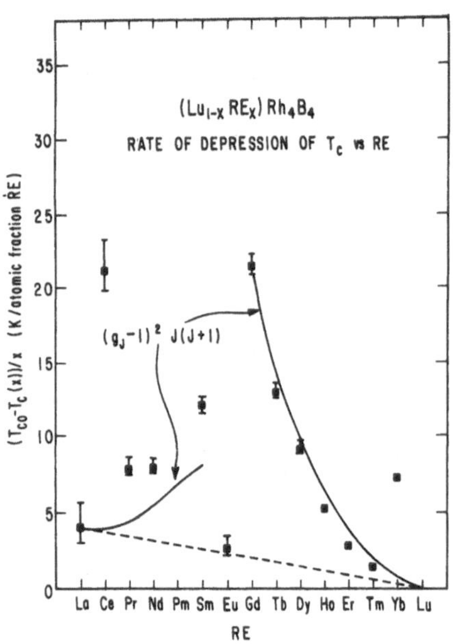

Fig. 4.3. Rate of depression of T_c vs RE for the $(Lu_{1-x}RE_x)Rh_4B_4$ system. The dashed line is the assumed baseline upon which the deGennes curve (solid line), normalized to the depression rate for RE = Gd, has been drawn. From [4.5]

The data shown in Fig.4.3 reveal several other interesting features. First, the depressions of T_c generally lie above the deGennes curve for the lighter RE ions to the left of Gd in the periodic table and slightly below the deGennes curve for the heavy RE ions to the right of Gd (except for Yb). This has been observed previously in the depression of T_c of La, $LaAl_2$ and other superconducting matrices by RE impurities [4.34] and appears to be a common behavior. Such a trend may be attributable to a general decrease in $|\mathscr{J}|$ with increasing RE atomic number and/or the splitting of the RE ion energy levels by the CEF, which is generally larger for the lighter RE ions. Second, the depressions of T_c for both the Ce and Yb members are anomalously large; this effect is frequently observed for the RE ion Ce and sometimes for the RE ions Pr, Sm, Eu, Tm and Yb and is associated with the Kondo effect [4.35]. For this case, the conduction electron spin-RE magnetic moment exchange interaction parameter $\mathscr{J}$ is negative due to the relatively strong hybridization of conduction electron and 4f electron states. Finally, the depression of T_c for Eu appears to be negligibly small. This suggests that Eu is trivalent when substituted for Lu in $LuRh_4B_4$ with a $4f^6$ configuration and corresponding nonmagnetic J = 0 Hund's rule ground state, leading to a substantial reduction in the magnitude of the exchange scattering rate.

4.2.3 Ferromagnetic Nonsuperconducting $RERh_4B_4$ Compounds: RE = Gd, Tb, Dy, Ho

Various properties of the $RERh_4B_4$ compounds with RE = Gd, Tb, Dy and Ho have been studied in order to more completely characterize the nature of CEF and magnetic interactions in this system. Displayed in Fig.4.4 are the low temperature magnetic heat capacities ΔC of these four compounds [4.4,5,36] obtained by subtracting the $LuRh_4B_4$ normal state specific heat data, which should be a reasonable estimate for the electronic and lattice contributions to the specific heat, from that of the magnetic compounds. Similar results have been obtained by KUMAGAI et al. [4.23]. All four magnetic compounds undergo remarkably sharp magnetic phase transitions at T_M with short-range correlation effects at temperatures above T_M notably absent.

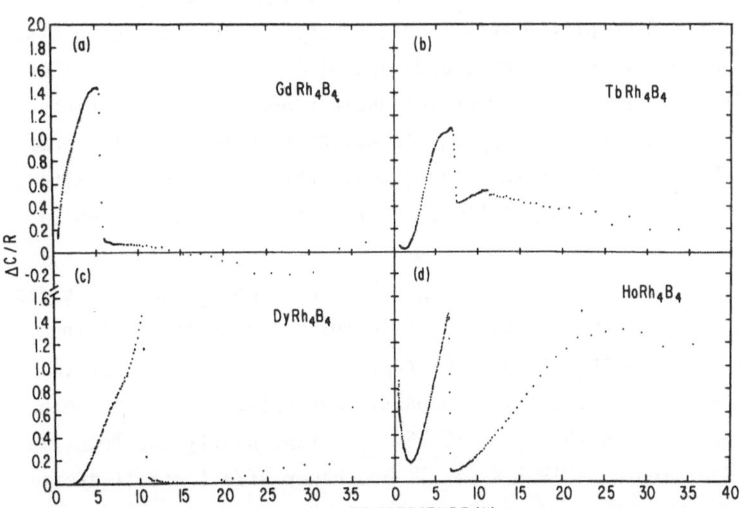

Fig.4.4. Magnetic heat capacity $\Delta C/R$ vs temperature for the ferromagnetic nonsuperconducting $RERh_4B_4$ compounds: (a) $GdRh_4B_4$, (b) $TbRh_4B_4$, (c) $DyRh_4B_4$ and (d) $HoRh_4B_4$. From [4.5]

As shown in Fig.4.4a, $GdRh_4B_4$ undergoes a jump in ΔC of about 3/2 R at its magnetic transition temperature of T_M = 5.8 K. In this case the value of the jump in ΔC is not in accord with mean field theory (MFT) [4.37] which predicts a discontinuity of 2.4 R for spin S = 7/2. Crystal field splitting of the Hund's rule ground state of the S-state Gd ion is expected to be negligible; indeed, the magnetic entropy ΔS below T_M is nearly 100% of R ln(8) [4.5]. Also, magnetic susceptibility data derived from Arrott plots at temperatures just above T_M follow a Curie-Weiss law with an effective magnetic moment μ_{eff} = 7.94 μ_B/Gd ion, the free ion value, and a Curie-Weiss temperature θ_p = 5.6 K [4.38].

At 1.6 K, the magnetization M in a magnetic field of 0.3 T is 86% of the saturation value of 7 μ_B/Gd ion [4.38]. This is to be contrasted with the values obtained in $HoRh_4B_4$, where M = 4.8 μ_B/Ho ion at T = 1.8 K and H = 10 T [4.36], and in $ErRh_4B_4$, where M = 7.65 μ_B/Er ion at T = 1.6 K and H = 10 T [4.39], both of which are reduced from the respective free ion values of 10 μ_B/Ho ion and 9 μ_B/Er ion. In addition, neutron diffraction measurements on $HoRh_4B_4$ and $ErRh_4B_4$ indicate values of the ground state magnetic moment of 8.7 μ_B/Ho ion [4.40] and 5.6 μ_B/Er ion [4.9], which are, respectively, greater than and less than the values derived from M above. These differences are presumably due, in part, to CEF effects and/or the presence of large anisotropy energies in these polycrystalline materials [4.36]. Nuclear magnetic resonance experiments on Gd in $GdRh_4B_4$ by KUMAGAI et al. [4.23] indicate that the magnitude of the hyperfine field at the Gd site is approximately the same as it is in pure Gd metal [4.41].

Three contributions to the heat capacity of $TbRh_4B_4$ are evident in Fig.4.4b [4.24]: (i) a nuclear Schottky anomaly, causing an upturn below 1.3 K, which arises from the hyperfine field splitting of the I = 3/2 ground state of the Tb nucleus into four equally spaced nuclear energy levels, (ii) an electronic Schottky anomaly peaking at 12 K due to the CEF-split energy levels of the Tb^{3+} J = 6 ground state, and (iii) an anomaly caused by the magnetic ordering at T_M = 7.4 K. The nuclear Schottky anomaly in $TbRh_4B_4$ can be fitted by the nuclear energy level spacings deduced for Tb metal. The magnetic entropy S_M associated with the magnetic phase transition at T_M for $TbRh_4B_4$ suggests that the ground state is a doublet, although the shape of the heat capacity anomaly follows MFT for spin S = 1/2 only up to 4 K [4.24].

At its magnetic transition temperature of T_M = 10.7 K, $DyRh_4B_4$ exhibits a discontinuity of ΔC = 3/2 R, as shown in Fig.4.4c. Both the MFT-like shape of the feature as well as the R ln 2 value of the magnetic entropy below T_M indicate that the ground state is a doublet. High temperature magnetic susceptibility data can be described by a Curie-Weiss law with μ_{eff} = 10.55 μ_B/Dy ion, nearly the free ion value of 10.63 μ_B/Dy ion, and θ_p = 19.3 K [4.5]. Mössbauer effect data indicate that the crystal field ground state magnetic moment is 9.2 μ_B/Dy ion at 4.2 K, corresponding to an almost pure $|\pm 15/2\rangle$ doublet [4.42]. Crystal field effects are

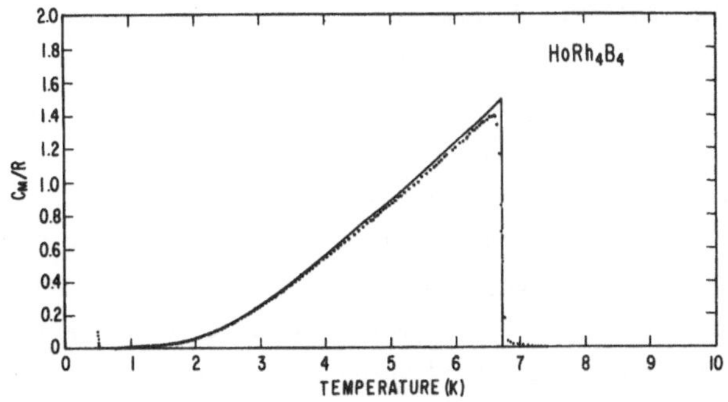

Fig.4.5. Temperature dependence of the specific heat anomaly of $HoRh_4B_4$ due to the ferromagnetic ordering. The solid line is the mean field result for an S = 1/2 ferromagnet. From [4.36]

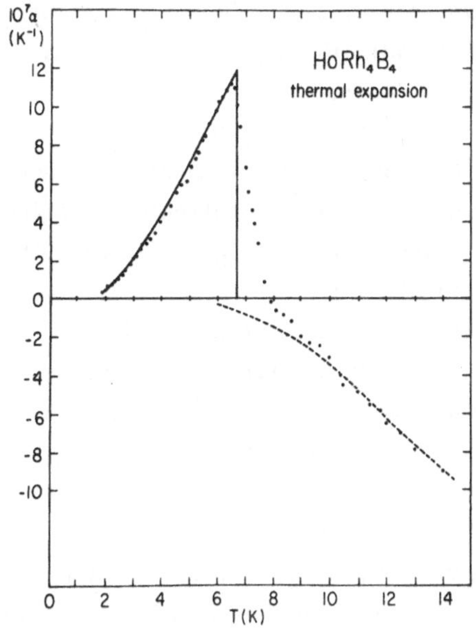

Fig.4.6. Linear thermal expansion coefficient α of polycrystalline $HoRh_4B_4$ vs temperature T. The solid line and the broken line are calculated from the specific heat using appropriate Grüneisen parameters. From [4.36]

also evident in heat capacity, magnetic susceptibility and Mössbauer effect measurements.

Three contributions to ΔC of $HoRh_4B_4$ in Fig.4.4d were identified as [4.4,36]: (i) A nuclear Schottky anomaly causing an upturn below 2 K which arises from the splitting of the Ho I = 7/2 nuclear ground state into eight equally spaced nuclear energy levels due to the effective hyperfine field, (ii) an electronic Schottky anomaly peaking at 26 K due to the CEF-split energy levels of the Ho^{3+} J = 8 Hund's rule ground state multiplet, and (iii) an anomaly C_M due to the spontaneous splitting of the ground state doublet (see below) upon magnetic ordering. The nuclear Schottky anomaly can be fitted by the nuclear energy level scheme previously derived for pure Ho metal. Thus, in at least $HoRh_4B_4$, $GdRh_4B_4$ and $TbRh_4B_4$ and probably the other $RERh_4B_4$ compounds, the effective hyperfine field acting at the site of the rare-earth nucleus is about the same in the $RERh_4B_4$ compounds as it is in *the pure rare-earth* metal. By subtracting fits of the nuclear and electronic

Schottky anomalies from the ΔC data, the data displayed in Fig.4.5 were derived. There is a pronounced discontinuity of 3/2 R in ΔC at T_M as well as an absence of a high temperature tail above T_M. Mean field theory for spin $S = 1/2$ and $T_M = 6.7$ K describes the C_M data remarkably well, as shown by the solid line in Fig.4.5. The entropy S_M between 0 and T_M is $S_M = R \ln 2$, confirming that the crystal field ground state is indeed a doublet. The temperature dependence of the linear thermal expansion coefficient α [4.36], shown in Fig.4.6, is also affected by the presence of magnetic ordering and CEF effects. The positive anomaly associated with the magnetic ordering has the same characteristic mean field shape as the specific heat while the negative feature above 8 K scales with the electronic Schottky anomaly. OTT et al. [4.36] have also found evidence for CEF effects in the magnetostrictive strain. The magnetic contribution to the electrical resistivity below T_M also behaves in accordance with the predictions of MFT [4.43]. The high temperature magnetic susceptibility may be described by a Curie-Weiss law with an effective magnetic moment μ_{eff} nearly equal to the free ion value of 10.60 μ_B/Ho ion and a paramagnetic Curie-Weiss temperature θ_p of 6.5 K [4.36]. At low temperatures, deviations from Curie-Weiss behavior occur because of CEF effects. Neutron diffraction experiments have shown that $HoRh_4B_4$ orders ferromagnetically at $T_M = 6.8$ K with the moment direction parallel to the tetragonal c-axis [4.40].

In addition, the temperature dependence of the spontaneous magnetization of $HoRh_4B_4$ is in excellent agreement with the theoretically derived function using MFT for spin $S = 1/2$ [4.43]. Thus, $HoRh_4B_4$ appears to be the first example of a metallic ferromagnet which exhibits ideal $S = 1/2$ mean field behavior.

4.2.4 Ferromagnetic Superconducting $RERh_4B_4$ Compounds: RE = Er

Much of the recent interest in the interplay between superconductivity and long-range magnetic order in the $RERh_4B_4$ compounds followed the discovery of re-entrant superconductive behavior due to the onset of magnetic order in $ErRh_4B_4$ by FERTIG et al. [4.8]. Typical AC magnetic susceptibility χ_{AC} and electrical resistance R_{AC} vs temperature data are shown in Fig.4.7 [4.44]. In zero magnetic field, the sample first becomes superconducting at an upper critical temperature $T_{c1} = 8.7$ K and then returns to a magnetically-ordered normal state at a lower critical temperature $T_{c2} \sim 0.9$ K. The thermal hysteresis that is evident in the χ_{AC} and R_{AC} vs temperature data indicates that a first-order transition occurs at T_{c2}. It has been suggested that the failure of the resistance below T_{c2} to attain its full normal state value may be due to the presence of filaments of another phase [4.45], although other explanations based on superconducting fluctuations have been suggested [4.8,46].

The first neutron scattering study on $ErRh_4B_4$ by MONCTON et al. [4.9] showed that the Er^{3+} magnetic moments ordered ferromagnetically, perpendicular to the unique tetragonal c-axis, in the neighborhood of T_{c2}, but with precursor scatter-

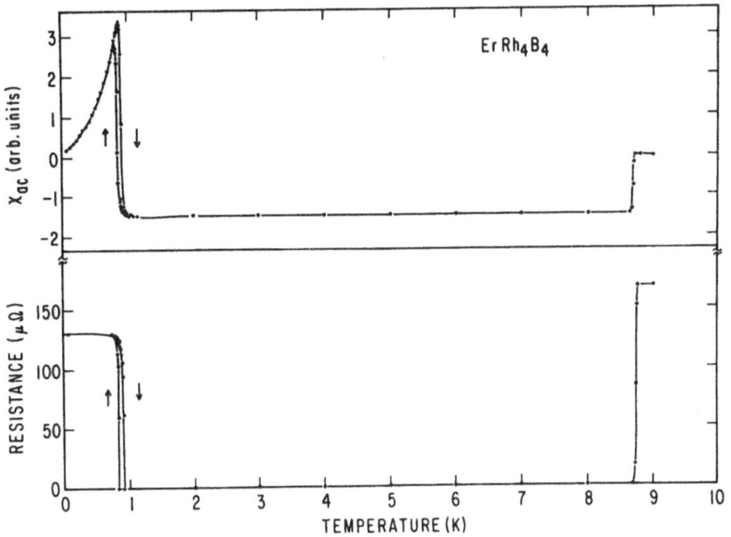

Fig.4.7. Typical AC magnetic susceptibility χ_{AC} and AC electrical resistance vs temperature data for $ErRh_4B_4$ in zero applied magnetic field. From [4.44]

ing that extended into the superconducting state up to ~1.4 K. Re-entrant superconductivity due to the onset of ferromagnetic order has also been observed in the compound $Ho_{1.2}Mo_6S_8$ [4.47,48] (Chaps.5,8).

The value of the saturation magnetization of $ErRh_4B_4$ determined from the neutron diffraction experiments was found to be only 5.6 μ_B/Er ion compared to the Hund's rule value of 9 μ_B expected for the Er^{3+} J = 15/2 ground state. Subsequently, SHENOY et al. [4.46] performed Er^{166} Mössbauer effect measurements on $ErRh_4B_4$ and inferred from the hyperfine field a value for the Er magnetic moment $(8.3 \pm 0.2)\mu_B$, considerably larger than the neutron scattering value [4.9], but still reduced from the free ion value. They suggested that the reason for this "dilemma" might be the existence of a disordered component of the Er magnetic moment that is not observed in the neutron diffraction experiments, which measure long-range ($\gtrsim$ 200 Å) magnetic moment correlations. However, the disordered component is detected, along with the ordered component, in the hyperfine technique that measures the single ion autocorrelation which, in the limit of slow electron relaxation, yields the value of the entire magnetic moment. It was conjectured that the disordered component of the Er magnetic moment might arise from superconducting fluctuations that were also present in the magnetically ordered state, as evidenced and noted before by the failure of the electrical resistance below T_{c2} to return to its full normal state value above T_{c1}, except in small applied magnetic fields. The 8.3 μ_B magnetic moment value for Er was attributed to a predominantly $|\pm 15/2\rangle$ ground state doublet as a result of the CEF.

Preliminary specific heat measurements appeared in the original report of FERTIG et al. [4.8] and were followed by more detailed measurements by WOOLF et al. [4.20].

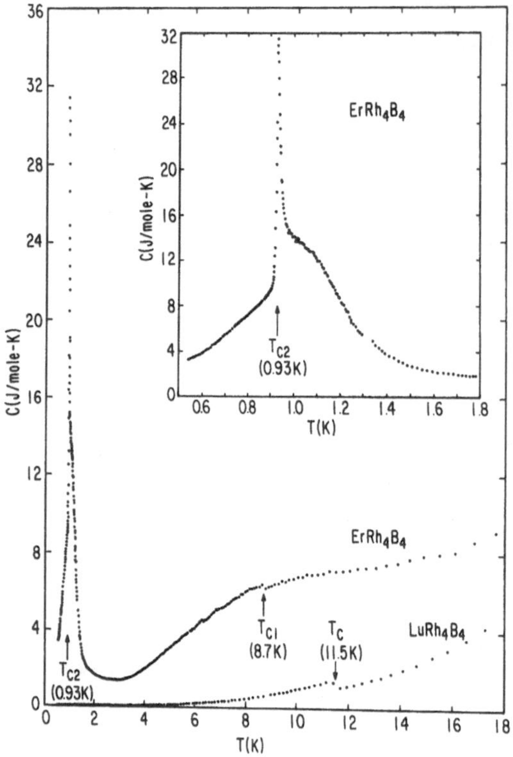

Fig.4.8. Specific heat C vs temperature T for $ErRh_4B_4$ and $LuRh_4B_4$. The inset shows a detailed plot of C vs T for $ErRh_4B_4$ in the vicinity of the re-entrant superconducting transition at T_{c2}. From [4.20]

Shown in Fig.4.8 is a plot of the heat capacity of $ErRh_4B_4$ and the isostructural nonmagnetic compound $LuRh_4B_4$ below 18 K [4.19]. The data reveal a jump in the heat capacity at the upper critical temperature T_{c1} = 8.7 K on a broad background with negative curvature, that is, a Schottky anomaly arising from the partial lifting by the CEF of the 16-fold degeneracy of the Er^{3+} J = 15/2 Hund's rule multiplet. At lower temperature (see inset), there is a spike-shaped feature at the lower superconducting critical temperature T_{c2} = 0.93 K (measured upon warming using AC magnetic susceptibility data) superimposed on another anomaly that is apparently associated with the long-range ferromagnetic ordering of the Er^{3+} magnetic moments in the vicinity of T_{c2}. Consistent with the thermal hysteresis in the physical properties at T_{c2}, the spike-shaped feature appears to represent a latent heat of transformation that is associated with a first-order transition at T_{c2} from the normal-ferromagnetic to the superconducting-paramagnetic state. There is also a shoulder in the heat capacity above T_{c2} whose origin may be attributable to a spatially modulated Er magnetization state, discussed below, that coexists with superconductivity. A thermal expansion anomaly near T_{c2} was also reported by OTT et al. [4.39].

Shown in Fig.4.9 is the excess heat capacity per mole $\Delta C/R$ vs temperature for $ErRh_4B_4$ that results after the electronic and lattice contributions, estimated from the normal state heat capacity of $LuRh_4B_4$, have been subtracted [4.20]. The

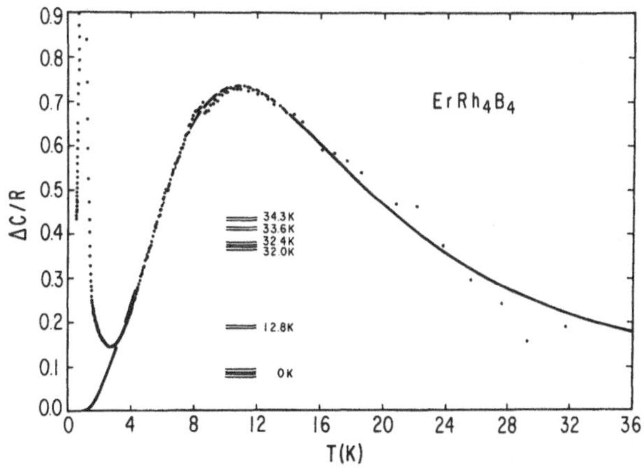

Fig.4.9. Magnetic heat capacity ΔC in units of the molar gas constant R vs temperature T for $ErRh_4B_4$. The solid line represents a calculated fit of the Schottky anomaly with a crystal field energy level scheme consisting of a ground state quartet and excited state doublets at temperatures of 12.8, 32,0, 32.4, 33.6, 34.3 and >100 K. From [4.20]

curve represents a Schottky anomaly which was fitted to the data and corresponds to the energy level scheme indicated in the figure. The ground state has been taken to be a quartet (or a ground state doublet separated from a low-lying excited state doublet by a few K) since the entropy associated with the magnetic anomaly is $\sim$R ln 4 per mole Er. Schottky anomalies are found in the heat capacity of other $RERh_4B_4$ compounds, as discussed herein, emphasizing that crystal field effects must be taken into account in order to achieve a detailed understanding of the extraordinary physical properties of the $RERh_4B_4$ compounds (see, for example, [4.44]). Recently, RADOUSKY et al. [4.49] measured the heat capacity of a series of $Y_{1-x}Er_xRh_4B_4$ compounds in an effort to separate crystal field effects from magnetic phenomena; they determined that there is a doublet 1.2 K higher in energy than the ground state doublet, consistent with Mössbauer effect data [4.42].

Measurements of the electrical resistance R vs temperature T in applied magnetic fields [4.8,39], isothermal magnetization M vs magnetic field H [4.39,50,51], specific heat [4.20,24], and ultrasonic attenuation [4.52,53] have been employed to determine the temperature dependence of the critical magnetic fields of polycrystalline samples of $ErRh_4B_4$. Generally, the resistively measured superconducting-normal transitions in a magnetic field are very broad, probably due to anisotropy of H_{c2}, and yield mean H_{c2} values that are more than twice as large as the values determined from the magnetization vs field measurements. The M vs H data on $ErRh_4B_4$ reported by OTT et al. [4.39] are shown in Fig.4.10. With decreasing temperature, the evolution of the M vs H curves suggests a change in the superconductive behavior from type II/2 to II/1. Such an effect has been predicted by MAEKAWA et al. [4.54] to occur as the temperature is decreased toward the Curie temperature in ferromagnetic superconductors as the interaction between vortices changes from repulsive to attractive due to the increase of the magnetization of the rare-earth ions. In this connection, ^{11}B NMR measurements on $ErRh_4B_4$ have been made between 1.2 and 4.2 K by KUMAGAI et al. [4.55]. In a field of 2.2 kG, the nuclear spin-

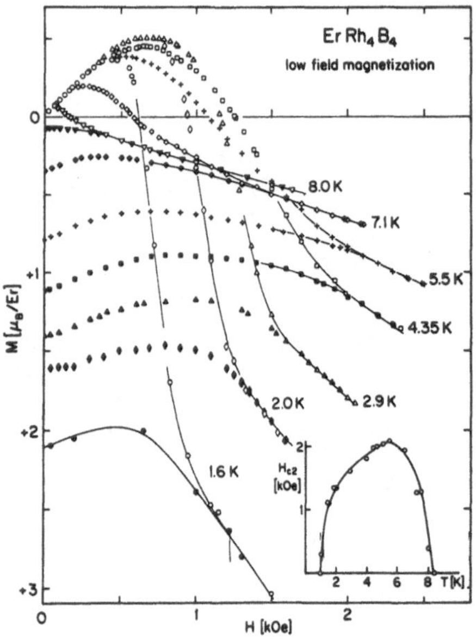

Fig.4.10. Low field magnetization M vs magnetic field H isotherms for $ErRh_4B_4$. The diamagnetic magnetization is plotted on the positive y-axis. The inset shows the $H_{c2}(T)$ curve as derived from magnetization and magnetostriction (not shown) measurements. From [4.39]

lattice relaxation time T_1 was found to exhibit BCS-like temperature dependence in the mixed state with $2\Delta = 3.52\ k_BT_{c1}$, where Δ is the superconducting energy gap. This was accompanied by an unusual deviation below 2.4 K where a positive magnetization occurs as evidenced by the behavior of the linewidth and the spin echo decay time T_2. With a further decrease in temperature, a stable Meissner state vanishes and a positive magnetization occurs even in zero magnetic field. KUMAGAI et al. [4.55] conjectured that this may be due to a transition near 2.4 K from type II/2 to type II/1 superconductive behavior accompanied by the onset of ferromagnetic order below 2.4 K.

CANTOR et al. [4.56] measured the electrical resistance of thin films of $ErRh_4B_4$ in applied magnetic fields and found that the perpendicular critical field $H_{c\perp}$ was larger than the parallel critical field $H_{c\parallel}$. This unusual result was found to be consistent with a simple model that included the effect of the magnetization induced in the film by the externally applied magnetic field. The $H_{c\perp}$ and $H_{c\parallel}$ vs T data obtained on a thin film $ErRh_4B_4$ specimen by CANTOR et al. are shown in Fig.4.11.

The resistively determined curve of H_{c2} vs T (defined from transition curve midpoints) reported for $ErRh_4B_4$ by OTT et al. [4.39] is displayed in Fig.4.16 along with resistively determined H_{c2} vs T data for the antiferromagnetic superconductors $NdRh_4B_4$, $SmRh_4B_4$ and $TmRh_4B_4$. An error that appeared in the H_{c2} vs T data for $ErRh_4B_4$ has been corrected in the data shown in Fig.4.16.

Magnetic susceptibility measurements on $ErRh_4B_4$ between T_{c1} and 300 K follow a Curie-Weiss law with an effective magnetic moment of 9.6 μ_B/Er ion, very close to the free ion value, and a low Curie-Weiss temperature $\theta_p \sim (1 \pm 1)$K [4.8,39]. The

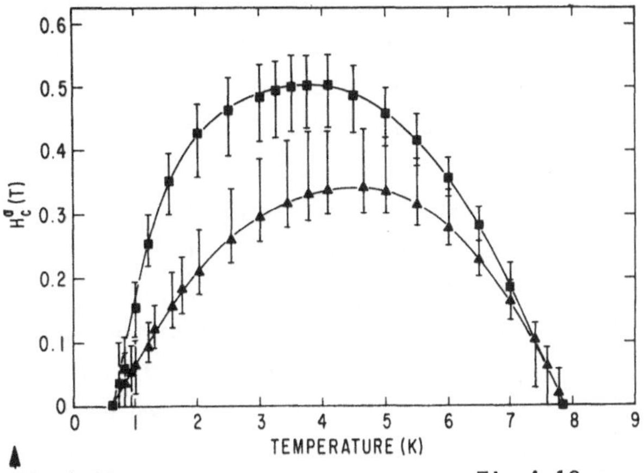

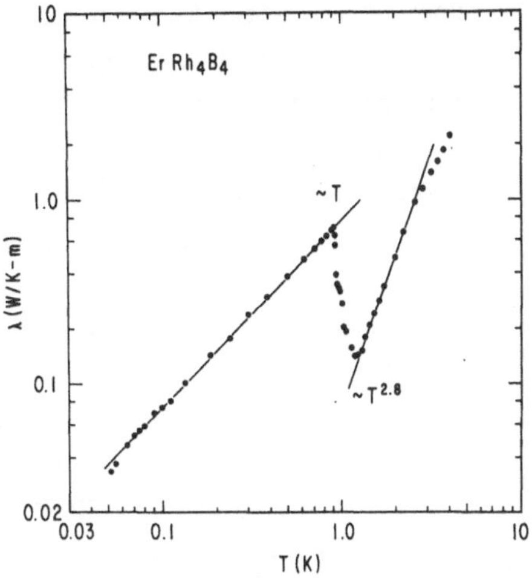

Fig.4.11 ▲ Fig.4.12 ►

Fig.4.11. Upper critical magnetic field vs temperature for a thin film of $ErRh_4B_4$ with the magnetic field applied parallel (triangles) or perpendicular (squares) to the surface of the film. The end points of the bars denote the fields at which the resistance is 10% or 90% of its normal state value. The dimensions of the film were approximately 2×0.1 cm by 4000 Å thick. From [4.56]

Fig.4.12. Thermal conductivity λ vs T for $ErRh_4B_4$ between 0.05 and 4 K, plotted on logarithmic scales. From [4.57]

magnetization at 1.6 K in a field of 10 T has not reached saturation and corresponds to 7.65 μ_B/Er ion [4.39].

The thermal hysteresis at T_{c2} has also been observed in the thermal conductivity [4.57] and heat capacity [4.58,59]. The thermal conductivity measurements of ODONI and OTT [4.57], shown in Fig.4.12, indicate that bulk superconductive behavior occurs in $ErRh_4B_4$ between T_{c2} and T_{c1} and disappears below T_{c2}.

SCHNEIDER et al. [4.60] reported anomalous temperature dependence with a peak near 5 K of the ultrasonic attenuation coefficient in $ErRh_4B_4$ at 15 MHz in zero field. Subsequently, measurements were reported of the attenuation coefficient at 15 MHz in magnetic fields up to 1.2 T between 1.5 K and 12 K [4.61]. The zero field maximum at 5 K was found to shift to lower temperatures with increasing field and to be riding on the high temperature tail of a maximum associated with the ferromagnetic transition. The magnitude of the attenuation at low temperatures increases as the magnetic field strength is increased, until at fields greater than 0.4 T, the low field maximum has disappeared and been replaced by another maximum in the attenuation which moves to higher temperatures as the field is increased to 1.2 T. The ultrasonic attenuation measurements of TOYOTA et al. [4.62] were carried out near 1 K and in magnetic fields up to 1.4 T. A broad, thermally hysteretic peak in the zero field attenuation was observed near T_{c2}. The attenuation anomaly was much stronger upon cooling, extended up to about 1.4 K, well above T_{c2}, and is probably associated with the sinusoidally modulated magnetic state that coexists with superconductivity. At large magnetic fields ($>H_{c2}$), a larger attenuation is observed when the propagation vector of the sound wave is perpendicular to the field direction than when it is parallel. The results were interpreted in terms of calculations by TACHIKI et al. [4.63] that suggested that ultrasonic attenuation caused by spin fluctuations in ferromagnetic superconductors would exhibit a characteristic dependence on temperature and magnetic field.

Electron tunneling measurements on $ErRh_4B_4$ have been carried out by ROWELL et al. [4.64] and UMBACH et al. [4.28] on polycrystalline thin film specimens through oxide barriers, and by POPPE [4.27] on single crystals through vacuum. The more recent studies by UMBACH et al. and POPPE indicate that $ErRh_4B_4$ is a strong coupled superconductor and yield values for $2\Delta/k_BT_{c1}$ of at least 4.2 and about 3.8, respectively. ROWELL et al. and POPPE both reported appreciable smearing of the energy gap below $\sim$1.5 K, which may be associated with the modulated magnetization state that coexists with superconductivity.

Following the initial neutron scattering experiments on $ErRh_4B_4$, MONCTON et al. [4.58] carried out a more complete investigation that included small angle scattering measurements. The neutron diffraction data revealed a broad ferromagnetic transition that extends up to $\sim$1.4 K, well above the temperature of the re-entrant superconducting transition of $T_{c2} \sim 0.9$ K, with significant thermal hysteresis between $\sim$0.8 and $\sim$1.4 K [Ref.4.19, Fig.1.13]. The width of the ferromagnetic transition has been attributed to a distribution of effective Curie temperatures within the material, while the hysteresis may be caused by the nucleation of normal ferromagnetic domains within the paramagnetic superconducting regions between T_{c2} and $\sim$1.4 [4.58,65,66]. The features in the neutron scattering vs temperature curve that apparently reflect the superconducting-normal transition at $T_{c2} \sim 0.9$ K include an inflection point and the disappearance of hysteretic behavior below $\sim T_{c2}$, measured upon cooling. The heat capacity data also reveal thermal hysteresis, part of which appears to be associated with the formation of ferromagnetic domains, discussed above in connection with the neutron scattering experiments (between $\sim$0.9 and $\sim$1.3 K), and part of which is due to the re-entrant superconducting transition at T_{c2}.

Perhaps the most provocative of the neutron diffraction results emerged from the small angle scattering studies of MONCTON et al. [4.58]. Shown in Fig.4.13 are their neutron intensity vs scattering angle data at various temperatures on a polycrystalline $ErRh_4B_4$ sample whose re-entrant superconducting-normal transition occurs at $\sim$0.7 K. A peak near 1° is seen to develop which grows in intensity as the temperature is decreased and then disappears abruptly when the sample becomes normal below T_{c2}. The data have been interpreted in terms of fluctuations into a state, proposed by BLOUNT and VARMA [4.67], in which the magnetization is sinusoidally modulated with a wavelength $\lambda \sim 100$ Å which was assumed to take the form of a spiral. These spatially modulated magnetic fluctuations, which coexist with and apparently

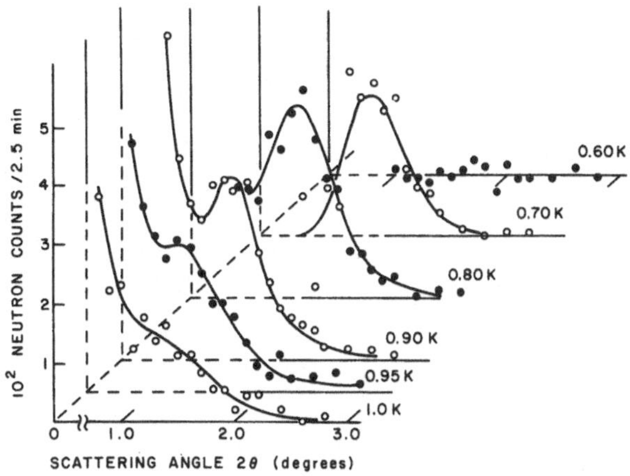

Fig.4.13. Small angle neutron scattering results on $ErRh_4B_4$ obtained at various temperatures. The peak at $2\theta=1.4°$ indicates an oscillatory magnetization with a wavelength of 100 Å. From [4.58]

owe their existence to the superconductivity, are reminiscent of the "cryptoferromagnetic" state suggested by ANDERSON and SUHL [4.68]. Several other theoretical investigations of spiral magnetic states with wavelengths $\sim 10^2$ Å that coexist with superconductivity in ferromagnetic superconductors above T_{c2} have recently appeared in the literature (for example, [4.69-72]). On the other hand, it has also been suggested that the periodic magnetic structure above T_{c2} may actually be a spontaneous vortex lattice [4.73,74]. The formation of a self-induced vortex lattice in a magnetic superconductor was first considered about a decade ago by KREY [4.75]. Recently, other possibilities have been examined such as a laminar structure, stabilized by the rare-earth magnetization in a self-consistent manner [4.76], and combined spiral magnetic and spontaneous vortex states [4.77]. Similar small angle scattering results have been obtained on polycrystalline specimens of the re-entrant ferromagnetic superconductor $HoMo_6S_8$ ([4.78]; also Chap.8).

MOOK and coworkers [4.79] performed neutron scattering measurements on $ErRh_4B_4$ in applied magnetic fields and observed long-range ferromagnetic order in fields greater than 1 kG at 1.6 K. Considerable hysteresis was found in the neutron scattering intensity vs magnetic field curve; long-range order with a small Er magnetic moment remained when the field was reduced to small values.

Quite recently, neutron diffraction and electrical resistivity measurements were carried out on single crystals of $ErRh_4B_4$ by SINHA et al. [4.80]. Their findings are consistent with the neutron scattering experiments on polycrystalline $ErRh_4B_4$ by MONCTON et al. [4.9,58], and revealed some important new features as well. Shown in Fig.4.14 are their results for the temperature dependence of (a) the ferromagnetic intensity from the (101) Bragg peak, (b) the satellite intensity, (c) the DC electrical resistivity, and (d) the ratio of the satellite to the ferromagnetic intensity for the (101) reciprocal lattice points. From these data, SINHA et al. concluded that superconductivity and long-range ferromagnetic order coexist, but in a spatially inhomogeneous manner between 0.71 K (T_{c2}, measured

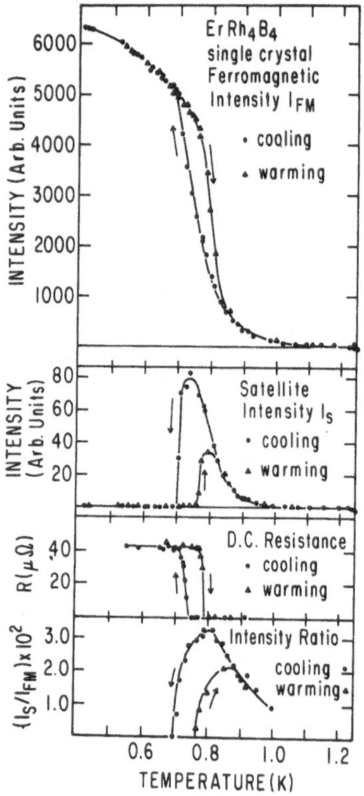

Fig.4.14. Neutron diffraction and electrical resistance data for an $ErRh_4B_4$ single crystal specimen. Temperature dependence of the ferromagnetic intensity [from the (101) Bragg peak], the satellite intensity, the DC electrical resistance and the ratio of the satellite to the ferromagnetic intensity for the (101) reciprocal lattice point. From [4.80]

upon cooling) and 1.2 K. The satellite peaks are associated with a transverse linearly polarized long-range magnetic structure with a wavelength of ~ 100 Å. The linearly polarized sinusoidal modulation lies along the [010] axis and the propagation directions are at 45° to the [001] and each of the [100] and [010] axes. With decreasing temperature, the modulated magnetic moment increases faster than the ferromagnetic moment and then disappears abruptly with the loss of superconductivity and transition to the normal ferromagnetic state. The higher temperature transition between the purely superconducting and magnetically ordered phases appears to be continuous. Weak scattering from small ferromagnetic regions of size ~ 100 Å and having a volume of about 5% of the total ferromagnetic volume is also seen and increases with the intensity of the sharp ferromagnetic peaks. GREENSIDE et al. [4.81] have shown that when the magnetic anisotropy is sufficiently strong, a linearly polarized sinusoidal state can be favored over both spiral and vortex states.

Measurements of the hyperfine magnetic field in $ErRh_4B_4$ in zero applied field in a range of temperatures around the re-entrant temperature T_{c2} have been reported by CORT et al. [4.82]. Using the Mössbauer effect of ^{57}Fe impurities as a hyperfine field microprobe, they found that the low temperature ferromagnetic order persists well above T_{c2} into the superconducting state in this compound. ROWELL et al. [4.63,83] have studied the effect of ion damage on the superconducting transition temperature, Curie temperature and electrical resistance of $ErRh_4B_4$. They also measured the critical current of $ErRh_4B_4$ and $Er_{0.43}Ho_{0.57}Rh_4B_4$ bridge-shaped samples prepared by photolithography and ion milling.

4.2.5 Antiferromagnetic Superconducting $RERh_4B_4$ Compounds: RE = Nd, Sm, Tm

In contrast to the re-entrant ferromagnetic superconductivity exhibited by $ErRh_4B_4$, antiferromagnetism coexists with superconductivity to temperatures below 0.07 K in $NdRh_4B_4$, $SmRh_4B_4$, and $TmRh_4B_4$. The similarity of the superconducting and magnetic properties of these three compounds ends there, however, as illustrated by the very different upper critical field curves, for example (see Fig.4.16). Presumably, factors such as the values of T_c and the Nèel temperature T_N, the magnetic anisotropy and the detailed magnetic structure contribute to these differences. The study of these materials has proven fruitful in providing tests of theoretical models describing how antiferromagnetic ordering influences the superconducting state.

MATTHIAS et al. [4.1] used ac magnetic susceptibility measurements to determine that $NdRh_4B_4$ becomes superconducting at T_c = 5.3 K. Since the $CeCo_4B_4$-type phase of this compound is metastable, additional rhodium and boron must be added to stabilize the primitive tetragonal phase. The resulting material, $NdRh_6B_6$, is composed primarily of $NdRh_4B_4$ and two impurity phases: RhB, which exhibits no magnetic or superconducting transitions between 0.06 and 20 K, and $NdRh_6B_4$, which becomes ferromagnetic at 4.9 K and typically accounts for approximately fifteen percent of the Nd^{3+} ions.

The AC electrical resistance R vs temperature T for a $NdRh_6B_6$ sample in various applied magnetic fields H is shown in Fig.4.15 [4.10]. For fields of 0.2 T or less, the sample remains superconducting below the critical temperature T_{c1}. In fields between 0.3 T and 0.6 T, however, the initial decrease in R is followed at lower temperature by a sharp increase in R and, as T is lowered further, R again rapidly decreases. Heat capacity measurements in zero applied magnetic field reveal two lambda-type anomalies that peak at T_{c2} = 1.31 K and T_{c3} = 0.89 K, respectively, indicating that two magnetic phase transitions are associated with the superconducting-to-normal state transitions below T_{c1} which are manifested in the resistance data [4.10]. The presence of thermal hysteresis in the R vs T data at T_{c3} suggests that the lower temperature transition is first order, whereas the absence of any such hysteresis at T_{c2} would imply that a second-order transition occurs at this temperature.

By defining the superconducting transition temperature as the temperature at which R is fifty percent of its normal state value, the upper critical field H_{c2} vs T data for $NdRh_6B_6$ shown in Fig.4.16 have been obtained [4.10]. The critical field curve shows an abrupt depression at T_{c2} followed by a rapid recovery below T_{c3}, although $H_{c2}(0)$ = 0.54 T remains below the value of $\sim$0.65 T expected from an

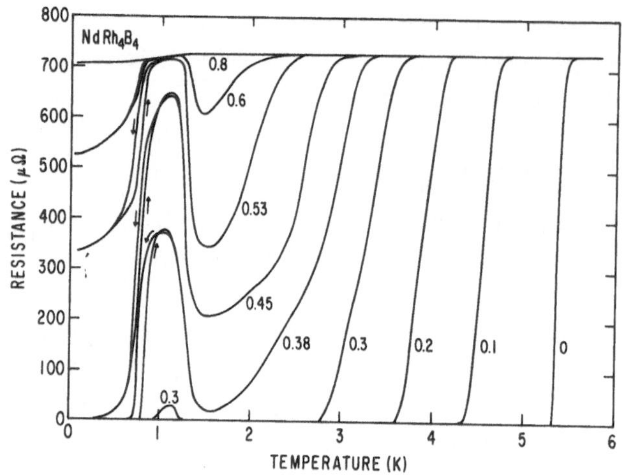

Fig.4.15. AC electrical resistance vs temperature for $NdRh_4B_4$ in various applied magnetic fields between 0 and 0.8 T. From [4.10]

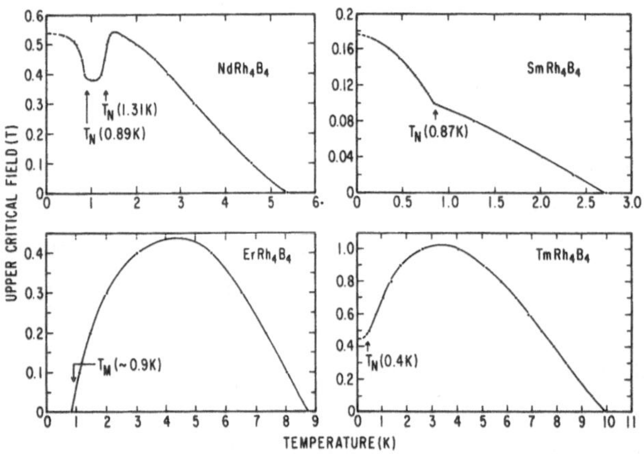

Fig.4.16. Upper critical magnetic field vs temperature for $NdRh_4B_4$ [4.10], $SmRh_4B_4$ [4.12], $ErRh_4B_4$ [4.39] and $TmRh_4B_4$ [4.13]

extrapolation of the data above 1.6 K. Therefore, the onset of magnetic order at T_{c2} causes additional pairbreaking, whereas the magnetic structure below T_{c3} is more compatible with superconductivity. Another unusual feature of the critical field curve is its positive curvature for $T > 3$ K.

The nature of the magnetic transitions at T_{c2} and T_{c3} appears to be rather complex. Between 20 and 300 K, the magnetic susceptibility χ follows a Curie-Weiss temperature dependence with an effective magnetic moment $\mu_{eff} = 3.58 \pm 0.05\ \mu_B$/Nd ion, close to the free ion value of 3.62 μ_B, and a Curie-Weiss temperature $\theta_p = -6.2 \pm 1.0$ K, which suggests that the Nd^{3+} ions interact antiferromagnetically [4.10]. However, isothermal magnetization M vs H curves above T_{c3} show a tendency to saturate with increasing $H > H_{c2}$, a behavior similar to that observed in $ErRh_4B_4$ [4.8] whereas below T_{c3}, the $NdRh_6B_6$ sample has a decreasing tendency to saturate [4.10]. Although the ferromagnetic ordering of the $NdRh_6B_4$ phase can, in part, account for the observed saturation, the change in behavior of the magnetization

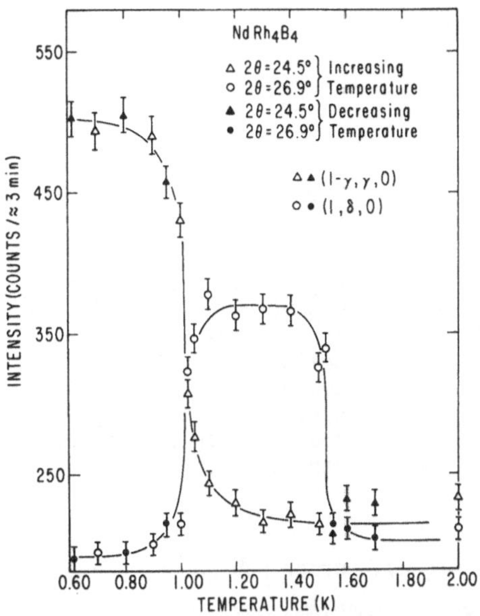

Fig.4.17. Temperature dependence of the neutron scattering intensity of a representative satellite from each of the two magnetic phases of $NdRh_4B_4$. From [4.11]

data below T_{c3} implies that the $NdRh_4B_4$ phase must contribute to the saturation as well. Neutron diffraction experiments [4.11] indicate that the magnetic phases of $NdRh_4B_4$ in zero applied field are body-centered tetragonal antiferromagnetic structures in which the Nd^{3+} moments are alternately parallel and antiparallel to the c-axis, with a sinusoidal modulation along the [100] direction with $\lambda = 46.5$ Å in the high temperature magnetic phase, and along the [110] direction with $\lambda = 45.2$ Å in the low temperature magnetic phase. The observed magnetic saturation above T_{c3} may indicate that applied magnetic fields that are sufficiently strong to completely quench superconductivity in this compound also induce ferromagnetic ordering of the Nd^{3+} ions. The temperature dependences of the neutron diffraction intensity of a representative satellite from each of the two magnetic phases of $NdRh_4B_4$ are shown in Fig.4.17 [4.11].

The H_{c2} vs T data for $SmRh_4B_4$, obtained from AC electrical resistance measurements, are shown in Fig.4.16 [4.12]. The critical temperature $T_c = 2.72$ K in zero applied field is slightly higher than the value of 2.5 K originally reported by MATTHIAS et al. [4.1]. At T = 0.85 K, a sharp discontinuity in the slope of the H_{c2} vs T curve occurs, and below this temperature H_{c2} is considerably larger than would be expected from an extrapolation of the high temperature data. This behavior can be explained well by MACHIDA's [4.84] theory of antiferromagnetic superconductors, in which the AG theory of superconductors with paramagnetic impurities is extended to correlated spin systems. In this model, the spin-fluctuations of the rare-earth magnetic moments have the dominant effect upon the superconducting properties, and the enhancement of H_{c2} below the Néel temperature T_N can be interpreted as resulting from the reduction of the net magnetization of the sample. The

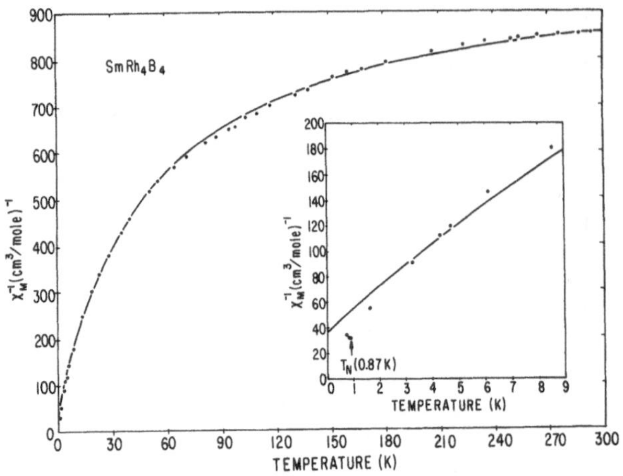

Fig.4.18. Inverse static magnetic susceptibility vs temperature for $SmRh_4B_4$. The solid line represents the sum of a Curie-Weiss law with μ_{eff}=0.632 μ_B and θ_p=-1.93 K and a temperature independent Van Vleck term. From [4.12]

modification of superconductivity by antiferromagnetic order has recently attracted considerable attention from theorists and several mechanisms have been proposed (Chap.9).

Heat capacity measurements on $SmRh_4B_4$ reveal a lambda-type anomaly peaking at T_N = 0.87 K, which supports the hypothesis given above that a phase transition occurs at the temperature at which the H_{c2} vs T curve changes slope [4.12]. Subtracting the electronic and lattice contributions to the specific heat of the isostructural nonmagnetic compound $LuRh_4B_4$ [4.44] results in a magnetic entropy between 0 and 16 K of $S_{mag} \simeq R \ln 2$. This indicates that the phase transition results from the antiferromagnetic ordering of the ground state doublet of the Sm^{3+} ions.

A plot of the reciprocal magnetic susceptibility χ^{-1} vs T for $SmRh_4B_4$ is shown in Fig.4.18 [4.12]. Since Sm^{3+} ions frequently have relatively low-energy angular momentum states above the Hund's rule ground state, a least squares fit of the data in the paramagnetic temperature range was made to the function

$$\chi_M = \frac{N}{k_B}\left(\frac{\mu_{eff}^2}{3(T-\theta_p)} + \frac{\mu_B^2}{\delta}\right) , \qquad (4.2)$$

where N is the number of Sm^{3+} ions per unit volume and k_B is Boltzmann's constant. The first term represents the Curie-Weiss expression for the J = 5/2 ground state, while the second term is the temperature independent Van Vleck correction arising from the accessible first excited J = 7/2 state. In the absence of crystal field effects, $\mu_{eff} = g_J\sqrt{J(J+1)}\mu_B = 0.85\ \mu_B$ and $\delta = 7\ \Delta E/20$, where ΔE is the energy difference between the J = 5/2 and J = 7/2 states. The χ vs T data are described well using the best fit parameters of $\mu_{eff} = 0.632\ \mu_B$, $\theta_p = -1.93$ K, and $\delta = 377$ K. Consistent with the specific heat data, the crystal field effects reduce the effective magnetic moment compared to its free ion value. These χ vs T data for $SmRh_4B_4$ have also been analyzed recently by STEWART [4.85] who deduced from his

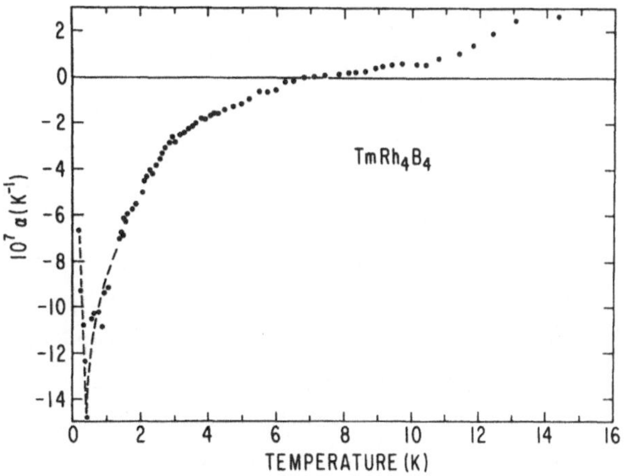

Fig.4.19. Linear thermal expansion coefficient α vs temperature for $TmRh_4B_4$ in zero applied magnetic field. From [4.13]

estimate of the reduction of the Sm effective magnetic moment by conduction electron polarization a relatively small value of $\mathcal{J} \sim 0.04$ eV-atom. Also, $\delta = 377$ K corresponds to $\Delta E = 1080$ K, which is somewhat less than, but comparable to, the value of ~1500 K estimated for free Sm^{3+} ions [4.86]. Finally, the χ^{-1} vs T curve exhibits a cusp near T_N which is characteristic of the antiferromagnetic transition at this temperature.

Thermal conductivity λ vs T measurements demonstrate conclusively that $SmRh_4B_4$ remains a bulk superconductor below its Néel temperature [4.87]. Whereas $ErRh_4B_4$ shows an abrupt increase in the thermal conductivity at T_M [4.57,88], λ monotonically decreases with decreasing T when the Sm^{3+} ions order magnetically. However, unlike an ordinary superconductor well below its critical temperature, in $SmRh_4B_4$ the thermal conductivity approaches a linear temperature dependence well below T_N. Thus, although superconductivity and antiferromagnetism coexist in this sample, the magnetic order seems to have some destructive effect upon the Cooper pairs in this material.

Considerable evidence exists that $TmRh_4B_4$, which becomes superconducting at 9.8 K [4.1], orders magnetically at 0.4 K. Shown in Fig.4.19 is the linear thermal expansion coefficient α vs T in zero applied field [4.13,89]. The data display a negative anomaly peaking at 0.4 K which is similar in character to the one observed in $ErRh_4B_4$ [4.39]. Specific heat measurements above 0.5 K show a sharp upturn in C concomitant with the α anomaly, which is indicative of a lambda-type anomaly [4.13, 89]. Since applied magnetic fields greatly enhance the α anomaly [4.89], it is probable that the features in the α and C data are associated with the long-range magnetic ordering of the Tm^{3+} magnetic moments. The negative sign of the α anomaly implies that applied pressure should depress the ordering temperature T_M. Neutron diffraction measurements indicate that the magnetic transition is at least partially antiferromagnetic in nature [4.14].

Low-frequency AC electrical resistance R vs T data for $TmRh_4B_4$ show that superconductivity persists to temperatures below 0.07 K in applied magnetic fields $H \leq 0.2$ T [4.13,89]. However, in H = 0.3 T, a thermally hysteretic, re-entrant transition occurs, although R at T = 0 remains considerably below its normal state value. In larger magnetic fields, R is constant below $T \sim 0.45$ K in any given field, with R(T = 0) increasing as H rises. This re-entrant behavior is very similar to that observed in $ErRh_4B_4$ [4.8,39] and suggests that in $H \geq 0.3$ T, $TmRh_4B_4$ orders ferromagnetically. The H_{c2} vs T curve obtained from these data, which is plotted in Fig. 4.16, shows that the upper critical field is a nonmonotonic function of temperature having a maximum of 1.0 T near T = 3 K. A rather similar H_{c2} vs T curve has been observed in another antiferromagnetic superconductor, $Er_{1.2}Mo_6S_8$ [4.90], where the decrease of H_{c2} has been attributed to the increasing paramagnetic susceptibility of the Er^{3+} magnetic moments. However, the behavior of the two compounds differ considerably below their magnetic ordering temperatures: whereas a sharp upturn in H_{c2} occurs when $Er_{1.2}Mo_6S_8$ orders antiferromagnetically, no such enhancement below the Néel temperature is apparent in $TmRh_4B_4$.

Low-field magnetization M vs H experiments also demonstrate that in applied fields, $TmRh_4B_4$ [4.91] and $ErRh_4B_4$ [4.39] behave similarly. In the temperature range $T_M \ll T < T_c$, both compounds are strongly irreversible type II/2 superconductors, but as T approaches T_M, the M vs H curves are more characteristic of type II/1 materials. This further indicates that $TmRh_4B_4$ becomes ferromagnetic in applied magnetic fields, since MAEKAWA et al. [4.54] and TACHIKI et al. [4.92] have predicted such a change in behavior near the Curie temperature of a ferromagnetic superconductor. High-field magnetization data for $H \leq 10$ T show that the saturation moment in $TmRh_4B_4$ is approximately 5 μ_B/Tm ion, which is well below the Hund's rule value of 7 μ_B/Tm ion [4.91]. An increasing tendency towards saturation with decreasing temperature is also revealed by magnetostriction measurements above 1 K [4.91].

Heat capacity measurements [4.13,89,93] show a Schottky anomaly peaking at approximately 20 K, indicating that the splitting of some of the 4f electron energy levels is relatively small. The magnetic susceptibility χ [4.89] follows a Curie-Weiss temperature dependence above T_c with $\mu_{eff} = 7.51 \pm 0.03$ μ_B/Tm ion, which is very close to the free ion value of 7.57 μ_B/Tm ion. This result implies that the crystal field ground state of the Tm^{3+} ions includes, to a large degree, the $|\pm 6\rangle$ eigenstates, thus supporting the conclusions of SHENOY et al. [4.42] in their Mössbauer studies on $TmRh_4B_4$. These χ vs T data also yield $\theta_p = -0.46 \pm 0.25$ K, which is consistent with the antiferromagnetic nature of the transition in zero field.

Further evidence that $TmRh_4B_4$ undergoes a metamagnetic transition from antiferromagnetism to ferromagnetism in $H \sim 0.3$ T has been obtained from thermal conductivity measurements [4.13,89]. In zero applied field, the λ vs T behavior for $TmRh_4B_4$ is very similar to that observed in $SmRh_4B_4$ [4.87], which confirms that bulk superconductivity and magnetic order coexist in the former compound. However, in $H = 60$ mT, the $TmRh_4B_4$ sample displays a thermally hysteretic transition near T_N, and λ, although remaining rather small below this temperature, follows a nearly linear T dependence. In $H = 0.1$ T, the sample reveals an extremely hysteretic transition near 1 K similar to that observed in $ErRh_4B_4$ [4.57,88], although λ still remains below its normal state value. Applied fields exceeding 0.2 T shift the transition to $T > 1$ K and the thermal conductivity is clearly electronic below the transition. In contrast, the electrical resistance of the sample remains zero below T_c for $H < 0.5$ T, and traces of superconductivity persist in fields exceeding 1.0 T. This discrepancy probably arises because some superconductivity exists in antiferromagnetic regions or domain walls [4.94,95] although in intermediate applied fields the bulk of the $TmRh_4B_4$ is normal and ferromagnetic.

The striking properties of the antiferromagnetic superconducting $RERh_4B_4$ compounds recounted here are quite similar to those of the other major class of antiferromagnetic superconductors, the $RE_xMo_6S_8$ and $RE_xMo_6Se_8$ compounds [4.96,97] (Chap.5).

4.3 Pseudoternary $RERh_4B_4$ Compounds

In order to investigate competing interactions not accessible in the pure ternary compounds, it is of significant interest to mix two ternary compounds together to form a pseudoternary system. Two types of $RERh_4B_4$ pseudoternaries have been formed, one in which a second RE element is substituted at the RE sites, and another in which another transition metal M element is substituted at the Rh sites. Not only is this an alternative method for studying the interaction between superconductivity and long-range magnetic order, but it also allows the effects of competing types of spin anisotropy and/or magnetic order to be explored. These systems have provided a critical test for theories involving the interplay between superconductivity and magnetism.

4.3.1 $(RE_{1-x}RE'_x)Rh_4B_4$ Compounds

JOHNSTON et al. [4.98] determined the composition dependencies of the superconducting critical temperatures T_{c1} and T_{c2} and the magnetic ordering temperature T_M for the system $(Er_{1-x}Ho_x)Rh_4B_4$ using χ_{AC} measurements in order to investigate the transition from the re-entrant superconductivity observed in $ErRh_4B_4$ to the ferromagnetism reported for $HoRh_4B_4$. The low temperature phase diagram for $(Er_{1-x}Ho_x)Rh_4B_4$, delineating the paramagnetic, superconducting and magnetically ordered phases, is displayed in Fig.4.20. We have added another phase boundary to Fig.4.20, discussed below, that separates ferromagnetic ordering of the Ho^{3+} magnetic moments along the tetragonal c-axis, and independent ferromagnetic ordering at a lower temperature of

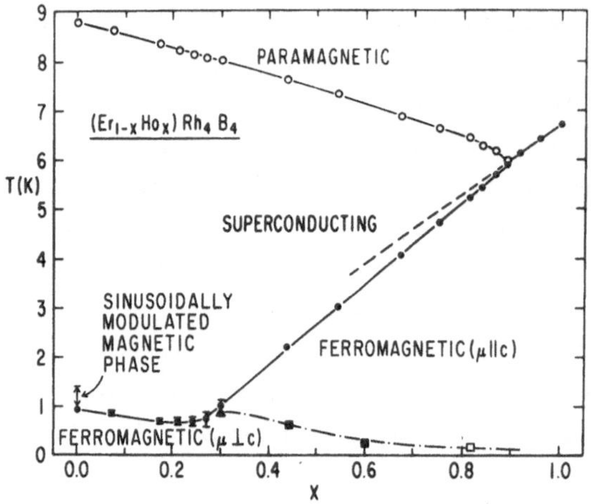

Fig.4.20. Low temperature phase diagram for the system $(Er_{1-x}Ho_x)Rh_4B_4$ determined from AC magnetic susceptibility and neutron diffraction measurements. The vertical bars on T_{c2} data points indicate the observed thermal hysteresis. The solid lines represent phase boundaries that are defined by χ_{AC} data (open and solid circles) from [4.98], while the dot-dashed line is based on χ_{AC} data from [4.98] (open squares) and [4.107] (solid squares) and neutron diffraction data from [4.103,104] (solid triangle). The temperature range of the sinusoidally modulated magnetic phase in $ErRh_4B_4$ [4.58,80] is also indicated in the figure

the Er^{3+} magnetic moments within the basal plane. The sinusoidally modulated magnetic phase above T_{c2} is also indicated in Fig.4.20 [4.58,80]. There is a tricritical point at the concentration $x_c = 0.89$ at which T_{c1}, T_{c2} and T_M become coincident. Also, the T_{c2} vs x phase boundary for $x < x_c$ is depressed relative to a linear extrapolation of T_M vs x from $x > x_c$ (dashed curve in Fig.4.20). This depression was ascribed to the decrease in the conduction electron spin susceptibility in the superconducting state relative to that in the normal state, which would depress $T_{c2} \approx T_M$ if the magnetic moments couple primarily via the RKKY interaction [4.98]. Finally, there is a minimum in the T_{c2} vs x phase boundary near $x = 0.25$. The effect of applied pressure p on a re-entrant $(Er_{0.11}Ho_{0.89})Rh_4B_4$ sample also shows a tricritical point [4.99] at a critical pressure p_c with a similar depression of the T_{c2} vs x phase boundary for $p < p_c$, indicating the similarity of the destruction of superconductivity either by presssure or chemical substitution. For $x = 0.89$, re-entrant superconductivity gives way to ferromagnetism above a critical pressure since $dT_M/dp > dT_{c1}/dp$.

Specific heat measurements performed on selected $(Er_{1-x}Ho_x)Rh_4B_4$ samples by MACKAY et al. [4.100] are presented in Fig.4.21. The MFT-like behavior in $HoRh_4B_4$ greatly simplifies the interpretation of these data. For $x = 0.912$, just above x_c, the anomaly associated with the magnetic transition has a shape similar to that of $HoRh_4B_4$. However, at $x = 0.813$, just below x_c, a spike-shaped feature appears to be superimposed on top of the MFT-like magnetic transition. Since the sample with $x = 0.813$ is re-entrant, whereas that with $x = 0.912$ is purely ferromagnetic, the spike-shaped feature is associated with the first-order re-entrant superconducting to normal-ferromagnetic transition at T_{c2}. For the four superconducting members of this system presented in Fig.4.21, the spike-shaped feature was coincident with the destruction of superconductivity as measured by χ_{AC}. The low temperature heat capacity of $(Er_{0.7}Ho_{0.3})Rh_4B_4$ shows two maxima, indicative of two distinct orderings.

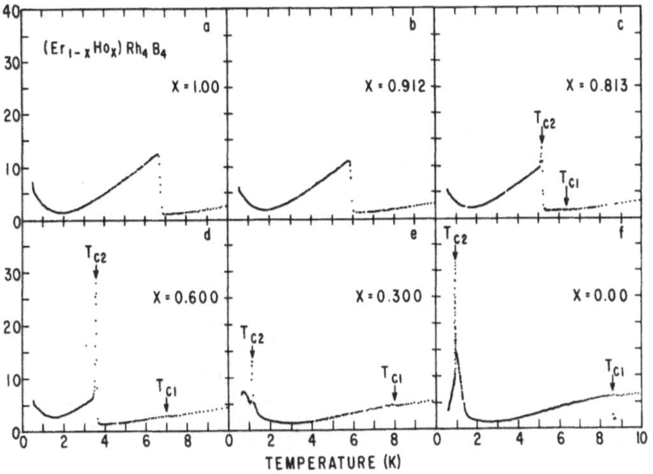

Fig.4.21. Heat capacity C vs temperature for certain members of the $(Er_{1-x}Ho_x)Rh_4B_4$ system. The arrows indicate the upper (T_{c1}) and lower (T_{c2}) superconducting-to-normal state transition temperatures as measured by AC magnetic susceptibility. From [4.100]

Also, Schottky anomalies present in each of the pseudoternary compounds could be described as the weighted sum of the respective Schottky anomalies for $ErRh_4B_4$ and $HoRh_4B_4$, indicating the additive nature of the crystal field effects [4.24,101].

Neutron diffraction experiments have shown that in the $RERh_4B_4$ tetragonal unit cell, the ferromagnetic alignment is in the basal plane for $ErRh_4B_4$ [4.9] but along the unique tetragonal c-axis for $HoRh_4B_4$ [4.40]. MOOK et al. [4.102,103] have performed neutron scattering experiments on $(Er_{0.4}Ho_{0.6})Rh_4B_4$ which suggest that only the Ho^{3+} ions order along the c-axis at T_M = 3.60 K. Recent neutron diffraction data on $(Er_{0.7}Ho_{0.3})Rh_4B_4$ [4.103,104] demonstrate that the Ho^{3+} ions order along the c-axis while at a lower temperature the Er^{3+} ions independently order in the basal plane, in agreement with the specific heat data. Thus, the T_M and T_{c2} vs x phase boundary of Fig.4.20 apparently arises from the ordering of the Ho^{3+} ions for $0.25 \lesssim x \lesssim 1.00$ and from the ordering of the Er^{3+} ions for $0 \leq x \lesssim 0.25$. The independent orderings of the two species of RE ion and consequent minimum in the T_{c2} phase boundary is typical for systems consisting of spins with competing anisotropies [4.105,106]. The specific heat [4.100] and neutron diffraction [4.103,104] data and a peak below T_{c2} in the $\chi_{AC}(T)$ curves [4.98,107] were used to construct the phase boundary (dot-dashed line) in Fig.4.20 separating independent $\mu(Ho^{3+})$ c-axis and $\mu(Er^{3+})$ a-axis ferromagnetic ordering.

Analysis [4.108] of the neutron diffraction data on $(Er_{0.4}Ho_{0.6})Rh_4B_4$ [4.102, 103] indicates that the actual T_M of 3.6 K is about 0.2 K less than would have occurred in the absence of superconductivity, in agreement with theoretical predictions [4.66,73]. Using a simple model to relate the spontaneous magnetization to the heat capacity, the spike-shaped feature in the heat capacity at T_{c2} was shown to be due primarily to the rapid onset of the spontaneous magnetization. Thus, the major contribution to the latent heat of transformation at the first-order re-entrant transition is magnetic in origin. The temperature dependence of the magnetization

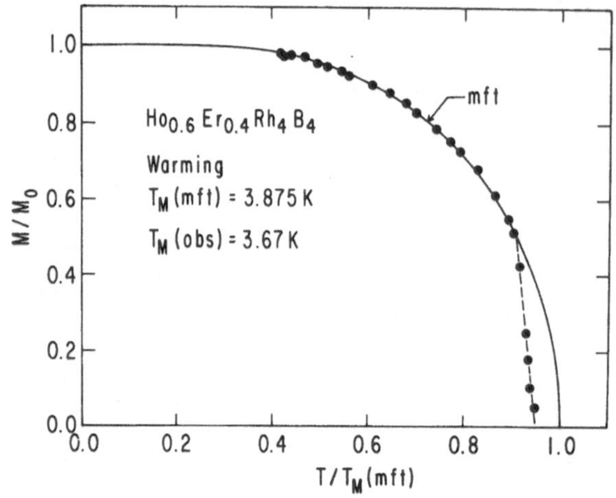

Fig.4.22. Temperature dependence of the magnetization M(T) derived from the (101) peak intensity for $Ho_{0.6}Er_{0.4}Rh_4B_4$. The solid curve is calculated from mean field theory. From [4.108]

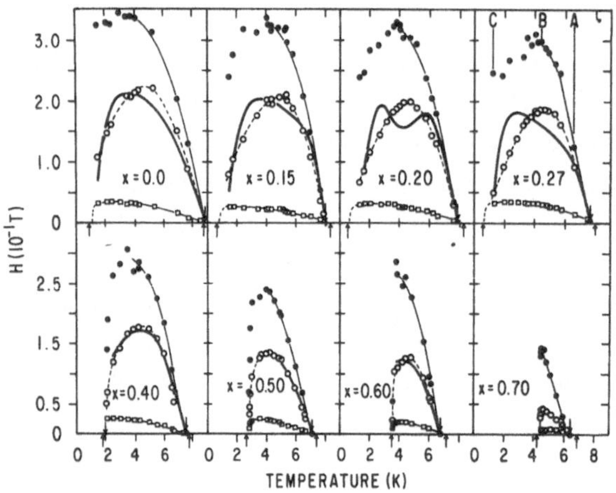

Fig.4.23. Lower critical field H_{c1} (open squares), upper critical field H_{c2} (open circles), and upper critical field of the superconducting subsystem $H_{c2}^{(0)}$ (full circles) vs temperature for the system $(Er_{1-x}Ho_x)Rh_4B_4$ obtained from magnetization measurements. The arrows pointing upward indicate T_{c1} and T_{c2} from [4.98]. The arrows pointing downward indicate T_{c1} and T_{c2} from [4.109]. From [4.109]

M(T) of $(Er_{0.4}Ho_{0.6})Rh_4B_4$ derived from the (101) peak intensity is shown in Fig. 4.22. The solid line in the figure is calculated from MFT.

The upper H_{c2} and lower H_{c1} critical fields shown in Fig.4.23 have been determined for various members of this system using isothermal magnetization curves [4.109,110]. As for $ErRh_4B_4$ and $TmRh_4B_4$, the H_{c2} vs T curves are bell shaped. However, for $x \gtrsim 0.40$, H_{c2} decreases sharply near T_{c2}. Considering the behavior of H_{c2} in $ErRh_4B_4$ near T_{c2}, it seems likely that a similar sharp depression would also be observed in H_{c2} near T_{c2} for the samples with $x < 0.40$ if that temperature range were accessible in this experiment. These abrupt drops in H_{c2} are probably associated with the onset of the magnetic ordering of the Ho^{3+} ions for $x > 0.25$ and with the Er^{3+} ions for $x < 0.25$. The competing superconducting and ferromagnetic interactions in the $(Er_{1-x}Ho_x)Rh_4B_4$ system have been the subject of several recent theoretical papers [4.111-113,116].

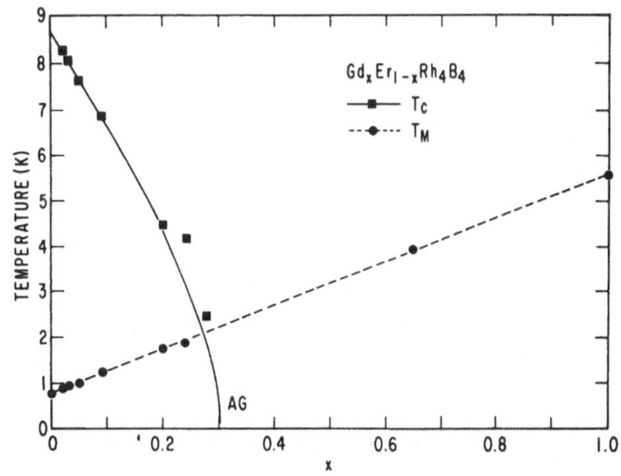

Fig.4.24. Low temperature phase diagram for the system $(Gd_xEr_{1-x})Rh_4B_4$. From [4.114, 115]

Another system studied which consists of a re-entrant superconductor mixed with a ferromagnet is $(Gd_xEr_{1-x})Rh_4B_4$. As shown in Fig.4.24, the phase diagram also shows a tricritical point near $x_c \approx 0.28$ at which T_{c1}, T_{c2} and T_M become coincident [4.114,115]. The more rapid depression of the T_c of $ErRh_4B_4$ by substitution of Gd for Er as compared to Ho for Er may be expected because of the larger deGennes factor for Gd as compared to Ho. In addition, T_c vs x varies according to AG theory except near x_c where small deviations occur. In contrast to the $(Er_{1-x}Ho_x)Rh_4B_4$ behavior, T_M is linear in x in $(Gd_xEr_{1-x})Rh_4B_4$, probably because Gd is an S state ion so that anisotropic crystal field effects arising from the Gd^{3+} ions should be absent. This implies that the ferromagnetic moment is in the basal plane for all values of x [4.116]. WANG et al. [4.114,117] also measured the upper critical field for various re-entrant members of this system and found that for small x, H_{c2} is enhanced over that of $ErRh_4B_4$. Measurements have also been made of the specific heat [4.118] and of T_c and T_m vs pressure [4.119]. Finally, unusual dependences of χ_{AC} and R vs T in various low magnetic fields for $(Gd_{0.28}Er_{0.72})Rh_4B_4$ were noted [4.115].

HUANG and coworkers [4.114,120] have investigated the upper critical field in $(Y_{1-x}Gd_x)Rh_4B_4$ compounds in order to determine the dominant physical mechanism governing H_{c2}. As x increases, they reported that the magnetic state of the samples changes from a dilute magnetic alloy for x = 0.01 to a spin glass for x = 0.1 to a re-entrant ferromagnet for x = 0.2. A phase diagram determined from their data also indicates a linear depression of T_M with x as well as AG-like behavior of T_c with x. Using the theory of MAEKAWA and TACHIKI [4.121], they find that H_{c2} is primarily determined by the exchange field of the Gd^{3+} spins. A more complete phase diagram was obtained for this system by ADRIAN et al. [4.122] which was in good agreement with that obtained by HUANG et al. Their magnetization, magnetic susceptibility and upper critical field measurements also revealed that macroscopic electromagnetic effects are of minor importance for the suppression of H_{c2} [4.122].

A phase diagram delineating the paramagnetic, superconducting and magnetically ordered regimes for $(Lu_{1-x}Ho_x)Rh_4B_4$ has been obtained by MAPLE et al. [4.4]. Again there is a critical concentration $x_c = 0.92$ at which T_{c1}, T_{c2} and T_M become coincident. Below $x = 0.28$, re-entrant superconductivity is observed to disappear. The behavior of the $(Er_{1-x}Ho_x)Rh_4B_4$ phase diagram for $x > 0.3$ is almost identical to that of the $(Lu_{1-x}Ho_x)Rh_4B_4$ system, which again indicates that the ordering of the Ho^{3+} ions determines the T_{c2} phase boundary for $x > 0.3$ in $(Er_{1-x}Ho_x)Rh_4B_4$. Heat capacity measurements on $(Lu_{0.50}Ho_{0.50})Rh_4B_4$ [4.4] also show a spike-shaped feature at T_{c2} superimposed on the sharp leading edge of the high temperature part of the magnetic ordering anomaly. This again indicates that the superconducting to normal-ferromagnetic state transition at T_{c2} is first order. A theory by SAKAI et al. [4.123] has reproduced the phase diagram of the $(Lu_{1-x}Ho_x)Rh_4B_4$ system quite accurately.

To observe the dependence of T_{c1} and T_{c2} on Er composition in $ErRh_4B_4$, OKUDA and coworkers [4.124,125] mapped out the phase diagram of the $(Er_xY_{1-x})Rh_4B_4$ system. They found that below $x = 0.43$, long-range magnetic ordering disappeared. Measurements of H_{c2} showed similar behavior to that described for $(Y_{1-x}Gd_x)Rh_4B_4$. In particular, a sudden decrease in H_{c2} for $0.8 \lesssim x \lesssim 1.0$ was attributed to the first-order transition at T_{c2}. SMITH et al. [4.116,126] performed a similar probe by constructing the paramagnetic, superconducting and re-entrant phase boundaries of $(Tm_xEr_{1-x})Rh_4B_4$. While re-entrant behavior was found to disappear for $x \gtrsim 0.4$, the χ_{AC} technique used could not observe the nonre-entrant magnetic ordering as x decreases from 1.00. TSE et al. [4.31] recently reported ^{11}B NMR and static magnetization measurements on $(Y_{1-x}Er_x)Rh_4B_4$ pseudoternary compounds. The results suggested a strong conduction electron-RE magnetic moment exchange interaction, which is surprising in view of the small depression of T_c with x. Large hyperfine interactions below T_c were also observed and attributed to an anomalously large uncompensated conduction electron spin polarization in the superconducting state (Chap.7).

The superconducting and magnetic transitions as a function of composition have also been determined in the superconducting system $(Sm_{1-x}Er_x)Rh_4B_4$ [4.127], wherein the nature of the magnetic ordering is altered from antiferromagnetism in $SmRh_4B_4$ to ferromagnetism in $ErRh_4B_4$. Both heat capacity and χ_{AC} data indicate that the nature of the magnetic ordering involving both the Er^{3+} and Sm^{3+} ions changes continously with x with the re-entrant character of the transition disappearing rapidly near $x = 0.2$. One of the unusual aspects of this system is that the magnetic ordering temperature T_M is enhanced between $x = 0.3$ and 0.9 over that of each of the ternary end members. It has been suggested [4.127] that the absence of the sinusoidally modulated magnetic state seen in $ErRh_4B_4$ might partially account for the enhancement of T_M as several theories suggest [4.66,73]. Measurements of H_{c2} vs T have also been made for various values of x in the $(Sm_{1-x}Er_x)Rh_4B_4$ system [4.128].

Table 4.3.

System		x_c^*	$x_s^\dagger$
$(Er_{1-x}Ho_x)Rh_4B_4$	[4.98]	0.89	
$(Gd_xEr_{1-x})Rh_4B_4$	[4.114]	0.28	
$(Y_{1-x}Gd_x)Rh_4B_4$	[4.122]	0.32	~ 0.16
$(Lu_{1-x}Ho_x)Rh_4B_4$	[4.4]	0.92	0.28
$(Er_xY_{1-x})Rh_4B_4$	[4.124]		0.43
$(Sm_{1-x}Er_x)Rh_4B_4$	[4.127]		0.18
$(Tm_xEr_{1-x})Rh_4B_4$	[4.116]		~ 0.35

*Critical concentration at which T_{c1}, T_{c2} and T_M coincide.
†Critical concentration at which $T_{c2} \to 0$.

VINING and SHELTON [4.129] found a nonlinear but smooth dependence of T_c on x in the superconducting pseudoternary system $(Th_{1-x}Y_x)Rh_4B_4$. In addition, they observed nonlinear behavior of T_c with pressure similar to that of $LuRh_4B_4$ [4.33,99] for samples with x > 0.80. In contrast, all magnetic $RERh_4B_4$ compounds displayed linear T_c (or T_M) vs pressure curves up to 20 kbar [4.99].

Some interesting conclusions can now be drawn by considering the pseudoternary systems that have been synthesized in conjunction with the $(Lu_{1-x}RE_x)Rh_4B_4$ rate of depression of T_c data [4.5]. For convenience, the critical concentration x_c at which T_{c1}, T_{c2} and T_M become coincident and the concentration x_s at which $T_{c2} \to 0$ for the various systems are summarized in Table 4.3. Since the substitution of Gd for Lu rapidly depresses the T_c of $LuRh_4B_4$, $x_c \sim 0.3$ is rather small and a considerable portion of the AG-like T_c vs x curve can be realized for $(Gd_xEr_{1-x})Rh_4B_4$ and $(Y_{1-x}Gd_x)Rh_4B_4$. In contrast, dT_c/dx for RE = Ho, Er and Tm is comparatively small for $(Lu_{1-x}RE_x)Rh_4B_4$ so that only the initial linear portion of the AG-like curve is apparent in the $(Er_{1-x}Ho_x)Rh_4B_4$, $(Lu_{1-x}Ho_x)Rh_4B_4$, $(Er_xY_{1-x})Rh_4B_4$ and $(Tm_xEr_{1-x})Rh_4B_4$ systems. This also leads to the relatively large values of $x_c \sim 0.9$ in $(Er_{1-x}Ho_x)Rh_4B_4$ and $(Lu_{1-x}Ho_x)Rh_4B_4$.

MACKAY et al. [4.5] also extrapolated their initial depression rate data for $(Lu_{1-x}RE_x)Rh_4B_4$ to x = 1.00 and found that the observed T_c values for $ErRh_4B_4$ and $TmRh_4B_4$ agreed with the extrapolated values. An expected value of T_c for $HoRh_4B_4$ (if it were not to order magnetically at a higher temperature) derived from the $(Er_{1-x}Ho_x)Rh_4B_4$ and $(Lu_{1-x}Ho_x)Rh_4B_4$ phase diagrams is also in agreement with the value predicted from the initial depression rate data for $(Lu_{1-x}Ho_x)Rh_4B_4$. This again suggests that for RE = Ho, Er and Tm in $(Lu_{1-x}RE_x)Rh_4B_4$, the T_c vs x curve between x = 0 and x = 1.0 is still linear. In contrast, the predicted values of T_c for $NdRh_4B_4$ and $SmRh_4B_4$ are lower than those actually observed. However, in these lighter compounds, CEF effects seem to be larger. Under certain conditions, then, the slope of the T_c vs x curve for these RE ions in $(Lu_{1-x}RE_x)Rh_4B_4$ would be expected to decrease in magnitude at higher concentrations of RE [4.130,131] which would lead to the initial depression rate data predicting too small a value for T_c. In fact, this is just the behavior of T_c observed by WOOLF and MAPLE [4.127] for the $(Sm_{1-x}Er_x)Rh_4B_4$ system.

In the above-mentioned systems consisting of two magnetic species, the magnetic ordering transitions have been mapped out in rather complete detail for $(Er_{1-x}Ho_x)Rh_4B_4$, $(Gd_xEr_{1-x})Rh_4B_4$ and $(Sm_{1-x}Er_x)Rh_4B_4$. In the first system, the magnetic ordering of the Er^{3+} and of the Ho^{3+} ions appear to occur independently of one another as is typical for mixtures of systems with competing spin anisotropies (and different Steven's parameters α_J). However, there seems to be a single magnetic transition for all x in the latter two systems. It is interesting to note that α_J for both Sm^{3+} and Er^{3+} is positive suggesting that Sm^{3+} also orders in the basal plane [4.24] while the spatially symmetric Gd^{3+} ion has no preferred direction of orientation. Thus it seems that in the absence of strong competing anisotropic crystal field forces, the magnetic ordering will involve both RE ions in a cooperative phase transition for all values of x.

The pseudoternary systems $(Nd_{1-x}Ho_x)Rh_4B_4$, $(Sm_{1-x}Ho_x)Rh_4B_4$ and $(Lu_{1-x}Dy_x)Rh_4B_4$ have also been investigated at low temperatures recently [4.101].

4.3.2 $RE(M_xRh_{1-x})_4B_4$ Compounds

Pseudoternary systems of the type $RE(M_xRh_{1-x})_4B_4$, in which a second transition metal element M is subtituted for the rhodium, have displayed many unusual superconducting and magnetic properties. By varying the concentration x, the influence of the transition metal upon the superconductivity in ternary compounds may be studied. Also, since the relative strengths of the magnetism and superconductivity change considerably with x, the interplay between these two competing orderings may be observed without disturbing the integrity of the rare-earth sublattice.

KU and coworkers have studied many systems in which iridium is substituted at the rhodium sites in the $CeCo_4B_4$-type structure [4.132-135]. The pure $REIr_4B_4$ compounds tend to be metastable, but superconductivity has been observed for RE = Ho, Er and Tm, with T_c = 2.0 K, 2.1 K and 1.5 K, respectively, and for various pseudoternary $(Y_xLu_{1-x})Ir_4B_4$ compounds [4.132]. The metastability of these compounds is somewhat surprising, since with Ir and Rh having nearly identical size and valence, an isomorphism between the two REM_4B_4 structures would be expected. Analyses of the lattice parameters determined from X-ray diffraction patterns for $Er(Ir_xRh_{1-x})_4B_4$ [4.132] and $Dy(Ir_xRh_{1-x})_4B_4$ [4.134] reveal that the c/a ratio abruptly decreases with increasing x for x > 0.6, whereas c/a remains comparatively constant below this concentration. The metastability of the iridium-rich compounds may therefore be caused by distortion of the shape of the Rh-Ir tetrahedral clusters.

Shown in Fig.4.25 is the low temperature phase diagram reported by KU et al. [4.133] for the pseudoternary system $Ho(Ir_xRh_{1-x})_4B_4$. As Ir is substituted for Rh,

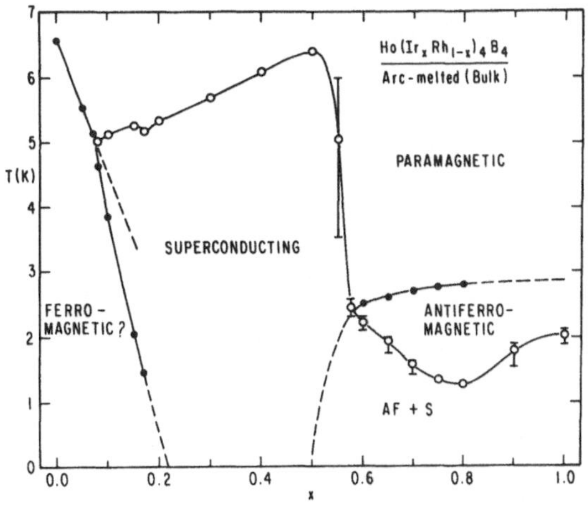

Fig.4.25. Low temperature phase diagram for the system $Ho(Ir_xRh_{1-x})_4B_4$ determined from AC and static magnetic susceptibility measurements. From [4.133]

T_M decreases linearly until a critical point similar to that observed in $(Er_{1-x}Ho_x)Rh_4B_4$ [4.98] is reached. Above this critical concentration, T_M decreases more rapidly and T_c increases, until near x = 0.5, a very abrupt decrease in T_c occurs. This sharp drop appears to be an intrinsic characteristic of the $RE(Ir_xRh_{1-x})_4B_4$ pseudoternary systems, since similar features are observed for RE = Er [4.132] and Dy [4.134] near x = 0.6, the concentration at which the c/a ratio begins to decrease. Another feature common to these three systems is the occurrence of a T_c minimum at x = 0.8. KU et al. have suggested that the abrupt change in T_c near x = 0.5 occurs because, for higher Ir concentrations, each unit cell contains, on the average, less than one Rh_4 tetrahedron, which appears to be primarily responsible for the high values of T_c observed in the $RERh_4B_4$ compounds [4.17,18]. They also interpreted the minima of T_c as an indication that the rhodium and iridium atoms are randomly distributed in only one of the two transition metal tetrahedra in each unit cell rather than in both [4.133].

Another unusual feature of the $Ho(Ir_xRh_{1-x})_4B_4$ pseudoternary system is that for $x \geq 0.6$, the samples first order antiferromagnetically and, upon further cooling, become superconducting. Recently, specific heat and critical field measurements have shown that antiferromagnetic order also occurs for $x < 0.6$ [4.136], so it is clear that superconductivity and antiferromagnetism can coexist in this system regardless of whether the ratio T_N/T_c is greater or less than one. A similar coexistence of superconductivity and long-range magnetic order with $T_M > T_c$ is observed in $Dy(Ir_xRh_{1-x})_4B_4$, although the nature of the magnetic order has not been unambiguously determined [4.134]. Since the T_M vs x and T_c vs x curves pass smoothly through each other, apparently the two types of ordering have comparatively little effect upon each other in these systems.

WOOLF et al. [4.137,138] have studied in depth the pseudoternary compound $Ho(Ir_{0.7}Rh_{0.3})_4B_4$, in which T_N = 2.6 K and $T_c \sim 1.4$ K. Heat capacity measurements

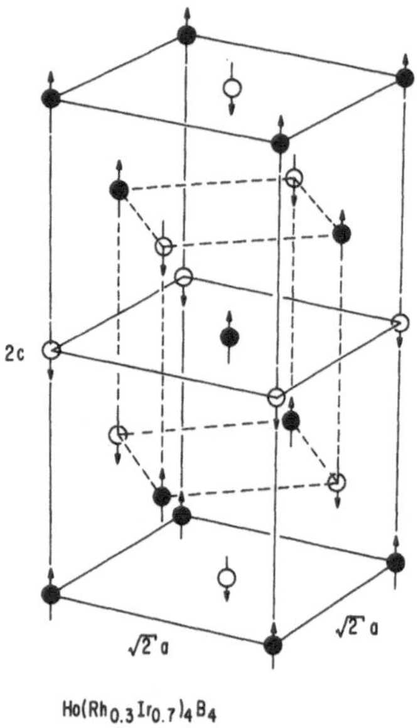

Fig.4.26. Proposed magnetic unit cell for $Ho(Rh_{0.3}Ir_{0.7})_4B_4$. The crystallographic unit cell is outlined in dashed lines and the Rh, Ir and B atoms have been removed for clarity. From [4.139]

have confirmed the bulk character of the magnetic order in this material and reveal that the ground state of the Ho^{3+} magnetic moments is a doublet. Thermal conductivity data indicate that the contribution to λ by phonon-electron scattering exceeds that due to the electrons, thereby preventing any definite conclusion regarding the bulk nature of the superconductivity from this measurement. However, M vs H data below T_c attest to the presence of bulk superconductivity [4.138]. Recently, HAMAKER et al. [4.139] have determined that $Ho(Ir_{0.7}Rh_{0.3})_4B_4$ orders antiferromagnetically at T_N = 2.75 K using elastic neutron scattering. The magnetic structure of this compound, shown in Fig.4.26, is body-centered tetragonal with magnetic lattice parameters $a_M = \sqrt{2}$ a and c_M = 2c and consists of stacks of antiferromagnetic sheets in the x-y plane. In this arrangement each of the Ho^{3+} magnetic moments, which are aligned along the c-axis, is parallel to the magnetic moments of half of its nearest Ho neighbors and antiparallel to those of its remaining Ho nearest neighbors. Such a configuration may represent a compromise between competing magnetic interactions, such as RKKY and dipole interactions, which may have different signs for the coupling between magnetic moments. The observed value of the saturation magnetic moment μ_s at 1.5 K is 9.6 ± 0.6 μ_B, nearly equal to the Hund's rule value of 10 μ_B, so the crystal-field ground state doublet must consist primarily of the $|\pm 8\rangle$ states.

The second class of transition metal pseudoternary systems which have been studied in detail consists of the $RE(Ru_xRh_{1-x})_4B_4$ compounds. JOHNSTON [4.140] first reported that the substitution of approximately five percent ruthenium for the rhodium stabilizes a body-centered tetragonal structure which is similar to the $CeCo_4B_4$-type phase. Furthermore, while the $RE(Ru_{0.15}Rh_{0.85})_4B_4$ compounds are superconducting (RE = Y, Pr, Eu, Dy, Ho, Er, Tm, and Lu), long-range magnetic order occurs in $RERu_4B_4$ (RE = Nd, Gd, Tb, Dy, Ho, and Er) except for $LuRu_4B_4$, which is superconducting. Studies of the effect of pressure upon the values of T_c and T_M show that lattice effects are not the principal cause of this dramatic change in behavior, but rather electronic changes resulting from the differing valences of the ruthenium and rhodium atoms appear to play a more important role [4.141].

The exact reasons for the strikingly different behavior of the $RE(Ru_{0.15}Rh_{0.85})_4B_4$ and $RERu_4B_4$ compounds are not yet clear, but a strong correlation between the c/a ratio and the superconducting properties has been observed. Whereas the rhodium-rich compounds all have $c/a < 2$, the $RERu_4B_4$ compounds exhibit $c/a > 2$ [4.140]. Furthermore, in both $Y(Ru_xRh_{1-x})_4B_4$ [4.140] and $Er(Ru_xRh_{1-x})_4B_4$ [4.142], a rapid depression of T_c is observed very near the concentration x at which $c/a = 2$, and a similar decrease in T_c is observed in $Dy(Ru_xRh_{1-x})_4B_4$ [4.143], although a detailed crystallographic study has not been made in this last system.

The magnetic order observed in the $RE(Ru_xRh_{1-x})_4B_4$ compounds is also strongly affected by the variation of x. Long-range magnetic order has been observed for all values of x in $Dy(Ru_xRh_{1-x})_4B_4$, even though no superconductivity occurs for $x > 0.35$ [4.143]. Upper critical field measurements indicate that antiferromagnetism coexists with the superconductivity in the Rh-rich samples, whereas magnetization data show that $DyRu_4B_4$ is ferromagnetic [4.144]. Similarly, magnetic susceptibility measurements suggest that $Gd(Ru_{0.15}Rh_{0.85})_4B_4$ is antiferromagnetic and that $GdRu_4B_4$ is ferromagnetic [4.145], while H_{c2} vs T curves for Rh-rich $Ho(Ru_xRh_{1-x})_4B_4$ and $Er(Ru_xRh_{1-x})_4B_4$ compounds reveal antiferromagnetism in these materials as well [4.146], whereas $ErRu_4B_4$ is ferromagnetic [4.147]. Therefore, the change from antiferromagnetism to ferromagnetism may be an intrinsic characteristic of the $RE(Ru_xRh_{1-x})_4B_4$ materials. Although the weakening of superconductivity appears to result inherently from the nature of the electronic band structure, it is not yet clear whether the change of the magnetic order is coupled in some way with the destruction of superconductivity in the systems containing rare earths with magnetic moments. Another problem which still requires further investigation is the detailed manner in which the antiferromagnetism changes to ferromagnetism as ruthenium is gradually substituted for rhodium in these materials.

4.4 Ternary REM_xSn_y Compounds

Re-entrant behavior has also been observed in some of the REM_xSn_y compounds. These materials have attracted a lot attention because of the relative ease with which large single crystals may be made for a considerable number of compounds. Since a detailed discussion of their composition and structure may be found elsewhere

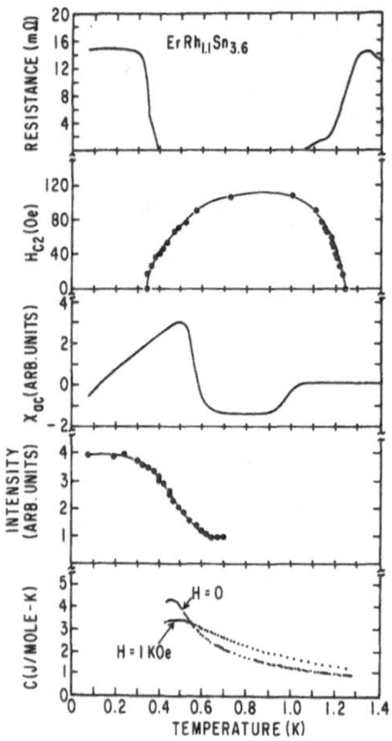

Fig.4.27. AC electrical resistance, upper critical field, AC magnetic susceptibility χ_{AC}, neutron scattering intensity of the (111) peak and heat capacity C vs temperature for $ErRh_{1.1}Sn_{3.6}$. From [4.147]

in this volume, we shall concentrate here upon the interplay of superconductivity and magnetism in these materials.

Measurements by REMEIKA et al. [4.148] have determined that $ErRh_{1.1}Sn_{3.6}$ is a re-entrant ferromagnetic superconductor. Their data, which are summarized in Fig.4.27, reveal that superconductivity is quenched, as indicated by the increase of both the resistance and AC magnetic susceptibility, upon the development of ferromagnetic order, which is evidenced by the lambda-type anomaly in specific heat data and the development of neutron scattering intensities only at the Bragg positions in the crystal lattice. Static magnetic susceptibility [4.149] and AC magnetic susceptibility [4.150] data indicate that in finite applied magnetic fields, the increasing paramagnetic induction of the sample is responsible for the destruction of superconductivity at T_{c2}, thereby accounting for the bell shape of the H_{c2} vs T curve.

In contrast to $ErRh_4B_4$ and $Ho_{1.2}Mo_6S_8$, however, $ErRh_{1.1}Sn_{3.6}$ exhibits no thermal hysteresis at the re-entrant transition either resistively or inductively. Also, the spike-shaped anomaly in the heat capacity data at T_{c2}, which is characteristic of the first-order coupled superconducting to normal-magnetic transition in previously reported re-entrant superconductors [4.20,100,151], is absent in $ErRh_{1.1}Sn_{3.6}$, as may be seen in Fig.4.27. Neutron diffraction results show that even at T = 0.07 K, true long-range order does not develop in this compound [4.148]. One explanation which may account for these properties is that partially occupied equivalent sites for the Sn and Er atoms within the crystal lattice cause disorder within the magnetic moment sub-structure [4.152]. This could also explain the sample dependence of the superconducting and magnetic behavior which was observed by OTT et al. [4.153], who reported that some $ErRh_xSn_y$ samples do not become superconducting at all, whereas others are superconducting without becoming re-entrant. In this last case, thermal conductivity measurements show that above T_c, λ depends linearly on T, whereas λ decreases more rapidly for $T < T_c$, as is expected for a bulk superconductor. Surprisingly, applied magnetic fields in excess of 0.1 T do not affect the temperature dependence of λ, even though in a magnetic field of 40 mT, electrical resistance measurements reveal no traces of superconductivity in this sample.

LAMBERT et al. [4.150] determined from AC magnetic susceptibility data that in H = 0, $ErOs_xSn_y$ also becomes re-entrant without any thermal hysteresis at T_{c2}. In contrast, $TmRh_{1.3}Sn_{4.0}$ becomes re-entrant only in applied magnetic fields of 0.12 T or more [4.150], and heat capacity measurements indicate that no magnetic order occurs above 0.1 K [4.154]. In this latter case, the increasing magnetization of the Tm^{3+} ions in applied magnetic fields is the sole cause of the destruction of superconductivity at low temperatures, which is similar to the behavior of $ErRh_{1.1}Sn_{3.6}$ in finite magnetic fields.

Recently, FISK et al. [4.155] reported data on the magnetic and superconducting properties of $REOs_xSn_y$ compounds with RE = Gd, Tb, Dy, Ho, Er and Tm. The compounds of Tb and Ho are superconducting only, those of Er and Tm are re-entrant superconductors, and those of Gd and Dy appear to exhibit some type of short-range magnetic order at low temperatures.

4.5 Summary

The $RERh_4B_4$ and REM_xSn_y compounds considered herein, as well as the $RE_xMo_6X_8$ compounds discussed in Chap.5 of this volume, have provided unique opportunities for studying the interplay between superconductivity and long-range magnetic order, a subject that has intrigued both experimentalists and theoreticians alike for more than two decades. A wide variety of experimental techniques have been employed to investigate superconducting-magnetic interactions in these ternary RE compounds, and a great deal of theoretical effort has been expended in developing an understanding of their remarkable physical properties.

The $RERh_4B_4$ compounds (as well as the $RE_xMo_6X_8$ compounds) have two special attributes: (i) the exchange interaction between the conduction electron spins and the RE magnetic moments is relatively weak, and (ii) the RE ions form an ordered sublattice. The small magnitude of the exchange interaction parameter $\mathcal{J}$ (~ 0.01 eV-atom) enables the ternary RE compounds to retain their superconductivity, even in the presence of relatively large concentrations of RE magnetic moments. It also results in ordering of the RE magnetic moments via the RKKY interaction at relatively low temperatures that are comparable to superconducting transition temperatures. The ordered sublattice of the RE ions leads to *long-range* magnetic ordering. characterized by sharp magnetic ordering temperatures T_M with well-defined features in the physical properties at T_M.

The small magnitude of $\mathcal{J}$ may be traced to the transition metal molecular units or "clusters" that, along with the RE ions, are the building blocks of the $RERh_4B_4$ (and $RE_xMo_6X_8$) crystal structure. The transition metal clusters also appear to play an important role in the superconductivity of the ternary RE compounds, several of which have high values of T_c and/or H_{c2}. The relative confinement of the transition metal 4d-electrons within the clusters produces narrow peaks in the density of states at the Fermi level, while the external and internal vibrational modes of the clusters themselves can affect the electron-phonon interaction.

Parallel investigations of the $RERh_4B_4$ compounds reviewed here and the $RE_xMo_6X_8$ compounds have revealed that long-range magnetic ordering of the RE ions can develop below T_c. Two different types of coupled superconducting and magnetic behavior have been observed, depending upon whether the nature of the magnetic ordering is ferromagnetic or antiferromagnetic.

When the ordering is ferromagnetic, superconductivity that occurs at an upper critical temperature T_{c1} is destroyed at a lower critical temperature $T_{c2} \sim T_M$, where T_M is the Curie temperature. Thermal hysteresis in various physical properties as well as a sharp feature in the heat capacity indicate that the transition at T_{c2} from the paramagnetic-superconducting state to the ferromagnetic-normal state is first order. Moreover, in the ferromagnetic superconductors $ErRh_4B_4$ and $HoMo_6S_8$, a sinusoidally modulated magnetic state with a wavelength $\sim$100 Å has been found to coexist with superconductivity within a narrow temperature interval above T_{c2}. In contrast, when the ordering is antiferromagnetic, it coexists with superconductivity in zero applied magnetic field, although it can modify superconducting properties by means of several proposed mechanisms. The most notable example is the curve of H_{c2} vs temperature which can be either enhanced or depressed below the Néel temperature T_N.

Experiments on pseudoternary $RERh_4B_4$ compounds have been very fruitful in elucidating superconducting-magnetic interactions and in studying the effects of competing types of spin anisotropy and/or magnetic order. Complex and interesting low temperature phase diagrams have been discovered and investigated and have yielded evidence for the suppression of the Curie temperature by superconductivity and the coexistence of superconductivity and antiferromagnetic order, but with $T_N > T_c$.

Finally, the physical properties of the $RERh_4B_4$ compounds that are only superconducting or magnetic are quite striking. The nonmagnetic $LuRh_4B_4$ and YRh_4B_4 compounds have relatively high T_c's of $\sim$11 K. Strong deviations of the Curie temperatures of the ferromagnetic $RERh_4B_4$ compounds from the deGennes' scaling that is expected for an isostructural series of RE compounds that order via the RKKY interaction suggest that CEF, magnetoelastic and possibly even dipolar interactions [4.156] are involved. Pronounced CEF effects have, in fact, been observed in the heat capacity, magnetization and thermal expansion of nearly all of the $RERh_4B_4$ compounds. A variety of magnetic structures, some unusual and complicated, have been found among the $RERh_4B_4$ ternary and pseudoternary compounds. The nearly perfect conformation of the ferromagnetic $HoRh_4B_4$ compound to the predictions of mean field theory appears to be unique.

Along with the $RE_xMo_6X_8$ compounds, the $RERh_4B_4$ and REM_xSn_y compounds discussed here have proven to be a reservoir of new and interesting physical phenomena and a rich testing ground for theoretical predictions.

References

4.1 B.T. Matthias, E. Corenzwit, J.M. Vandenberg, H. Barz: Proc. Natl. Acad. Sci. USA *74*, 1334 (1977)
4.2 J.M. Vandenberg, B.T. Matthias: Proc. Natl. Acad. Sci. USA *74*, 1336 (1977)
4.3 Yu.B. Kuz'ma, N.S. Bilonizhko: Soviet Phys.-Cryst. *16*, 897 (1972)
4.4 M.B. Maple, H.C. Hamaker, D.C. Johnston, H.B. MacKay, L.D. Woolf: J. Less-C. Metals *62*, 251 (1978)
4.5 H.B. MacKay, L.D. Woolf, M.B. Maple, D.C. Johnston: J. Low Temp. Phys. *41*, 639 (1980)
4.6 M.B. Maple: Solid State Commun. *12*, 653 (1973)
4.7 M.A. Ruderman, C. Kittel: Phys. Rev. *96*, 99 (1954);
T. Kasuya: Progr. Theoret. Phys. (Kyoto) *16*, 45 (1956);
K. Yosida: Phys. Rev. *106*, 893 (1957)
4.8 W.A. Fertig, D.C. Johnston, L.E. DeLong, R.W. McCallum, M.B. Maple, B.T. Matthias: Phys. Rev. Lett. *38*, 987 (1977)
4.9 D.E. Moncton, D.B. McWhan, J. Eckert, G. Shirane, W. Thomlinson: Phys. Rev. Lett. *39*, 1164 (1977)
4.10 H.C. Hamaker, L.D. Woolf, H.B. MacKay, Z. Fisk, M.B. Maple: Solid State Commun. *31*, 139 (1979)
4.11 C.F. Majkrzak, D.E. Cox, G. Shirane, H.A. Mook, H.C. Hamaker, H.B. MacKay, Z. Fisk, M.B. Maple: To appear in Phys. Rev. B
4.12 H.C. Hamaker, L.D. Woolf, H.B. MacKay, Z. Fisk, M.B. Maple: Solid State Commun. *32*, 289 (1979)
4.13 H.C. Hamaker, H.B. MacKay, M.S. Torikachvili, L.D. Woolf, M.B. Maple, W. Odoni, H.R. Ott: J. Low Temp. Phys. *44*, 553 (1981)
4.14 C.F. Majkrzak, S.K. Satija, G. Shirane, H.C. Hamaker, Z. Fisk, M.B. Maple: To be published
4.15 A.A. Abrikosov, L.P. Gor'kov: Sov. Phys. JETP *12*, 1243 (1961)
4.16 P.G. deGennes: J. Phys. Rad. *23*, 510 (1962)
4.17 J.M. Vandenberg, B.T. Matthias: Science *198*, 194 (1977)
4.18 T. Jarlborg, A.J. Freeman, T.J. Watson-Yang: Phys. Rev. Lett. *39*, 1032 (1977)
4.19 Ø. Fischer, M.B. Maple (eds.): *Superconductivity in Ternary Compounds I*, Topics in Current Phys., Vol.32 (Springer, Berlin, Heidelberg, New York 1982)
4.20 L.D. Woolf, D.C. Johnston, H.B. MacKay, R.W. McCallum, M.B. Maple: J. Low. Temp. Phys. *35*, 651 (1979)
4.21 L.D. Woolf, H.C. Hamaker, S.E. Lambert, M.B. Maple, H.R. Ott: To be published
4.22 S.D. Bader, G.S. Knapp, S.K. Sinha, P. Schweiss, B. Renker: Phys. Rev. Lett. *37*, 344 (1976)
4.23 K. Kumagai, Y. Inoue, K. Asayama: J. Phys. Soc. Jpn. *47*, 1363 (1979)
4.24 L.D. Woolf: Ph. D. Thesis, University of California, San Diego (1980), unpublished
4.25 H.R. Ott, A.M. Campbell, H. Rudigier, H.C. Hamaker, M.B. Maple: Physica *108*B, 751 (1981)
4.26 A.J. Freeman, T. Jarlborg, T.J. Watson-Yang: J. Mag. Magn. Materials *7*, 296 (1978)

4.27 U. Poppe: Physica *108*B, 805 (1981)
4.28 C.P. Umbach, L.E. Toth, E.D. Dahlberg, A.M. Goldman: Physica *108*B, 803 (1981)
4.29 Y. Kuwasawa, L. Rinderer, B.T. Matthias: J. Low Temp. Phys. *37*, 179 (1979)
4.30 C.Y. Huang, S.E. Kohn, S. Maekawa, J.L. Smith: Solid State Commun. *32*, 929 (1979)
4.31 P.K. Tse, A.T. Aldred, F.Y. Fradin: Phys. Rev. Lett. *43*, 1825 (1979)
4.32 W. Odoni, G. Keller, H.R. Ott, H.C. Hamaker, D.C. Johnston, M.B. Maple: Physica *108*B, 1227 (1981)
4.33 L.E. DeLong, H.B. MacKay, M.B. Maple: In *Ternary Superconductors*, ed. by G.K. Shenoy, B.D. Dunlap, F.Y. Fradin (North-Holland, Amsterdam 1981) pp.193-196
4.34 M.B. Maple: In *MAGNETISM: A Treatise on Modern Theory and Materials*, Vol.V, ed. by H. Suhl (Academic, New York 1973) and references cited therein
4.35 M.B. Maple, L.E. DeLong, B.C. Sales: In *Handbook on the Physics and Chemistry of Rare Earths*, ed. by K.A. Gschneidner, Jr., L. Eyring (North-Holland, Amsterdam 1978), and references cited therein
4.36 H.R. Ott, L.D. Woolf, M.B. Maple, D.C. Johnston: J. Low Temp. Phys. *39*, 383 (1980)
4.37 D.C. Mattis: *The Theory of Magnetism* (Harper and Row, New York 1965) p.229
4.38 F. Acker, H.C. Ku: J. Low Temp. Phys. *42*, 449 (1981)
4.39 H.R. Ott, W.A. Fertig, D.C. Johnston, M.B. Maple, B.T. Matthias: J. Low Temp. Phys. *33*, 159 (1978)
4.40 G.H. Lander, S.K. Sinha, F.Y. Fradin: J. Appl. Phys. *50*, 1990 (1979)
4.41 Y. Seiwa, J. Tsuda, A. Hirai, C.W. Searle: Phys. Lett. *43*A, 23 (1973)
4.42 G.K. Shenoy, P.J. Viccaro, D. Niarchos, J.D. Cashion, B.D. Dunlap, F.Y. Fradin: In *Ternary Superconductors*, ed. by G.K. Shenoy, B.D. Dunlap, F.Y. Fradin (North Holland, Amsterdam 1981) pp.163-167
4.43 H.R. Ott, G. Keller, W. Odoni, L.D. Woolf, M.B. Maple, D.C. Johnston, H.A. Mook: Phys. Rev. B*25*, 477 (1982)
4.44 M.B. Maple, H.C. Hamaker, L.D. Woolf, H.B. MacKay, Z. Fisk, W. Odoni, H.R. Ott: In *Crystalline Electric Field and Structural Effects in f-Electron Systems*, ed. by J.E. Crow, R.P. Guertin, T.W. Mihalisin (Plenum, New York 1980) pp.533-543
4.45 M.B. Maple: In *Science and Technology of Rare Earth Materials*, ed. by E.C. Subbarao, W.E. Wallace (Academic, New York 1980) pp.167-193
4.46 G.K. Shenoy, B.D. Dunlap, F.Y. Fradin, S.K. Sinha, C.W. Kimball, W. Potzel, F. Pröbst, G.M. Kalvius: Phys. Rev. B*21*, 3886 (1980)
4.47 M. Ishikawa, Ø. Fischer: Solid State Commun. *23*, 37 (1977)
4.48 J.W. Lynn, D.E. Moncton, W. Thomlinson, G. Shirane, R.N. Shelton: Solid State Commun. *26*, 493 (1978)
4.49 H.B. Radousky, G.S. Knapp, J.S. Kouvel, T.E. Klippert, J.W. Downey: In *Ternary Superconductors*, ed. by G.K. Shenoy, B.D. Dunlap, F.Y. Fradin (North Holland, Amsterdam 1981) pp.151-154
4.50 F. Behroozi, G.W. Crabtree, S.A. Campbell, M. Levy, D. Snider, D.C. Johnston, B.T. Matthias: In *Ternary Superconductors*, ed. by G.K. Shenoy, B.D. Dunlap, F.Y. Fradin (North Holland, Amsterdam 1981) pp.155-158
4.51 F. Behroozi, M. Levy, D.C. Johnston, B.T. Matthias: Solid State Commun. *38*, 515 (1981)
4.52 S.C. Schneider, M. Levy, M. Tachiki, D.C. Johnston: Physica *108*B, 807 (1981)
4.53 S.C. Schneider, M. Levy, R. Chen, M. Tachiki, D.C. Johnston, B.T. Matthias: Solid State Commun. *40*, 61 (1981)
4.54 S. Maekawa, M. Tachiki, S. Takahashi: J. Mag. Magn. Mat. *13*, 324 (1979)
4.55 K. Kumagai, Y. Inoue, K. Asayama: Solid State Commun. *35*, 531 (1980)
4.56 R.H. Cantor, E.D. Dahlberg, A.M. Goldman, L.E. Toth, G.L. Christner: Solid State Commun. *34*, 485 (1980)
4.57 W. Odoni, H.R. Ott: Phys. Lett. *70*A, 480 (1979)
4.58 D.E. Moncton, D.B. McWhan, P.H. Schmidt, G. Shirane, W. Thomlinson, M.B. Maple, H.B. MacKay, L.D. Woolf, Z. Fisk, D.C. Johnston: Phys. Rev. Lett. *45*, 2060 (1980)
4.59 H.B. MacKay: Ph. D. Thesis, University of California, San Diego (1979), unpublished

4.60 S.C. Schneider, M. Levy, D.C. Johnston, B.T. Matthias: Phys. Lett. *80*A, 72 (1980)
4.61 S.C. Schneider, R. Chen, M. Levy, D.C. Johnston, B.T. Matthias: In *Ternary Superconductors*, ed. by G.K. Shenoy, B.D. Dunlap, F.Y. Fradin (North Holland, Amsterdam 1981) pp.147-150
4.62 N. Toyota, S.B. Woods, Y. Muto: Solid State Commun. *37*, 547 (1981)
4.63 M. Tachiki, T. Koyama, H. Matsumoto, H. Umezawa: Solid State Commun. *34*, 269 (1980)
4.64 J.M. Rowell, R.C. Dynes, P.H. Schmidt: In *Superconductivity in d- and f-Band Metals*, ed. by H. Suhl, M.B. Maple (Academic, New York 1980) pp.409-418
4.65 D.E. Moncton: J. Appl. Phys. *50*, 1880 (1979)
4.66 G. Shirane, W. Thomlinson, D.E. Moncton: In *Superconductivity in d- and f-Band Metals*, ed. by H. Suhl, M.B. Maple (Academic, New York 1980) pp.381-389
4.67 E.I. Blount, C.M. Varma: Phys. Rev. Lett. *42*, 1079 (1979)
4.68 P.W. Anderson, H. Suhl: Phys. Rev. *116*, 898 (1959)
4.69 H. Suhl: J. Less-C. Metals *62*, 225 (1978)
4.70 L.N. Bulaevski, A.I. Rusinov, M. Kulîc: Solid State Commun. *30*, 59 (1979)
4.71 H. Matsumoto, H. Umezawa, M. Tachiki: Solid State Commun. *31*, 157 (1979)
4.72 K. Machida, T. Matsubara: Solid State Commun. *31*, 791 (1979)
4.73 M. Tachiki, H. Matsumoto, T. Koyama, H. Umezawa: Solid State Commun. *34*, 19 (1980)
4.74 C.G. Kuper, M. Revzen, A. Ron: Phys. Rev. Lett. *44*, 1545 (1980)
4.75 U. Krey: Int. J. Magnetism *3*, 65 (1972); Int. J. Magnetism *4*, 153 (1973)
4.76 M. Tachiki: Proc. 16th Intern. Conf. Low Temp. Phys., Part III, to appear in Physica B + C (1982)
4.77 C.R. Hu, T.E. Ham: Physica *108*B, 1041 (1981)
4.78 J.W. Lynn, G. Shirane, W. Thomlinson, R.N. Shelton, D.E. Moncton: Phys. Rev. B*24*, 3817 (1981)
4.79 H.A. Mook, M.B. Maple, Z. Fisk, D.C. Johnston: Solid State Commun. *36*, 287 (1980)
4.80 S.K. Sinha, G.W. Crabtree, D.G. Hinks, H.A. Mook: Proc. 16th Intern. Conf. Low Temp. Phys., Part III, to appear in Physica B + C (1982)
4.81 H.S. Greenside, E.I. Blount, C.M. Varma: Phys. Rev. Lett. *46*, 49 (1981)
4.82 G. Cort, R.D. Taylor, J.O. Willis: Physica *108*B, 809 (1981)
4.83 J.M. Rowell, R.C. Dynes, P.H. Schmidt: Solid State Commun. *30*, 191 (1979)
4.84 K. Machida: J. Low Temp. Phys. *37*, 583 (1979)
4.85 A.M. Stewart: Phys. Rev. B*24*, 4080 (1981)
4.86 J.H. Van Vleck: *The Theory of Electric and Magnetic Susceptibilities* (Oxford, London 1932) pp.245-256
4.87 H.R. Ott, W. Odoni, H.C. Hamaker, M.B. Maple: Phys. Lett. *75*A, 243 (1980)
4.88 H.R. Ott, W. Odoni: In *Superconductivity in d- and f-Band Metals*, ed. by H. Suhl, M.B. Maple (Academic, New York 1980) pp.403-408
4.89 H.C. Hamaker, H.B. MacKay, L.D. Woolf, M.B. Maple, W. Odoni, H.R. Ott: Phys. Lett. *81*A, 91 (1981)
4.90 M. Ishikawa, Ø. Fischer: Solid State Commun. *24*, 747 (1977)
4.91 H.R. Ott, H. Rudigier, H.C. Hamaker, M.B. Maple: In *Ternary Superconductors*, ed. by G.K. Shenoy, B.D. Dunlap, F.Y. Fradin (North-Holland, Amsterdam 1981) pp.159-162
4.92 M. Tachiki, H. Matsumoto, H. Umezawa: Phys. Rev. B*20*, 1915 (1979)
4.93 J.L. Smith, C.Y. Huang, J.J. Tsou, J.C. Ho: J. Appl. Phys. *50*(3), 2330 (1979)
4.94 B.T. Matthias, H. Suhl: Phys. Rev. Lett. *4*, 51 (1960)
4.95 M. Tachiki, A. Kotani, H. Matsumoto, H. Umezawa: Solid State Commun. *32*, 599 (1979)
4.96 M. Ishikawa, Ø. Fischer, J. Müller: J. Physique *39*, C6-1379 (1978), and references cited therein
4.97 M.B. Maple: J. Physique *39*, C6-1374 (1978), and references cited therein
4.98 D.C. Johnston, W.A. Fertig, M.B. Maple, B.T. Matthias: Solid State Commun. *26*, 141 (1978)
4.99 R.N. Shelton, D.C. Johnston: In *High Pressure and Low Temperature Physics*, ed. by C.W. Chu, J.A. Woollam (Plenum, New York 1978) pp.409-416
4.100 H.B. MacKay, L.D. Woolf, M.B. Maple, D.C. Johnston: Phys. Rev. Lett. *42*, 918 (1979); Erratum-Phys. Rev. Lett. *43*, 89 (1979)

4.101 M.B. Maple, H.C. Hamaker, D.C. Johnston, H.B. MacKay, M.S. Torikachvili, L.D. Woolf: To be published
4.102 H.A. Mook, W.C. Koehler, M.B. Maple, Z. Fisk, D.C. Johnston: In *Superconducitvity in d- and f-Band Metals*, ed. by H. Suhl, M.B. Maple (Academic, New York 1980) pp.427-432
4.103 H.A. Mook, W.C. Koehler, M.B. Maple, Z. Fisk, D.C. Johnston, L.D. Woolf: Phys. Rev. B*25*, 372 (1982)
4.104 H.A. Mook, M.B. Maple, Z. Fisk, D.C. Johnston, L.D. Woolf: In *Ternary Superconductors*, ed. by G.K. Shenoy, B.D. Dunlap, F.Y. Fradin (North Holland, Amsterdam, 1981) pp.179-180
4.105 S. Fishman, A. Aharony: Phys. Rev. B*18*, 3507 (1978), and references cited therein
4.106 S. Maekawa, C.Y. Huang: In *Crystalline Electric Field and Structural Effects in f-Electron Systems*, ed. by J.E. Crow, R.P. Guertin, T.W. Mihalisin (Plenum, New York 1980) pp.561-568
4.107 S.E. Lambert, L.D. Woolf, M.B. Maple: Unpublished results
4.108 L.D. Woolf, D.C. Johnston, H.A. Mook, W.C. Koehler, M.B. Maple, Z. Fisk: Proc. 16th Intern. Conf. Low Temp. Phys., Part III, to appear in Physica B + C (1982)
4.109 H. Adrian, K. Müller, G. Saemann-Ischenko: Phys. Rev. B*22*, 4424 (1980)
4.110 M. Ishikawa: Phys. Lett. *74*A, 263 (1979)
4.111 R.M. Hornreich, H.G. Schuster: Phys. Lett. *70*A, 143 (1979)
4.112 C. Balseiro, L.M. Falicov: Phys. Rev. B*19*, 2548 (1979)
4.113 B. Schuh, N. Grewe: Solid State Commun. *37*, 145 (1981)
4.114 R.H. Wang, R.J. Laskowski, C.Y. Huang, J.L. Smith, C.W. Chu: J. Appl. Phys. *49*, 1392 (1978)
4.115 S. Kohn, R.H. Wang, J.L. Smith, C.Y. Huang: J. Appl. Phys. *50*, 1862 (1979)
4.116 S. Maekawa, J.L. Smith, C.Y. Huang: Phys. Rev. B*22*, 164 (1980)
4.117 R.H. Wang, C.Y. Huang, J.L. Smith: J. Physique *39*, C6-373 (1978)
4.118 J.C. Ho, C.Y. Huang, J.L. Smith: J. Physique *39*, C6-381 (1978)
4.119 C.W. Chu, C.Y. Huang, S. Kohn, J.L. Smith: J. Less. C. Metals *62*, 245 (1978)
4.120 C.Y. Huang, S.E. Kohn, S. Maekawa, J.L. Smith: Solid State Commun. *32*, 929 (1979)
4.121 S. Maekawa, M. Tachiki: Phys. Rev. B*18*, 4688 (1978)
4.122 H. Adrian, R. Müller, R. Behrle, G. Saemann-Ischenko, G. Voit: Physica *108*B, 1281 (1981)
4.123 O. Sakai, M. Tachiki, H. Matsumoto, H. Umezawa: Solid State Commun. *39*, 279 (1981)
4.124 K. Okuda, Y. Nakakura, K. Kadowaki: Solid State Commun. *32*, 185 (1979)
4.125 K. Okuda, Y. Nakakura, K. Kadowaki: J. Mag. Magn. Mat. *15-18*, 1575 (1980)
4.126 J.L. Smith, R.B. Roof, V.O. Struebing: Bull. Am. Phys. Soc. *23*, 322 (1978)
4.127 L.D. Woolf, M.B. Maple: In *Ternary Superconductors*, ed. by G.K. Shenoy, B.D. Dunlap, F.Y. Fradin (North-Holland, Amsterdam 1981) pp.181-184
4.128 S.E. Lambert, L.D. Woolf, M.B. Maple: To be published
4.129 C.B. Vining, R.N. Shelton: In *Ternary Superconductors*, ed. by G.K. Shenoy, B.D. Dunlap, F.Y. Fradin (North Holland, Amsterdam 1981) pp.189-192
4.130 P. Fulde, L.L. Hirst, A. Luther: Z. Physik *230*, 155 (1970)
4.131 J. Keller, P. Fulde: J. Low Temp. Phys. *4*, 289 (1971)
4.132 H.C. Ku, B.T. Matthias, H. Barz: Solid State Commun. *32*, 937 (1979)
4.133 H.C. Ku, F. Acker, B.T. Matthias: Phys. Lett. *76*A, 399 (1980)
4.134 H.C. Ku, F. Acker: Solid State Commun. *35*, 937 (1980)
4.135 H.C. Ku, H. Barz: In *Ternary Superconductors*, ed. by G.K. Shenoy, B.D. Dunlap, F.Y. Fradin (North-Holland, Amsterdam 1981) pp.209-212
4.136 K.N. Yang, S.E. Lambert, H.C. Hamaker, M.B. Maple, H.A. Mook, H.C. Ku: To appear in Proc. IV Conf. Superconductivity in d- and f-Band Metals, Karlsruhe, FRG, June 28-30, 1982
4.137 L.D. Woolf, S.E. Lambert, M.B. Maple, H.C. Ku, W. Odoni, H.R. Ott: Physica *108*B, 761 (1981)
4.138 L.D. Woolf, S.E. Lambert, M.B. Maple, F. Acker, H.C. Ku, W. Odoni, H.R. Ott: To be published
4.139 H.C. Hamaker, H.C. Ku, M.B. Maple, H.A. Mook: To appear in Solid State Commun.

4.140 D.C. Johnston: Solid State Commun. *24*, 699 (1977)
4.141 R.N. Shelton, C.U. Segre, D.C. Johnston: In *Ternary Superconductors*, ed. by G.K. Shenoy, B.D. Dunlap, F.Y. Fradin (North Holland, Amsterdam 1981) pp.205-208
4.142 H.E. Horng, R.N. Shelton: In *Ternary Superconductors*, ed. by G.K. Shenoy, B.D. Dunlap, F.Y. Fradin (North Holland, Amsterdam 1981) pp.213-216
4.143 H.C. Hamaker, M.B. Maple: In *Ternary Superconductors*, ed. by G.K. Shenoy, B.D. Dunlap, F.Y. Fradin (North Holland, Amsterdam 1981) pp.201-204
4.144 H.C. Hamaker, M.B. Maple: Physica *108*B, 757 (1981)
4.145 D.C. Johnston: Solid State Commun. *42*, 453 (1982)
4.146 Y. Muto, H. Iwasaki, T. Sasaki, N. Kobayashi, M. Ikebe, M. Isino: In *Ternary Superconductors*, ed. by G.K. Shenoy, B.D. Dunlap, F.Y. Fradin (North Holland, Amsterdam 1981) pp.197-200
4.147 H. Iwasaki, M. Isino, Y. Muto: Physica *108*B, 759 (1981)
4.148 J.P. Remeika, G.P. Espinosa, A.S. Cooper, H. Barz, J.M. Rowell, D.B. McWhan, J.M. Vandenberg, D.E. Moncton, Z. Fisk, L.D. Woolf, H.C. Hamaker, M.B. Maple, G. Shirane, W. Thomlinson: Solid State Commun. *34*, 923 (1980)
4.149 K. Andres, J.P. Remeika, G.P. Espinosa, A.S. Cooper: Phys. Rev. B*23*, 1179 (1981)
4.150 S.E. Lambert, Z. Fisk, H.C. Hamaker, M.B. Maple, L.D. Woolf, J.P. Remeika, G.P. Espinosa: In *Ternary Superconductors*, ed. by G.K. Shenoy, B.D. Dunlap, F.Y. Fradin (North Holland, Amsterdam 1981) pp.247-250
4.151 L.D. Woolf, M. Tovar, H.C. Hamaker, M.B. Maple: Phys. Lett. *71*A, 137 (1979)
4.152 J.M. Vandenberg: Mat. Res. Bull. *15*, 835 (1980)
4.153 H.R. Ott, W. Odoni, Z. Fisk, J.P. Remeika: In *Ternary Superconductors*, ed. by G.K. Shenoy, B.D. Dunlap, F.Y. Fradin (North Holland, Amsterdam 1981) pp.251-254
4.154 J.M. Dundon: Private communication
4.155 Z. Fisk, S.E. Lambert, M.B. Maple, J.P. Remeika, G.P. Espinosa, A.S. Cooper, H. Barz, S. Oseroff: Solid State Commun. *41*, 63 (1982)
4.156 M. Redi, P.W. Anderson: Proc. Natl. Acad. Sci. USA *78*, 27 (1981)

5. Superconductivity and Magnetism in $(RE)Mo_6X_8$ Type Compounds

M. Ishikawa, Ø. Fischer, and J. Muller

With 17 Figures

The problem of the coexistence between superconductivity and magnetism has been revived following the discovery of superconductivity in the ternary rare-earth (RE) compounds of the $(RE)Mo_6X_8$-type (X = S, Se) [5.1,2] and $(RE)Rh_4B_4$-type [5.3]. The renewed interest in the subject has indeed increased with the discovery of reentrant superconductivity at the ferromagnetic transition in two of these compounds, $HoMo_6S_8$ [5.4] and $ErRh_4B_4$ [5.5]. This was immediately followed by the discovery of the coexistence of antiferromagnetism and superconductivity in several other compounds [5.6,7]. After these novel experimental findings [5.8,9], a large amount of both experimental and theoretical work has been carried out [5.10] and we have now gained considerable insight into the problem of the interplay between superconductivity and magnetism.

First, we would like to point out that the difference between these new ternary compounds and the pseudobinary compounds investigated earlier, resides in two factors.

(i) The ternary compounds have a *regular* lattice of magnetic ions. They are, therefore, much less sensitive to inhomogeneities than the pseudobinaries as far as magnetic ordering is concerned.

(ii) The exchange interaction between the rare-earth ions and the conduction electrons in ternary rare-earth compounds is rather weak owing to the particular crystal structure of these materials. Thus, in spite of the relatively high concentration of magnetic ions, most of these ternary rare-earth compounds are superconducting with only a small influence by the magnetic ions on the paramagnetic superconducting state.

In this chapter we shall restrict our discussion to the ternary rare-earth molybdenum chalcogenides: $(RE)Mo_6X_8$ (X = S, Se). As described in [5.11], these compounds may be characterized as Mo_6-cluster compounds where the conduction electrons have mainly d-character; also the large Mo-RE distance of about 6.5 Å is responsible for the low value of the exchange interaction. Most of the experimental results described here have been obtained from samples prepared by powder-metallurgical methods. Very little work has been done on melted samples and so far it has not been possible to grow single crystals of any reasonable size for this type of ternary compound (see also [Ref.5.11, Chap.4]). Unfortunately, as it turned out, the superconductivity is rather sensitive to small changes of composition and/or lattice parameters and thus to homogeneity. For example, the T_c of $PbMo_6S_8$ has been reported to vary between 11 K and 15 K, depending on the preparation technique used. For the same reasons the width of the superconducting transition is usually found to be rather large. The observed correlation between T_c and lattice parameters suggests that these variations are associated with changes in composition, for example, by the exact amount of Pb present in the compound. For these reasons and also because the relevant portions of the ternary phase diagrams are not yet known, detailed analyses and comparisons of results from different research groups on certain properties, especially the critical temperature, should be made with care.

This chapter is organized as follows: in Sect.5.1.1 the results for the re-entrant superconductor $HoMo_6S_8$ are summarized and the relationship between ferromagnetism and superconductivity is discussed. Several of the characteristic features of antiferromagnetic superconductors are described in Sect.5.1.2. The last part of Sect.5.1 is devoted to the "anomalous" $(RE)Mo_6X_8$ (RE = Ce, Eu, Tm, and Yb) compounds. Several pseudoternary systems have been investigated and some have revealed unexpected behavior. Section 5.2 outlines many of their most striking features. Much valuable information about the coexistence problem can be extracted from H_{c2} versus T curves. Analyses of such curves are presented in Sect.5.3 with particular emphasis on antiferromagnetic superconductors. The estimate of the magnitude of the exchange interaction and the mechanisms for the magnetic ordering are reviewed in Sect.5.4. Finally, Sect.5.5 summarizes the essential results obtained so far for this class of ternary compounds. At the end of the chapter, an extensive list of references to experimental work is given. References to theoretical work, however, are limited to representative ones only.

5.1 Ternaries

5.1.1 Re-Entrant Superconductors

Among the Chevrel-type superconductors, only $HoMo_6S_8$ and its pseudoternaries [5.8] belong to this class. They become superconducting at T_{c1} but at a lower temperature T_{c2}, they re-enter the normal state at the onset of long-range ferromagnetic order. In Fig.5.1, the DC resistance of $HoMo_6S_8$ is shown as a function of temperature for several applied DC magnetic fields. The particular sample shown in this figure becomes superconducting at about 1.3 K and returns to the normal state at 0.6 K in zero field. At this point it is worthwhile to remark on the sensitivity of T_{c1} to heat treatments, etc., for these Chevrel phases including the nonmagnetic ones; T_{c1} often varies by several tenths of a degree from one sample

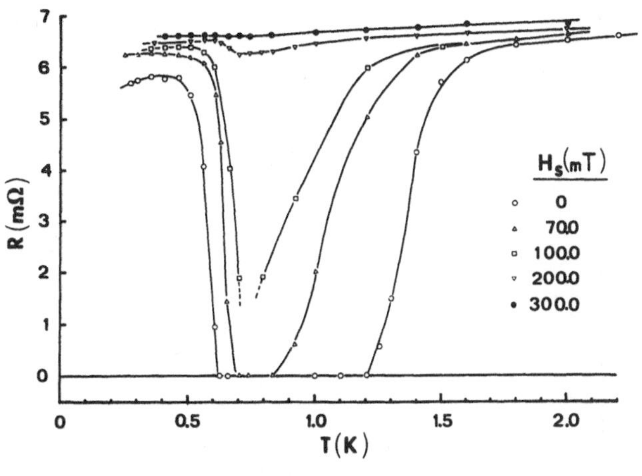

Fig.5.1. DC resistance vs temperature of $HoMo_6S_8$ in DC magnetic fields, 0,70,100,200 and 300 mTesla [5.4]

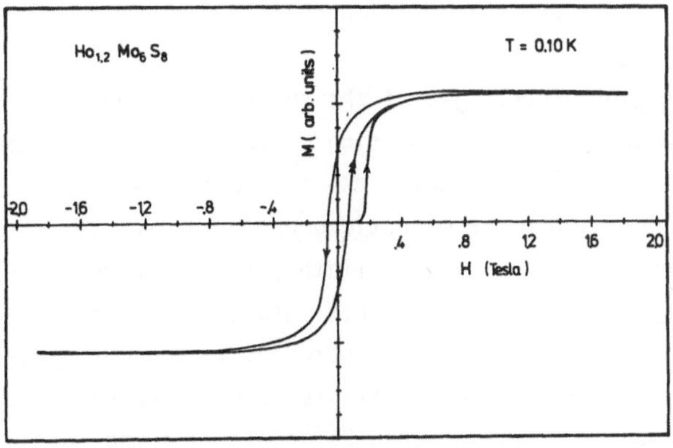

Fig.5.2. Magnetization vs applied magnetic field at T=0.1 K for $HoMo_6S_8$ [5.12]

to another, depending on the temperature and the duration of the heat treatments. In fact, a careful study of T_{c1} [5.12] showed that the onset of T_{c1} for $HoMo_6S_8$ may be as high as 2.15 K. The width of the transition at T_{c1} is also relatively large, typically a few tenths of a degree wide. Contrary to this, the transition at T_{c2} is usually very sharp and rather insensitive to details in the preparation. However, the latter transition is extremely sluggish just above T_{c2} on both heating and cooling. The time constant was found to be even longer in the applied magnetic fields [5.13].

A magnetic hysteresis curve for $HoMo_6S_8$ below T_{c2} is shown in Fig.5.2, which indicates that the samples becomes ferromagnetic below T_{c2}. The ferromagnetism has been confirmed by neutron diffraction [5.14], as discussed in Chap.8. A specific heat experiment on $HoMo_6S_8$ down to 0.5 K by WOOLF et al. [5.15] revealed a spike superimposed on a sawtooth-shaped feature at T_{c2}, which closely resembles the one found for another re-entrant superconductor, $ErRh_4B_4$ (Chap.4).

A particularly important question about this re-entrant behavior is how the uniform ferromagnetic phase grows in the superconducting phase, and whether or

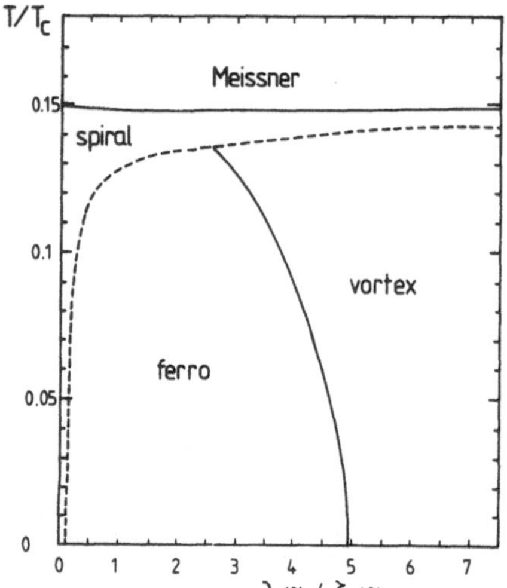

Fig.5.3. Phase diagram of a re-entrant superconductor as a function of $\lambda_L(0)/\xi(0)$. The dashed line denotes the first-order transition [5.18]

not there is a narrow coexistence region around T_{c2}. If the superconducting phase is not destroyed precisely at the magnetic phase transition T_m but remains down to T_{c2} ($< T_m$), then, in principle, the competition between the two collective phenomena leaves us with two possibilities: either the magnetic order is uniformly ferromagnetic and the superconducting order parameter is oscillatory (self-induced vortex state), or the superconductivity forces the ferromagnet into an oscillatory state (spiral or linearly polarized state) leaving the superconducting order parameter uniform. The competition between these two distinct states has been analyzed theoretically by several groups [5.16,17] (see also Chap.9) and it is found that the oscillatory magnetic state is favored for large exchange interaction and small values of $\kappa = \lambda_L(0)/\xi(0)$, where $\lambda_L(0)$ and $\xi(0)$ are the penetration and coherence length, respectively.

In order to illustrate how these proposed states appear one after another as a function of temperature, with other parameters fixed, a phase diagram derived by SAKAI et al. [5.18] is reproduced in Fig.5.3. However, the temperature range where these states are stable for the re-entrant superconductors discovered so far, seems to be so limited that it would be very difficult to experimentally confirm such a phase diagram at present. Recent neutron diffraction experiments on a powder sample of $HoMo_6S_8$ by LYNN et al. [5.19] show that a sinusoidally modulated magnetization with a period of about 250 Å appears about 0.1 K above T_{c2} upon cooling. This seems to confirm that the two phenomena really coexist in just such a narrow temperature interval. However, a more precise determination of the nature of this magnetic order necessitates single crystals. The field dependence of the observed neutron peak and the temperature dependence of the period appear to rule out the possibility of a vortex state. A similar situation appears to be true for $ErRh_4B_4$ [5.20]. The first order nature of the phase transition in the vicinity of T_{c2} has been confirmed by measuring the latent heat with good accuracy for $(Er_{1-x}Ho_x)Rh_4B_4$ [5.21]. However, it was not possible to discern a predicted second-order transition either above or below the first-order transition.

Although the experiments on $HoMo_6S_8$ support an oscillatory magnetic state, we believe that the relevant parameters of this system are not very far from those suitable for a self-induced vortex state with a larger temperature interval of coexistence. For instance, an analysis of existing data on Chevrel compounds shows that they have a relatively low exchange constant and a value of about 5 for κ. Therefore, we believe that it would be interesting to look for such a state in systems having a lower exchange constant and a larger value of κ.

5.1.2 Antiferromagnetic Superconductors

Most of the remaining Chevrel-type ternary RE-compounds belong to this class in which coexistence of long-range antiferromagnetic (AF) order and superconductivity has been found. In other words, the RE magnetic moments order antiferromagnetically in the superconducting state without completely destroying superconductivity. The possibility of this type of coexistence was demonstrated theoretically many years ago by BALTENSPERGER and STRÄSSLER [5.22]. Due to the absence of an average polarization one could crudely state that the coexistence for the antiferromagnetic case is more easily attained than for the ferromagnetic case. However, when the magnetic subsystem orders antiferromagnetically, the superconducting state is nevertheless considerably modified and this is readily reflected as strong anomalies in the superconducting properties around T_m. These anomalies in several systems are illustrated below as well as in Sect.5.3.

The coexistence of antiferromagnetism and superconductivity was first discovered in $TbMo_6S_8$, $DyMo_6S_8$, $ErMo_6S_8$ [5.6] and $ErMo_6Se_8$ [5.7], and subsequently in many other ternary compounds, such as $REMo_6X_8$ [5.8] and $RERh_4B_4$ [5.9]. Hence, this type of coexistence is now believed to be a rather common phenomenon in magnetic superconductors, confirming the theoretical prediction [5.22]. In this section, characteristic features of this type of coexistence are illustrated in detail by the data of different kinds of experiments for some typical compounds. In the following, we shall generally refer to the influence of magnetism on superconductivity as "pairbreaking effects", although this is not always the appropriate term. For a more detailed discussion, Sect.5.3 should be consulted.

The pairbreaking effect due to the ordering of magnetic moments manifests itself, for example, as an anomaly around the magnetic ordering temperature T_m in the electrical resistance R, or AC susceptibility χ_{AC} in low applied DC magnetic fields. In Figs.5.4,5, R and χ_{AC} of $GdMo_6S_8$ are shown as a function of temperature for several applied magnetic fields. The data were taken on a sintered cylindrical sample of 4mm in diameter and about 12 mm long and the resistance ratio $R_{300}/R_{4.2}$ was typically 10 to 20 for such a sintered sample. Without a magnetic field its

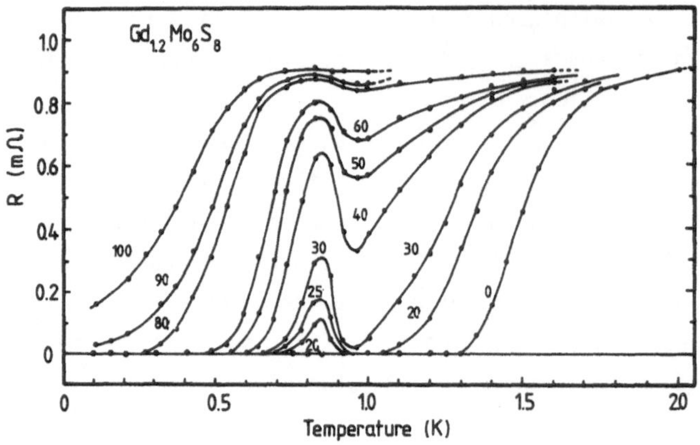

Fig.5.4. DC resistance vs temperature of $GdMo_6S_8$ in DC magnetic fields (indicated in mTesla)

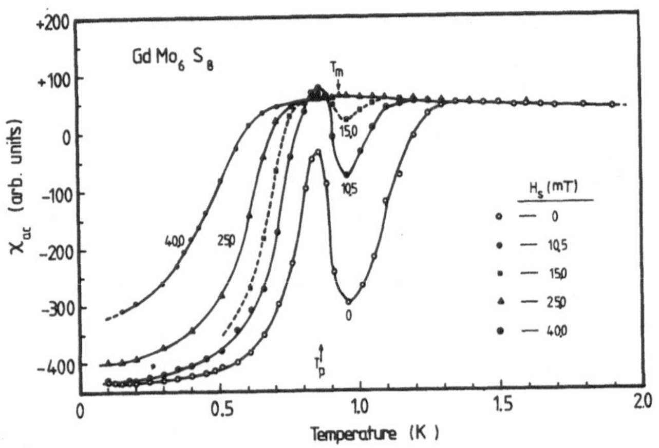

Fig.5.5. AC susceptibility vs temperature of $GdMo_6S_8$ in DC magnetic fields (indicated in mTesla)

resistance remains zero below 1.3 K, but in low fields a finite resistance appears around its T_m of 0.85 K and drops again to zero at lower temperatures. The corresponding variation of χ_{AC} on the same sample is shown in Fig.5.5. The superconducting critical temperature T_c determined by χ_{AC} is considerably lower than that determined by R, with the onset temperature indicated by χ_{AC} roughly corresponding to the temperature where R becomes zero. Such a large difference in inductively and resistively determined T_c's is commonly found in these ternary superconductors which indicates a considerable sensitivity of these materials to inhomogeneities. As the figure shows, χ_{AC} in zero field reveals a big peak at T_m for $GdMo_6S_8$ and this peak increases with fields up to about 25 mT. Among this class of superconductors such a peak in χ_{AC} for a zero field has to date only been found for $GdMo_6S_8$. This fact indicates that the pairbreaking effect is most important for $GdMo_6S_8$ which has the largest spin moment, i.e., the largest deGennes factor among the RE ions. Hence, it might be possible to find re-entrant behavior even in an antiferromagnetic superconductor with a still larger exchange constant. The second example is given in

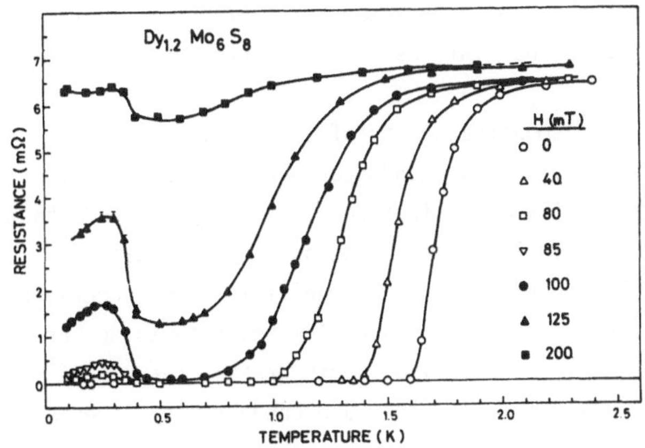

Fig.5.6. DC resistance vs temperature of $DyMo_6S_8$ in DC magnetic fields (indicated in mTesla) [5.6]

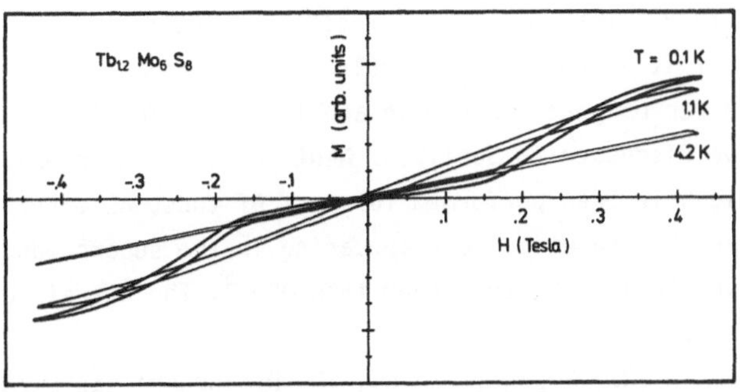

Fig.5.7. Magnetization vs applied magnetic field at T=0.1, 1.1 and 4.2 K for $TbMo_6S_8$ [5.12]

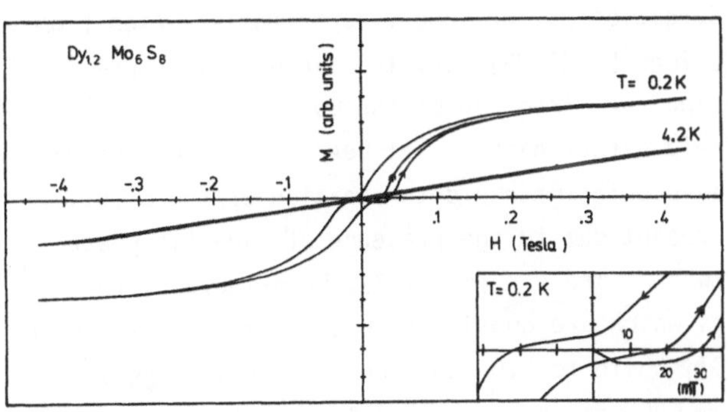

Fig.5.8. Magnetization vs applied magnetic field at T = 0.2 and 4.2 K for $DyMo_6S_8$ [5.12]

Fig.5.6 for $DyMo_6S_8$. The rise of R at T_m is more abrupt here than for the previous case in $GdMo_6S_8$. The physical meaning of this difference in the R-T characteristics will become more evident when their upper critical field (H_{c2}-T) curves are deduced from the R-T curves shown later in Sect.5.4. In Fig.5.7, the magnetic hysteresis (M-H) curves taken by a conventional integration technique at three different temperatures are shown for another AF superconductor, $TbMo_6S_8$. Above its T_m (~ 0.9 K), it shows paramagnetic behavior but below T_m, an antiferromagnetic curve with a spin flipping field of about 0.15 Tesla is found. Below T_m and in low fields

(< 5 mT), a diamagnetic contribution has been detected but is not shown in the figure [5.12]. A slightly more complicated but interesting M-H curve was found for $DyMo_6S_8$ which is shown in Fig.5.8.

The curve at 0.2 K ($< T_m$ = 0.45 K) demonstrates one of the peculiar magnetic properties of $DyMo_6S_8$. Below its T_m, the AF superconductor has an extremely low spin flipping field of about 20 mT and the magnetization has already reached about 80% of its saturation value at H = 0.1 Tesla, which is below its upper critical field value [5.12]. This has also been confirmed by neutron diffraction experiments in magnetic fields [5.23]. The neutron experiments have also shown that the ferromagnetic order at a field of 60 mT has a correlation length greater than 300 Å. Therefore, below 0.45 K and around H = 0.1 Tesla, the superconductivity "coexists" with ferromagnetism in $DyMo_6S_8$ although we do not know how the superconducting and ferromagnetic components are distributed in the sample; the ferromagnetic order is possibly confined along flux lines in the vortex state, as studied in detail by KRZYSZTÓN [5.24]. Even if it is so, $DyMo_6S_8$ would be the first superconductor to sustain a long-range ferromagnetic order in applied magnetic fields. More concerning this point will be discussed in Sect.5.3. Neutron diffraction experiments in a zero magnetic field have been performed for most of these AF superconductors and they confirmed that the magnetic order appearing in the superconducting state is long range with a correlation length longer than 300 Å. The details of these experiments are given in Chap.8.

The antiferromagnetic structure in (RE) Mo_6S_8 (RE = Gd, Tb, Dy and Er) has been determined as A-1 type, i.e., AF consisting of {100} planes with the moments alternating between planes, while the direction of the moments apparently varies from [111] to [100] depending on the RE ions [5.23,25]. Crystal field effects presumably play an important role in determining the direction of the moments in each of these compounds. However, the crystal field effect has not yet been extensively studied for Chevrel compounds, partly because calorimetric determination of the crystal field splittings is precluded at present due to the presence of magnetic phase transitions of impurity phases between 3 and 5 K [5.26,27]. Nevertheless, for selenides, crystal field splittings which are consistent with cubic symmetry were found in neutron experiments [5.28], while for the sulfides a fairly large axial crystalline field was inferred from EPR measurements [5.29]. From the neutron diffraction experiments on the sulfides (RE) Mo_6S_8, the saturation moments are found to be very close to their free ion values except for $ErMo_6S_8$ [5.23]. Hence, the crystal field effect seems to be particularly strong in $ErMo_6S_8$. The magnetic structure of the selenides is still not well known because of complex spectra arising from the presence of unidentified impuritiy phases [5.30,31].

Low temperature specific heat (C_p) measurements have been performed [5.7,15 31-33] on some of these AF superconductors and revealed a pronounced lambda type anomaly at the magnetic phase transition, as shown for $GdMo_6S_8$ in Fig.5.9. Accord-

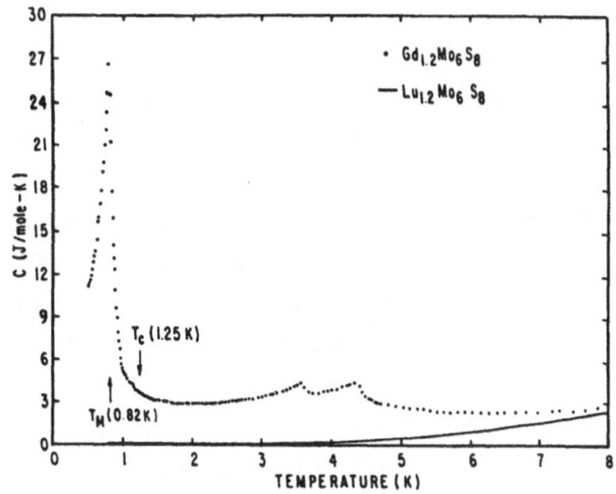

Fig.5.9. Specific heat vs temperature for $GdMo_6S_8$ and nonmagnetic $LuMo_6S_8$ [5.26]

ing to the analysis of these data, almost the full magnetic entropy of Gd^{3+} was found for $GdMo_6S_8$ [5.26] and $GdMo_6Se_8$ [5.32], while for the other RE ions, the magnetic entropy is much less than the Hund's rule value of R ln(2J + 1). This again confirms the importance of the crystal field effect in these compounds. The electronic density of states at the Fermi level N(0) and the Debye temperature θ_D are important parameters which are usually obtained from C_p measurements. However, for these AF superconductors a precise determination of these parameters is difficult because of the intrinsic AF phase transition at T_m and additional magnetic transitions of impurity phases at higher temperatures, as can be seen for $GdMo_6S_8$ in Fig.5.9. For our subsequent analyses, we may take N(0) = 0.2 and 0.43 states/eV-atom-spin and θ_D = 220 K and 200 K for all the heavy rare-earth (except Yb) sulfides [5.27,33] and selenides [5.33], respectively.

The susceptibility χ of these AF superconductors has been measured [5.33-35] with a Faraday magnetometer between 1.5 and 300 K. These measurements show that χ follows a Curie-Weiss law above about 50 K with an effective moment for free trivalent ions and a negative paramagnetic Curie temperature. At lower temperatures, an appreciable amount of deviation from the Curie-Weiss law has been observed which is due to crystal field and magnetic correlation effects.

5.1.3 Other Ternaries with Ce, Eu, Tm and Yb

Among the $REMo_6X_8$-type compounds there are several particular compounds which do not belong to either of the two classes discussed in the previous sections, but do deserve some comments because of their peculiar physical properties. These are the compounds with an ambivalent RE ion such as Ce, Eu, Tm and Yb. This is the subject of the present section.

Neither sulfides nor selenides of Ce and Eu are superconducting. It is, however, not very surprising that $CeMo_6X_8$ is nonsuperconducting since superconducting compounds with trivalent Ce ions are believed to be the exception. On the other hand, it was really surprising that $EuMo_6X_8$ with divalent Eu ions is not superconducting under normal conditions because its supposedly isoelectronic and isostructural counterparts $PbMo_6X_8$ and $SnMo_6X_8$ are good superconductors. MAPLE et al. reported an anomalous increase of resistance at low temperatures of $EuMo_6X_8$ and characterized the behavior as valence fluctuations [5.36]. This explanation has, however, been questioned by the recent results of Mössbauer experiments [5.37]. More recently, BAILLIF et al., using high quality bulk samples of $EuMo_6S_8$, discovered a structural transformation around 110 K from the room-temperature rhombohedral to a low temperature triclinic structure [5.38]. The phase transition, confirmed by X-ray diffraction, is accompanied by a sharp heat capacity peak yielding a latent heat of 46 J/gm at. A small hysteresis of the resistance anomaly further ascertains the first-order nature of the transition. Since the electronic specific heat coefficient for the low temperature phase was found to be vanishingly small (about a tenth of that of copper) and the substantial resistance increase sets in precisely at the transformation temperature, it was concluded that the low temperature phase of $EuMo_6S_8$ is a small gap semiconductor. Even if the causal relationship between the unusual transport behavior and the structural instability remains to be elucidated, one has to admit that the rare-earth nature of the divalent cation is not essential for the 110 K phase transition. Indeed, the isoelectronic compound $BaMo_6S_8$ exhibits an entirely analogous instability [5.39].

High pressure experiments on $EuMo_6S_8$ revealed superconductivity with T_c up to 11 K above 7 Kbar [5.40,41]. These investigations also showed that the pressure removes the anomalous resistance and restores normal metallic behavior. Although the observed value of T_c corresponds well to what one would expect for $EuMo_6S_8$ from the behavior of other Chevrel compounds, doubt has been cast concerning the origin of the observed superconductivity. MCCALLUM et al. failed to observe bulk superconductivity in their samples under pressure and suggest that the superconductivity observed by other groups is due to an impurity phase [5.42]. On the other hand, $EuMo_6Se_8$ has not yet been forced into the superconducting state under pressure, although the temperature dependence of its resistance at normal pressure is similar to that of $EuMo_6S_8$ under pressure [5.43]. These interesting questions concerning both $EuMo_6S_8$ and $EuMo_6Se_8$ remain to be clarified. Since the valence states of the RE ions in $CeMo_6X_8$ and $EuMo_6X_8$ are trivalent and divalent [5.37], respectively, the RE ions in these compounds should order magnetically at lower temperatures. In fact, their magnetic transitions have been confirmed; for example, $EuMo_6S_8$ orders antiferromagnetically below 0.25 K [5.44] and $CeMo_6S_8$ orders at about 2 K, but its magnetic state is a very complex one, being neither AF nor F [5.34,35,45]. However, it is noted here that in view of the picture given above for $EuMo_6S_8$, the magnetic ordering in this compound cannot be analyzed in the same way as for the other compounds in Sect.5.4, i.e., in the framework of the RKKY interaction.

For $YbMo_6S_8$, magnetic ordering has been reported at about 2.7 K ($< T_c \sim 9$ K) which is considerably higher than T_m of the other $REMo_6S_8$ compounds discussed in the previous sections. From calorimetric and Mössbauer experiments [5.46,47], it has been claimed that this magnetic order, due to the trivalent Yb ions, really occurs in the Chevrel phase and not in any spurious impurity-phases. Accordingly, a coexistence of the magnetic order with superconductivity was also proposed in this compound. However, its high T_c of about 9 K with respect to other $REMo_6S_8$ compounds should clearly be attributed to the divalent Yb ions or at least to a mixed configuration (a static mixing of Yb^{2+} and Yb^{3+}). Hence, more experiments on this interesting compound will be required to study whether the magnetic order is of long range and whether the magnetic order and the superconductivity truly take place in the same phase.

$TmMo_6S_8$ is superconducting below about 2 K like other $REMo_6S_8$ compounds containing trivalent RE ions, but magnetic ordering has not been detected above 70 mK [5.13]. Thus, the trivalent Tm ions in this compound are probably characterized by a non-Kramers ion with a nonmagnetic ground state.

5.2 Pseudoternaries

Pseudoternary systems have turned out to be very interesting in the study of the interplay of magnetism and superconductivity. Indeed, new phenomena, often unexpected from the behavior of the ternaries, have been found both in the Chevrel phases and in the rhodium borides. However, pseudoternaries are, in general, even more susceptible to the actual metallurgical state of the samples than the ternaries are. Thus, the results on pseudoternaries should be examined with even more care [5.48]. Mainly due to the lack of detailed knowledge of the phase diagrams do we feel strongly that the techniques of sample preparation and analysis must be improved further in the future in order to understand the details of the experimental results correctly. With these facts in mind, we shall discuss in this section some typical examples of pseudoternary molybdenum chalcogenides.

One of the most interesting pseudo-ternary systems is $(Ho_{1-x}Eu_x)Mo_6S_8$, which is the first complete system that has been investigated with respect to the coexistence problem [5.44]. Both $HoMo_6S_8$ and $EuMo_6S_8$ have unique low temperature properties within the series of $(RE)Mo_6S_8$. $HoMo_6S_8$ is the only re-entrant superconductor and $EuMo_6S_8$ is not superconducting, although it should be so from simple considerations of the other ternaries, as mentioned in the preceding sections. The results are summarized by a plot of T_c and T_m as a function of the Eu concentration x in Fig. 5.10. In the investigated temperature range (above 70 mK), three distinct regions of concentration regarding superconductivity and magnetic transitions exist: on the Ho rich side, a re-entrant superconducting region, and in the middle, a supercon-

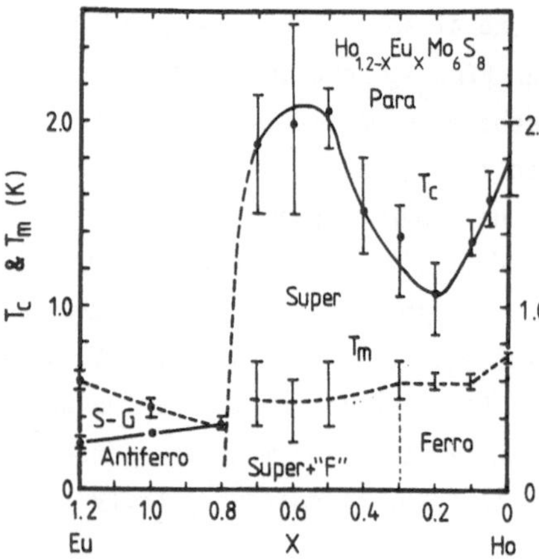

Fig.5.10. Low temperature phase diagram of $(Ho_{1-x}Eu_x)Mo_6S_8$ [5.44]

ducting region coexisting with some sort of ferromagnetic order whose precise nature is not yet understood. However, from magnetization measurements a finite remanent moment was confirmed in the magnetic state. Therefore, this would be the first case where superconductivity and net zero-field magnetization coexist in a wide temperature range and is hence more suitable for testing proposed phase diagrams such as the one shown in Fig.5.3. More detailed experiments, in particular neutron diffraction similar to those done for $HoMo_6S_8$ [5.19], should be performed in order to see, for example, in what manner the two cooperative phenomena coexist. Of particular importance would be to check if in the region of $x = 0.3 \sim 0.8$ an oscillatory magnetic state (or a vortex state) occurs in a temperature interval wider than in $HoMo_6S_8$. On the Eu rich side there is a nonsuperconducting region of the spin glass-type at intermediate temperatures and an antiferromagnetic one at lower temperatures. The strong increase in T_c in the middle of the present phase diagram is consistent with the expectation that $EuMo_6S_8$ should have been a good high T_c superconductor if it were not for the anomalous decrease of T_c around $x = 0.8$. As to the origin of the observed behavior, an anomaly in the lattice parameters [5.44] suggests that a high temperature phase transition may take place, similar to the one in $EuMo_6S_8$ which was discussed in Sect.5.1.3. It should also be noted that there is no significant correlation between T_c and T_m in this system, i.e., T_c varies considerably with x, while T_m remains relatively constant.

Another pseudoternary system $(Ho_{1-x}Lu_x)Mo_6S_8$ was investigated in the dilute region ($0 \leq x \leq 0.2$) and rather peculiar results were found [5.8]. Resistance measurements suggest that this system may indeed undergo a third transition from the ferromagnetic state back into another superconducting state at the lowest temperatures. This point certainly deserves more detailed study.

Another group of pseudoternary systems which have extensively been investigated so far are $(Pb_{1-x}Eu_x)Mo_6S_8$ and $(Sn_{1-x}Eu_x)Mo_6S_8$. Soon after the publication of the very peculiar but nevertheless interesting concentration dependence of T_c [5.1] and the H_{c2} versus T curves [5.49] (see also Chapt.3), a number of investigations with various experimental methods were undertaken [5.45,50-54]. At low x-values, an extremely small depression of T_c for Eu^{2+} ions was found [5.49]. However, at higher x-values, T_c abruptly drops and superconductivity disappears above $x \simeq 0.9$, as we have just seen above for $(Ho_{1-x}Eu_x)Mo_6S_8$. This anomalous behavior must be correlated with the unusual properties of $EuMo_6S_8$ itself, as discussed in Sect. 5.1.3. From Mössbauer experiments on the system $(Sn_{1-x}Eu_x)Mo_6S_8$, a weak Korringa relaxation was reported from which a very weak exchange constant was estimated [5.51]. However, for $(Pb_{1-x}Eu_x)Mo_6S_8$ [5.55] and $(Yb_{1-x}Eu_x)Mo_6S_8$ [5.37], the Korringa relaxation was not confirmed and more complicated relaxation behavior was suggested instead. EPR experiments on dilute $[Pb_{1-x}(RE)_x]Mo_6S_8$ and $[Sn_{1-x}(RE)_x]Mo_6S_8$ systems (RE: Gd and Eu) have confirmed the low values for the exchange constant [5.29]. Also it should be noted here that a more pronounced depression of T_c by Eu in the range of $0 < x < 0.5$ was reported for $(Sn_{1-x}Eu_x)Mo_6S_8$ and $(Yb_{1-x}Eu_x)Mo_6S_8$ by MCCALLUM et al. [5.37]. Such diverse results on these pseudoternaries of $EuMo_6S_8$ could be a consequence of an appreciable amount of inhomogeneity if the samples were single-phased. A particular difficulty with these pseudoternaries is that the lattice parameters of $PbMo_6S_8$ and $SnMo_6S_8$ are very close to the ones for $EuMo_6S_8$ so that if a two-phase region exists in the pseudoternary phase diagram, it may not be easily determined with X-ray diffraction.

Let us make another remark on the $[Pb_{1-x}(RE)_x]Mo_6S_8$ and $[Sn_{1-x}(RE)_x]Mo_6S_8$ systems. For these systems, an anomalous variation of the lattice parameters and T_c has been found at low x-values when RE = trivalent rare earth [5.56]. These anomalies are probably due to a possible nonstoichiometry on either the RE or Pb(Sn) sites in these materials.

5.3 Analysis of Upper Critical Field Curves

For the purpose of numerical analyses of the effects of the long-range magnetic order on superconductivity, upper critical field curves (H_{c2} vs T) can be conveniently extracted from the results of resistance measurements in applied magnetic fields such as those shown in Figs.5.1,4,6. It is, however, important to assure oneself that the H_{c2} - T curve does not depend much on the definition of H_{c2} because the resistive transitions are especially wide in magnetic fields for the ternary compounds under consideration, as mentioned in the previous sections. H_{c2} is usually determined from a midpoint of the resistive transition and thus, defined H_{c2} - T curves seem to represent an average physical quantity, at least for our compounds presented in this section [5.4]. For certain compounds, however, the general feature of the H_{c2} - T curve depends considerably on the definition of H_{c2} [5.54]. Accordingly, more careful analyses are required in such cases. Our

H_{c2} - T curves are shown in Figs.5.11,12,13 for RE: Gd, Tb; Dy, Ho; and Er, respectively. The anomalous behavior of the H_{c2} - T curves down close to the magnetic ordering temperature T_m can be interpreted [5.4] as a result of spin polarization in the framework of the additive pairbreaking theory. Below T_m, however, a sudden decrease of H_{c2} is found which cannot be explained with the ordinary pairbreaking theory. This decrease around T_m appears to be especially abrupt in the Dy and Ho compounds, and it seems to be somehow correlated with their strongly temperature-dependent staggered magnetization deduced from neutron diffraction experiments in a zero field [5.14,23,57], as shown in Fig.5.14. Both the magnetization [5.13] and neutron diffraction experiments in magnetic fields [5.23] on $DyMo_6S_8$, have revealed that the magnetic moments are ferromagnetically aligned in fields higher than about 0.1 Tesla which is below its H_{c2}, as mentioned in Sect.5.1.2. Hence, a ferromagnetic interaction induced by the applied magnetic field may be at least in part responsible for the anomalous H_{c2} - T curve of this compound [5.12]. On the other hand, in the other AF superconductors like Tb, Gd and Er compounds, there is a broad peak around T_m (onset temperature is indicated by an arrow in Fig.5.11) where the moments are still aligned antiferromagnetically in the corresponding fields [5.12,23]. The broad peak is then followed by a sharp rise at lower temperatures. This apparent difference between the two groups of AF superconductors is probably due to the difference in the value of T_m/T_c. For example, in the case of $DyMo_6S_8$, the magnetic transition happens to be in a region where H_{c2} in the paramagnetic state is nearly temperature independent. According to our analysis in terms of the multiple pairbreaking theory, $TbMo_6S_8$ and $DyMo_6S_8$ appear to be very similar, whereas $GdMo_6S_8$ behaves rather differently, as we will see below.

It should be noted here that the anomalous behavior of H_{c2} at the magnetic transition is not a particular one to these compounds but is also found in $(RE)Rh_4B_4$ compounds, as discussed in the previous chapter. It seems to be a general feature at the AF transition and therefore one would expect a rather universal explanation for it.

The H_{c2} - T curves of $(RE)Mo_6S_8$ were first analyzed by means of the additive pairbreaking theory and it was found that some additional temperature-dependent pairbreaking parameters are necessary to fully explain the anomalies [5.6] which we have just described above.

The procedures of the analysis are briefly described here. For more details, see [5.6,58] and Chap.3.

The upper critical field $H_{c2}(T)$ which we find experimentally, is expressed in Tesla by

$$H_{c2}(T) = H^*_{c2}(T) - M(H_{c2},T) - 3.56H^*_{c2}(0)\rho$$

$$- 0.22\frac{\alpha}{\lambda_{so}T_{co}}[H_{c2}(T) + H_J(H_{c2},T)]^2 \quad , \tag{5.1}$$

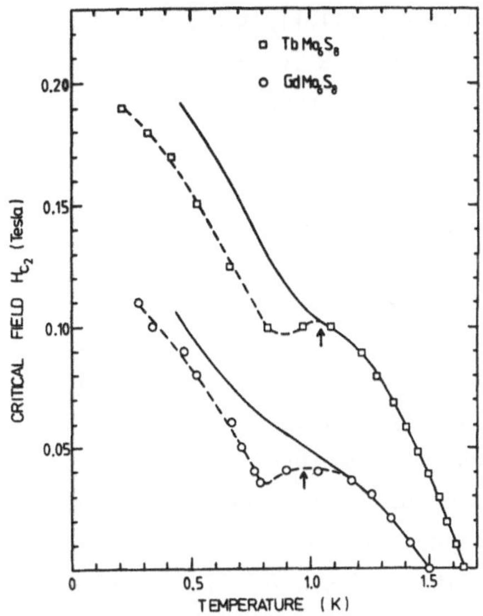

Fig.5.11

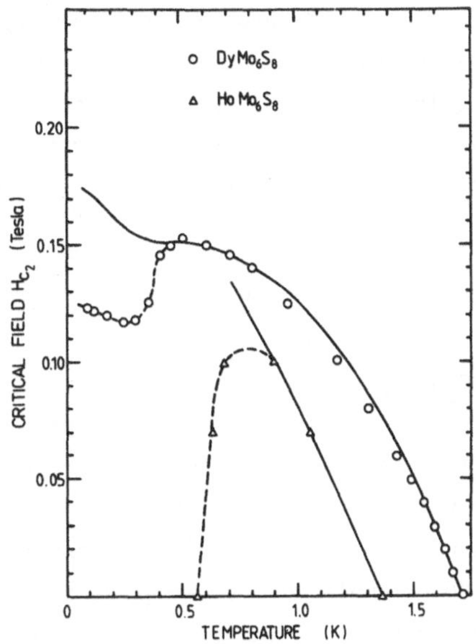

Fig.5.12

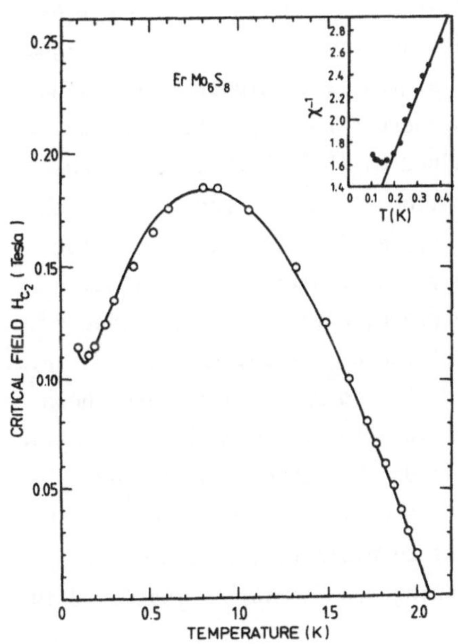

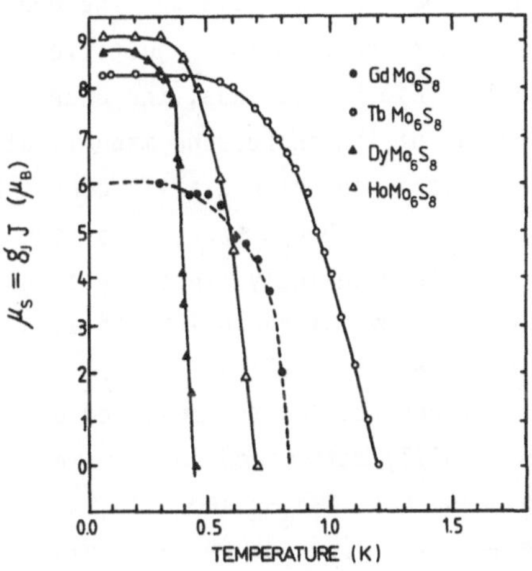

Fig.5.14 ▲

◄ Fig.5.13

Fig.5.11. Upper critical field curves of $GdMo_6S_8$ and $TbMo_6S_8$ (for details, see text)

Fig.5.12. Upper critical field curves of $DyMo_6S_8$ and $HoMo_{6\ 8}$ (for details, see text)

Fig.5.13. Upper critical field curve for $ErMo_6S_8$. The full line is the theoretical curve as explained in the text. The inset shows the inverse susceptibility determined in 80 mTesla

Fig.5.14. Temperature dependence of the magnetization for $(RE)Mo_6S_8$ (RE: Gd, Tb, Dy and Ho) as determined by neutron diffraction [5.14,23,57]

where $H^*_{c2}(T)$ is the orbital critical field which is essentially the critical field the sample would have in the absence of magnetic moments; α is the Maki parameter and λ_{so} the spin-orbit scattering parameter. The last term represents the spin-polarization effect due to an effective field H_J of the magnetic moments. H_J may be estimated from the experimental values of M(H/T) or $\chi(T)$ H using the relationship

$$H_J(H,T) = \frac{g-1}{2gN\mu_B^2}\,\Gamma M(H,T) \quad . \tag{5.2}$$

Here, g is the Landé g-factor, N the total number of atoms per unit volume and Γ $(= 2\sqrt{\langle J^2(q)\rangle})$ the exchange constant as it is usually defined [5.6]. ρ, in the third term, is a sum of the scattering pairbreaking parameters in the present scheme which may be temperature dependent. In order to see the effect of magnetic order, we first assume ρ to be temperature independent and equal to ρ_{AG}, the pair-breaking parameter due to exchange scattering in the Abrikosov-Gor'kov approximation [5.59].

The high temperature part (above T_m) of the H_{c2} - T curves was fitted with Γ^2/λ_{so} as a fitting parameter. The curves obtained in this fashion are shown by the full lines in Figs.5.11-13. The good fit obtained for $ErMo_6S_8$ in such a wide temperature range down to T_m justifies the application of (5.1) for the paramagnetic region (Fig.5.13). Thus, the decrease of H_{c2} at low temperatures is explained as a result of the increasing magnetization with decreasing temperature. However, the rapid variation of H_{c2} at T_m cannot be explained by (5.1) because this scheme would require an abrupt increase of the magnetization at the AF transition which seems extremely unlikely. In order to gain more information about this anomalous behavior, let us introduce into (5.1) an additional temperature-dependent pair-breaking parameter $\rho'(T)$ as $\rho = \rho_{AG} + \rho'(T)$. Therefore, $\rho'(T)$ can be obtained from the difference between the measured and the calculated H_{c2} values assuming $\rho = \rho_{AG}$. Curves of $\rho'(T)$ determined in this way for the Gd, Tb and Dy compounds are shown in Fig.5.15. For $TbMo_6S_8$ and $DyMo_6S_8$ the $\rho'(T)$ curves closely resemble the behavior of the magnetic-order parameter shown in Fig.5.14. On the other hand, for $GdMo_6S_8$ there appears, in addition, a prominent peak just below T_m. This may indicate the more important effect of spin fluctuations in this compound than in the others which in turn reflects that $GdMo_6S_8$ is close to being re-entrant, as mentioned in Sect.5.1.2.

This preliminary analysis with the aid of the multiple pairbreaking theory certainly demonstrates a need for a more careful theoretical treatment of magnetic ordering in the superconducting state. So far several attempts using quite different approaches have been made in order to explain the anomalous H_{c2} - T curves [5.60-63]. MACHIDA et al. [5.60] were the first to fit these curves by taking into account the effects of the energy gaps of spin density waves on the Fermi sur-

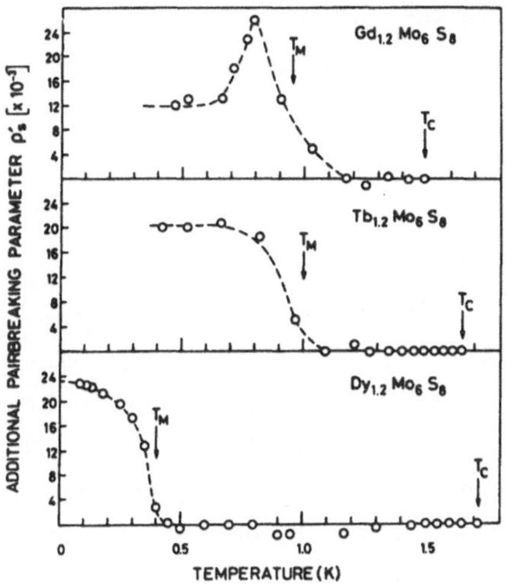

Fig.5.15. Additional pairbreaking parameter vs temperature for $GdMo_6S_8$, $TbMo_6S_8$ and $DyMo_6S_8$ (for details, see text)

face and spin fluctuations near the Néel temperature. In their model, a new order parameter $\Delta_{\pm\vec{Q}}$ was assumed to be stable, where $\vec{Q}$ is the wave vector of the antiferromagnetic order. However, this approach was subsequently criticized by others. NASS et al. [5.61] found with the use of a model calculation that the stable-order parameter should be of the usual BCS type rather than $\Delta_{\pm\vec{Q}}$. Recently, ZWICKNAGL and FULDE [5.62] presented a more detailed theoretical work where the original idea of BALTENSPERGER and STRÄSSLER [5.22] was incorporated into the strong-coupling theory of Eliashberg. According to their model, the pairing in AF supercondcutors is no longer the usual BCS time-reversed type, but is formed between the two states related by a time-reversal plus a lattice translation. The anomalous H_{c2} - T curves are interpreted as a result of the influence of the staggered magnetization on the phonon-mediated quasiparticle attraction. As expected, in their model, the reduction of H_{c2} below T_m roughly follows the sublattice magnetization found by neutron experiments [5.23]. However, a considerable discrepancy still remains between the calculated and the experimental H_{c2} - T curves near T_m. As suggested in their paper, careful treatment of magnetic interactions in an applied magnetic field would be required for an improvement. At this point, it may be worth noting that if the idea of ZWICKNAGL and FULDE turned out to be correct, these materials would be the first superconducting metals to show a non-BCS type pairing.

On the other hand, a more phenomenological viewpoint was taken by SAKAI et al. [5.63]. Considering only the first two terms of (5.1) in the framework of the boson theory [5.64], they interpreted the apparently different types of anomalies as being essentially due to the different temperature dependence of the magnetization in magnetic fields which results from the difference in the ratio T_m/T_c.

From the above discussion, it may be concluded that a new phenomenon takes place at the antiferromagnetic transition in an AF superconductor. It is probably associated with changes in pairing and/or the effective interaction between electrons. However, to fully understand this new phenomenon, more theoretical as well as experimental studies are needed.

5.4 Overview of Exchange Interactions in (RE)Mo_6X_8 Compounds

In the preceding sections, superconducting and magnetic properties of (RE)Mo_6X_8-type compounds have been reviewed by emphasizing the specifics of each compound. In this section, their general behavior across the rare-earth series in the Periodic Table is discussed with special references to the results of the sulfides with heavy rare-earth ions. In Fig.5.16 the magnetic ordering temperature T_m and the superconducting transition temperature T_c of (RE)Mo_6X_8 type compounds are summarized.

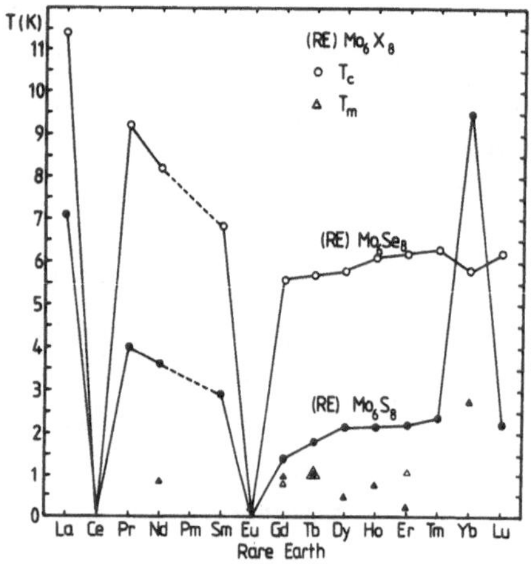

Fig.5.16. Superconducting (O,●) and magnetic (Δ,▲) transition temperatures of (RE)Mo_6X_8 for the rare-earth series

From their band structure calculations, FREEMAN and JARLBORG [5.65] (see also Chap.6) found that a large charge transfer from both the RE and Mo sites to the chalcogens explains most of their unusual superconducting and magnetic properties, i.e., the charge transfer of Mo electrons to the chalcogen sites yields the high Mo d-band density of states at the Fermi level by shifting E_F towards that of Nb, thereby favoring superconductivity. On the other hand, the charge transfer from the RE sites to the chalcogen sites yields the low conduction electron density at the RE sites which is responsible for the very weak exchange interactions between conduction electrons and the RE ions as well as between the RE ions themselves. These weak exchange interactions are evidenced by their low magnetic-ordering temperature and the low value of the exchange constant Γ as estimated in the following.

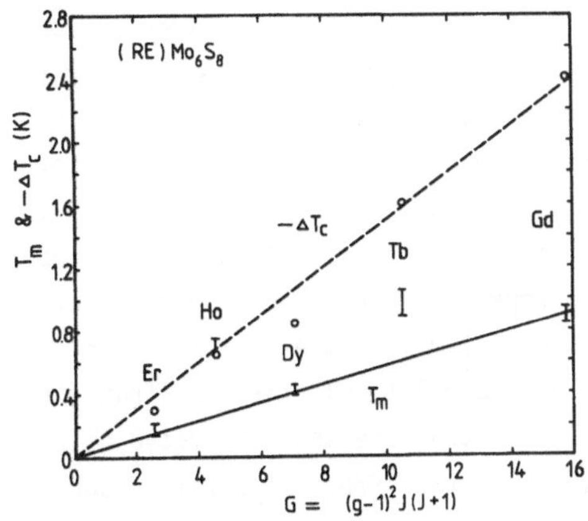

Fig.5.17. T_m and $-\Delta T_c$ vs de Gennes factor for the heavy rare-earth molybdenum sulfides [5.48]

In general, magnetic ordering in metallic substances, results from a competition and a complex combination of interactions of various origins [5.66] such as (i) indirect exchange interaction mediated by conduction electrons, (ii) a purely magnetic dipolar interaction and (iii) an anisotropic interaction caused by crystal fields to which magnetic ions are subjected. In addition to our still limited knowledge of the crystal field effects in these complex ternary compounds as mentioned in the previous section, the extremely small magnetic interaction energies involved in the magnetic transition makes a detailed analysis difficult. For the purpose of our qualitative discussion in this section, we will assume in the following that the RKKY-type indirect exchange interaction is responsible for the magnetic order and that it is not modified greatly by the presence of superconductivity in these ternary superconductors because of the strong spin-orbit interaction.

In the mean field approximation, T_m is proportional to $N(0)G\Gamma^2$ where G is the de Gennes factor, defined as $(g - 1)^2J(J + 1)$ [5.67a,b]. However, within the series of $(RE)Mo_6S_8$ compounds, T_m scales approximately with G as can be seen from Fig.5.17. The deviation from the proportionality may be attributed to the crystal field effect as the proportionality constant of the straight line in the figure yields about 55 meV for Γ [5.12].

Γ can, alternatively, be estimated from the decrease of T_c, $-\Delta T_c$ due to spin exchange scattering by using the A-G formula

$$k_B\Delta T_c = \frac{\pi^2}{8} cN(0)\Gamma^2G \quad , \tag{5.3}$$

where c is the concentration of magnetic ions.

As long as the magnetic moments are not correlated, this formula may also be used for a dense system. For this calculation, T_c's of hypothetical nonmagnetic compounds are required. However, it is rather difficult to unambiguously separate the spin exchange effect from other chemical effects, etc., for these complex ternary compounds. In fact, the estimated values of $-\Delta T_c$ for (RE)Mo_6S_8 vary within a fairly wide range, depending on the nonmagnetic compounds employed. However, we should like to make here a moderate estimate by taking $(La_{1-x}Lu_x)Mo_6S_8$ [5.33] as nonmagnetic reference compounds [5.12]. Also shown in Fig.5.17, $-\Delta T_c$ is proportional to G with a slope yielding about 25 meV for Γ. This value of Γ is close to those obtained using a similar analysis [5.36,58] and appreciably smaller than those of other RE binary compounds [5.68], as would be expected from the fact that the superconducting d-electrons on Mo sites are located relatively far away from the magnetic RE ions in our ternary compounds. It is noted here that the value of Γ estimated above from T_m is considerably higher than that from $-\Delta T_c$. This discrepancy may indicate a deficiency of the simple mean field approximation for T_m or that a multiband picture prevails so that the density of states in the expression for T_m is not necessarily the same as the one in (5.3) [5.48]. If the RKKY interaction is mediated only by a restricted number of conduction electrons such as (6sp + 5d) - electrons on RE atoms, our choice of the average total density of states estimated from the specific heat data and the use of an isotropic formula like ours would be inappropriate. The fact that no significant correlation exists between T_c and T_m when varied by alloying or applying high pressure [5.69] may suggest this kind of almost independent two-band model for magnetism and superconductivity. Furthermore, our calculation of the Ruderman-Kittel sum [5.70] for $HoMo_6S_8$ shows that only about 10 percent of the total conduction electrons participate in the magnetic interaction [5.13,48].

On the other hand, the value of Γ estimated from the A-G formula is in good agreement with the values found in the EPR and Mössbauer relaxation rate experiments [5.29,51,71]. This small value of Γ has also been substantiated by a recent band structure calculation [5.65].

A further check on the validity of the A-G formula can be obtained by substituting one RE ion for another. In this case, T_c should scale with the difference in the de Gennes factors as [5.48]

$$k_B\Delta T_c = \frac{\pi^2}{8} c x_B N(0) \Gamma^2 (G_B - G_A) \quad , \tag{5.4}$$

where ΔT_c is the change of T_c relative to T_c for A and x_B is the fraction of the RE ion B. In fact, it is found that T_c varies almost linearly with x_B in a wide range of x_B in most pseudoternary systems. This is true especially for (RE)Rh_4B_4 compounds (Chap.4). In the case of the Chevrel phases, however, it is difficult to verify this formula since ΔT_c is rather small and depends very much on nonmagnetic effects. Again, it is worth pointing out that it is of particular importance to carefully check the homogeneity and the composition in the application of the A-G formula to the Chevrel phases.

5.5 Summary

As we have seen in the preceding sections, a number of experimental and theoretical investigations have already revealed many exciting features of the new ternary superconductors.

One compound in this class of materials, $HoMo_6S_8$, shows the behavior of re-entrant superconductivity at the ferromagnetic transition, with a narrow temperature interval where an oscillatory magnetization develops as a result of the competition between the two collective phenomena. A number of other compounds order antiferromagnetically in the superconducting state. At the antiferromagnetic transition, an anomalous reduction of H_{c2} has been found for these compounds. This reduction of H_{c2} may indicate that when the system enters the AF state, the superconducting state is modified to a non-BCS type due to a possible change in pairing.

Pseudoternary systems have revealed an unexpected richness of phenomena. In particular it may be possible in systems such as $(Ho_{1-x}Eu_x)Mo_6S_8$ to find conditions where ferromagnetism and superconductivity coexist in a larger temperature interval. However, in order to properly study such pseudoternary systems it is crucial to make a careful diagnosis of the metallurgical state of the samples. Further development in sample preparation techniques in the future will assure us of even more fascinating research with these interesting Chevrel-type compounds.

References

5.1 Ø. Fischer, A. Treyvaud, R. Chevrel, M. Sergent: Solid State Commun. *17*, 721 (1975)
5.2 R.N. Shelton, R.W. McCallum, H. Adrian: Phys. Lett. *56*A, 213 (1976)
5.3 B.T. Matthias, E. Corenzwit, J.M. Vandenberg, H. Barz: Proc. Natl. Acad. Sci. USA *74*, 1334 (1977)
5.4 M. Ishikawa, Ø. Fischer: Solid State Commun. *23*, 37 (1977)
5.5 W.A. Fertig, D.C. Johnston, L.E. DeLong, R.W. McCallum, M.B. Maple, B.T. Matthias: Phys. Rev. Lett. *38*, 987 (1977)
5.6 M. Ishikawa, Ø. Fischer: Solid State Commun. *24*, 747 (1977)
5.7 R.W. McCallum, C.D. Johnston, R.N. Shelton, W.A. Fertig, M.B. Maple: Solid State Commun. *24*, 501 (1977)
5.8 M. Ishikawa, Ø. Fischer, J. Muller: J. de Physique *39*, C6-1379 (1978)
5.9 M.B. Maple: J. de Physique *39*, C6-1374 (1978)
5.10 G.K. Shenoy, B.D. Dunlap, F.Y. Fradin (eds.): *Ternary Superconductors* (North Holland, Amsterdam 1981)
5.11 Ø. Fischer, M.B. Maple (eds.): *Superconductivity in Ternary Compounds I*, Topics in Current Physics, Vol.32 (Springer, Berlin, Heidelberg, New York 1982)
5.12 M. Ishikawa, J. Muller: Solid State Commun. *27*, 761 (1978)
5.13 M. Ishikawa: Unpublished results (1977)
5.14 J.W. Lynn, D.E. Moncton, W. Thomlinson, G. Shirane, R.N. Shelton: Solid State Commun. *26*, 493 (1978)
5.15 L.D. Woolf, M. Tovar, H.C. Hamaker, M.B. Maple: Phys. Lett. *71*A, 137 (1979)
5.16 E.I. Blount, C.M. Varma: Phys. Rev. Lett. *42*, 1079 (1979);
L.N. Bulaevski, A.I. Rusinov, M. Kulič: Solid State Commun. *30*, 59 (1979;
H. Matsumoto, H. Umezawa, M. Tachiki: Solid State Commun. *31*, 157 (1979)

5.17 M. Tachiki, H. Matsumoto, T. Koyama, H. Umezawa: Solid State Commun. *34*, 19 (1980);
C.G. Kuper, M. Revzen, A. Ron: Phys. Rev. Lett. *44*, 1545 (1980);
see also [ref.5.10, pp.261,267,285]
5.18 O. Sakai, M. Tachiki, H. Matsumoto, H. Umezawa: Solid State Commun. *39*, 279 (1981)
5.19 J.W. Lynn, G. Shirane, W. Thomlinson, R.N. Shelton: Phys. Rev. Lett. *46*, 368 (1981);
J.W. Lynn, A. Raggazoni, R. Pynn, J. Joffrin: J. Physique Lett. *42*, L-45 (1981)
5.20 D.E. Moncton, D.B. McWhan, P.H. Schmidt, G. Shirane, W. Thomlinson, M.B. Maple, H.B. MacKay, L.D. Woolf, Z. Fisk, D.C. Johnston: Phys. Rev. Lett. *45*, 2060 (1980);
S.K. Sinha, G.W. Crabtree, D.G. Hinks, H. Mook: Phys. Rev. Lett. *48*, 950 (1982)
5.21 B. Lachal, M. Ishikawa, A. Junod, J. Muller: J. Low Temp. Phys. *46*, 467 (1982)
5.22 W. Baltensperger, S. Strässler: Phys. Kondens. Mater. *1*, 20 (1963)
5.23 W. Thomlinson, G. Shirane, D.E. Moncton, M. Ishikawa, Ø. Fischer: Phys. Rev. B*23*, 4455 (1981)
5.24 T. Krzysztoń: J. Mag. Mag. Materials, *15-18*, 1572 (1980)
5.25 S. Quézel, F. Tchéou, J. Rossat-Mignod, R. Chevrel, M. Sergent: Solid State Commun. *38*, 1003 (1981)
5.26 L.D. Woolf, M. Tovar, H.C. Hamaker, M.B. Maple: Phys. Lett. *74*A, 363 (1979)
5.27 A. Junod, M. Ishikawa: Unpublished data (1978)
5.28 J.W. Lynn, R.N. Shelton: J. Mag. Mag. Materials *15-18*, 1577 (1980);
J.W. Lynn: "Crystal Field Effects in Magnetic Superconductors" in *Crystalline Electric Field and Structural Effects in f-Electron Systems*, ed. by J.E. Crow, R.P. Guertin, T.W. Mihalisin (Plenum, New York 1980) pp.547-560
5.29 R. Odermatt, M. Hardiman, J. van Meijel: Solid State Commun. *32*, 1227 (1979);
R. Odermatt: Ph. D. Thesis, Université de Genève (1981)
5.30 J.W. Lynn, D.E. Moncton, G. Shirane, W. Thomlinson, J. Eckert, R.N. Shelton: J. Appl. Phys. *49*, 1389 (1978)
5.31 M.B. Maple, L.D. Woolf, C.F. Majkrzak, G. Shirane, W. Thomlinson, D.E. Moncton: Phys. Lett. *77*A, 487 (1980)
5.32 L.J. Azevedo, W.G. Clark, C. Murayama, R.W. McCallum, D.C. Johnston, M.B. Maple, R.N. Shelton: J. de Physique *39*, C6-365 (1978)
5.33 R.W. McCallum: Ph. D. Thesis, University of California (1977)
5.34 M. Pelizzone, A. Treyvaud, P. Spitzli, Ø. Fischer: J. Low Temp. Phys. *29*, 453 (1977)
5.35 D.C. Johnston, R.N. Shelton: J. Low Temp. Phys. *26*, 561 (1977)
5.36 M.B. Maple, L.E. DeLong, W.A. Fertig, D.C. Johnston, R.W. McCallum, R.N. Shelton: In *Valence Instabilities and Related Narrow Band Phenomena*, ed. by R.D. Parks (Plenum, New York 1977) p.17
5.37 R.W. McCallum, F. Claesen, F. Pobell: In [Ref.5.10, p.99]
5.38 R. Baillif, A. Junod, B. Lachal, J. Muller, K. Yvon: Solid State Commun. *40*, 603 (1981)
5.39 R. Baillif, A. Dunand, J. Muller, K. Yvon: Phys. Rev. Lett. *47*, 672 (1981)
5.40 C.W. Chu, S.Z. Huang, J.H. Lin, R.L. Meng, M.K. Wu, P.H. Schmidt: Phys. Rev. Lett. *46*, 276 (1981)
5.41 D.W. Harrison, K.C. Lim, J.D. Thompson, C.Y. Huang, P.D. Hambourger, H.L. Luo: Phys. Rev. Lett. *46*, 280 (1981)
5.42 R.W. McCallum, W. Kalsbach, T.S. Radhakrishnan, F. Pobell, R.N. Shelton, P. Klavins: Solid State Commun. *42*, 819 (1982
5.43 C.W. Chu, S.Z. Huang, J.H. Lin, R.L. Meng, M.K. Wu, P.H. Schmidt: In [Ref. 5.10, p.103]
5.44 M. Ishikawa, M. Sergent, Ø. Fischer: Phys. Lett. *82*A, 30 (1981)
5.45 G. Chouteau, R. Tournier, R. Chevrel, M. Sergent: In [Ref.5.10, p.107]
5.46 N.E. Alekseevskii, G. Wolf, N.M. Dobrovolskii, C. Hohlfeld: Phys. Status Solidi *44*(a), K79 (1977); Phys. Status Solidi *51*(a), 399 (1979)
5.47 P. Bonville, J.A. Hodges, P. Imbert, G. Jehanno, R. Chevrel, M. Sergent: Revue Phys. Appl. *15*, 1139 (1980)
5.48 M. Ishikawa: In [Ref.5.10, p.43]

5.49 Ø. Fischer, M. Decroux, R. Chevrel, M. Sergent: In *Superconductivity in d- and f-band Metals*, ed. by D.H. Douglass (Plenum, New York 1976) p.175
5.50 F.Y. Fradin, G.K. Shenoy, B.D. Dunlap, A.T. Aldred, C.W. Kimball: Phys. Rev. Lett. *38*, 719 (1977)
5.51 B.D. Dunlap, G.K. Shenoy, F.Y. Fradin, C.D. Barnet, C.W. Kimball: J. Mag. Mag. Mat. *13*, 319 (1979)
5.52 J. Bolz, G. Crecelius, H. Maletta, F. Pobell: J. Low Temp. Phys. *28*, 61 (1977)
5.53 J.O. Willis, J.D. Thompson, C.Y. Huang, H.L. Luo: J. Appl. Phys. *52*, 2174 (1981)
5.54 M. Isino, N. Kobayashi, Y. Muto: In [Ref.5.10, p.95]
5.55 T.S. Radhakrishnan, M.P. Janawadkar, S.H. Devare, H.G. Devare, R.G. Pillay, R. Janaki, A.M. Umarji, G.V. Subba Rao: In [Ref.5.10,p.91]
5.56 M. Sergent, R. Chevrel, C. Rossel, Ø. Fischer: J. Less-Comm. Metals *58*, 179 (1978)
5.57 C.F. Majkrzak, G. Shirane, W. Thomlinson, M. Ishikawa, Ø. Fischer, D.E. Moncton: Solid State Commun. *31*, 773 (1979)
5.58 Ø. Fischer, M. Ishikawa, M. Pelizzone, A. Treyvaud: J. Physique *40*, C5-89 (1979)
5.59 A.A. Abrikosov, L.P. Gorkov: Sov. Phys. JETP *12*, 1243 (1961)
5.60 K. Machida, K. Nokura, T. Matsubara: Phys. Rev. B*22*, 2307 (1980)
5.61 M.J. Nass, K. Levin, G.S. Grest: Phys. Rev. Lett. *46*, 614 (1981)
5.62 G. Zwicknagl, P. Fulde: Z. Phys. B*43*, 23 (1981)
5.63 O. Sakai, M. Tachiki, T. Koyama, H. Matsumoto, H. Umezawa: Phys. Rev. B*24*, 3830 (1981)
5.64 M. Tachiki, H. Matsumoto, H. Umezawa: Phys. Rev. B*20*, 1915 (1979); H. Matsumoto, R. Teshima, H. Umezawa, M. Tachiki: "Mixed States in Ferromagnetic Superconductors" (Preprint)
5.65 T. Jarlborg, A.J. Freeman: Phys. Rev. Lett. *44*, 178 (1980)
5.66 See, for example, R.J. Elliott (ed.): *Magnetic Properties of Rare-Earth Metals* (Plenum, New York 1972) Chaps.2,6
5.67a P.G. de Gennes: C.R. Hebd. Séan. Acad. Sci. *247*, 1836 (1958)
5.67b A.A. Abrikosov, L.P. Gor'kov: Sov. Phys. JETP *16*, 1575 (1963)
5.68 M.B. Maple: Appl. Phys. *9*, 179 (1976)
5.69 R.N. Shelton, C.U. Segre, D.C. Johnston: Solid State Commun. *33*, 843 (1980)
5.70 D.C. Mattis: *The Theory of Magnetism I*, Springer Series in Solid-State Sci., Vol.17 (Springer, Berlin, Heidelberg, New York 1981)
5.71 S. Oseroff, R. Calvo, D.C. Johnston, M.B. Maple, R.W. McCallum, R.N. Shelton: Solid State Commun. *27*, 201 (1978)

6. Electronic Structure and Superconductivity/ Magnetism in Ternary Compounds

A. J. Freeman and T. Jarlborg

With 11 Figures

The remarkable physical properties of the ternary compounds are a continuing subject of intense investigation – as is apparent from the various contributions to these volumes. A prodigious amount of experimental data has now been amassed, particularly regarding their superconducting and/or magnetic properties – and their coexistence – which have raised fundamental questions regarding their origin and have challenged long held traditional, theoretical understandings of these phenomena.

In this chapter, we describe the electronic structure of the rare-earth (RE) ternary borides and the Chevrel phase molybdenum chalcogenide compounds and use the results obtained to discuss the origin of their magnetic and superconducting properties. This work is derived from extensive theoretical studies of the self-consistent energy band structure of these compounds using state-of-the-art theoretical models (self-consistent local density and local spin density functional formalisms) on realistic representations of their crystalline structure in which all electrons and all (18 or 15) atoms per unit cell are included in a self-consistent scheme [6.1-6]. Since much of this work has not been submitted for publication it is fully described here in order to provide a self-contained discussion. As will be demonstrated, because of the vital role played by the charge transfer in the resulting properties of these systems, self-consistent calculations are essential for understanding the ternary compounds. A complete description of the earlier pioneering nonself-consistent cluster/tight binding calculations on the Chevrel compounds [6.7-9] is given in [Ref.6.10, Chap.6]; hence, we confine ourselves here to the work we have carried out on both the borides and Chevrel phase compounds.

To provide a framework and perspective on this work, we first describe the local density and local spin-density formalisms and the linearized muffin-tin orbital energy band method in Sect.6.1. The self-consistent energy band results on a number of the Chevrel compounds are presented in Sect.6.2 and used to discuss their superconducting/magnetic properties. In this work (and in that for the ternary borides discussed in Sect.6.2), all electrons and all atoms/cell are included with the core electrons (including the 4f's), recalculated in each iteration in a fully relativistic representation, and the conduction electrons treated semi-relativistically (all relativistic terms except spin-orbit). For the Chevrels, superconductivity is found to be due to the high Mo d-band density of states (DOS) at the Fermi energy (E_F) resulting from the unusual large charge transfer of Mo electrons to the chalcogen sites. There is also a large charge transfer from the metal site to the cluster (~ 2 electrons in Sn and Eu) giving essentially no occupied conduction bands, for example, at the Eu site and a divalent ion isomer shift, in very good agreement with the Mössbauer effect experiments of FRADIN et al. [6.11,12]. The conduction-electron DOS at the Eu site is found to be reduced by an order of magnitude from its metallic state value—in close agreement with their spin-lattice relaxation rate measurements. This low conduction-electron DOS yields very weak coupling of the 4f electrons to the conduction electrons and only a very weak Ruderman-Kittel-Kasuya-Yosida magnetic interaction, showing why all the Chevrel RE compounds—except Ce and Eu—are superconducting despite their having large local magnetic moments. The unusually high upper critical fields H_{c2} in these materials is found to be due to the unusually flat energy bands near E_F.

We are particularly concerned in Sect.6.3 with understanding the observations that the addition of magnetic RE impurities like Eu (in concentration up to 50% Eu) causes a large increase in the critical field of $SnMo_6S_8$ while T_c is hardly changed, with the depression occurring abruptly only at high concentration [6.13]. This behavior is contrary to (i) observations on all other materials with the addition of a local magnetic moment and (ii) the theory of ABRIKOSOV and GOR'KOV [6.14]. Our ferromagnetic (spin polarized) results for the Eu and Gd-compounds show a net, small but positive, magnetic moment on the metal site and a small but *negative* induced spin magnetic moment on the Mo site in the Eu compound. Fermi-contact contributions to the hyperfine field are calculated and found to be in good agreement with the Eu Mössbauer results and the negative NMR Knight shift results of FRADIN et al. [6.11,12]. These results demonstrate theoretically for the first time the validity of the FISCHER et al. [6.13] and FRADIN et al. [6.12] conclusion that the Jaccarino-Peter [6.15] mechanism is responsible for the large increase in the H_{c2} when large concentrations of Eu magnetic impurities are added to $SnMo_6S_8$. Finally, calculated Stoner factors for the paramagnetic phase and spin magnetization densities for the ferromagnetic phase are used to discuss qualitatively the origin of the different behavior observed for $GdMo_6S_8$ and $EuMo_6S_8$.

For the RE ternary borides discussed in Sect.6.3, we focus on the observation of magnetism and superconductivity, most notably the excitement generated by the discovery of re-entrant magnetism (elsewhere called re-entrant superconductivity) accompanying a transition to a normal metallic state at $T < T_c$ (the superconducting transition temperature) [6.16]. We discuss the origin of magnetism and superconductivity in the MRh_4B_4 compounds and re-entrant magnetism in $ErRh_4B_4$ using results of *ab initio* self-consistent LMTO energy band calculations. Here, the total and separate ℓ-decomposed contributions to the DOS arising from the two M, eight Rh and eight B atoms per unit cell are used to estimate their various contributions to magnetic ordering (via the 4f-5d RKKY interaction) and/or superconductivity (using a GASPARI-GYORFFY [6.17] model to obtain the electronic contribution to λ, the electron-phonon coupling parameter, and MCMILLAN strong-coupling theory [6.18] to obtain T_c). Comparisons are made with recent experimental results and several experimental speculations about the re-entrant magnetism state are presented including the possibility of a mixed state which is either simultaneously superconducting and ferromagnetic, or a magnetic field induced mixed state with regions which are normal but magnetic and interspersed within the superconducting regions.

6.1 Theoretical Approach

We are seeking accurate solutions of the Schrödinger or Dirac equation for the ternary compounds with their rather complex crystalline structures. It is, therefore, all the more important to recall that even for simple homonuclear metals, one faces the still very complex problem of obtaining solutions for what is after all, a many-body problem for a crystal with 10^{23} nuclei and electrons. There are a number of simplifying assumptions and approximations, such as the Born-Oppenheimer approximation, which reduce the many-body problem, involving the interactions between all the particles (nuclei and electrons) in the system, to a one-electron or independent electron model. This section briefly describes the underlying theoretical basis of the approach followed in obtaining self-consistent energy band results which are used to understand the superconducting/magnetic properties of the ternary compounds.

6.1.1 Local Density and Local Spin-Density Formalism

The local density functional (LDF) theory of HOHENBERG et al. [6.19] has provided a rigorous basis and justification of the single particle energy band description of the many-electron ground state properties of materials and has spurred the development of accurate, tractable , computational schemes for describing them from first principles. Briefly, the LDF formalism is based on the fundamental theorem that the ground state properties of an inhomogeneous interacting electron system are functionals of the electron density, $\rho(\underline{r})$, and that in the presence of an external potential $V_{ext}(\underline{r})$, the total ground state energy in its lowest variational state can be written as

$$E[\rho(\underline{r})] = \int V_{ext}(\underline{r})\rho(\underline{r})d\underline{r} + G[\rho(\underline{r})] \quad , \tag{6.1}$$

where $G[\rho(\underline{r})]$ is a universal functional of $\rho(\underline{r})$ and is *independent of the external potential* $V_{ext}(\underline{r})$. This theorem forms the basis of the approach to the electronic structure problem in that it provides an effective one-particle equation relating self-consistently the ground state wave functions to the energy functionals (i.e., potential) of the electronic system. Identifying the external potential for a polyatomic system as the electron-nuclear and internuclear interactions of the many-electron Hamiltonian and varying $E[\rho(\underline{r})]$ with respect to $\rho(\underline{r})$, one obtains an effective one-particle equation of the form

$$\left(-\frac{1}{2}\nabla^2 + \sum_m \frac{Z_m}{|\underline{R}_m - \underline{r}'|} + \int \frac{\rho(\underline{r}')}{|\underline{r} - \underline{r}'|}\, d\underline{r}' + \frac{\delta E_{xc}[\rho(\underline{r})]}{\delta\rho(\underline{r})}\right)\psi_j(\underline{r}) = \varepsilon_j\psi_j(\underline{r}) \quad . \qquad (6.2)$$

Here Z_m denotes the nuclear charge of the nucleus at site $\underline{R}_m$ and $E_{xc}[\rho(\underline{r})]$ denotes the total exchange and correlation energy of the interacting (inhomogeneous) electron system (square brackets are used to denote functional dependence). The eigenfunctions $\psi_j(\underline{r})$ are simply related to the total ground state charge density of the occupied one-particle states which, in turn, determines self-consistently the local density functional. The total ground state energy is then given by

$$E_{tot} = \sum_{j=1}^{\sigma_{oc}} \langle\psi_j(\underline{r})| -\frac{1}{2}\nabla^2|\psi_j(\underline{r})\rangle + \int \rho(\underline{r})\left(\sum_m \frac{Z_m}{|\underline{r} - \underline{R}_m|} + \frac{1}{2}\int \frac{\rho(\underline{r}')}{|\underline{r} - \underline{r}'|}\, d\underline{r}'\right) dr$$

$$+ \sum_{\substack{n,m \\ n \neq m}} \frac{Z_n Z_m}{\underline{R}_n - \underline{R}_m} + E_{xc}[\rho(\underline{r})] \quad , \qquad (6.3)$$

where the first term represents the kinetic energy, the second and third terms are the total electrostatic potential energy and the last term is the exchange and correlation energy. Note that the LDF formalism in the form described above makes no claim on the physical significance of the eigenvalues ε_j or the ψ_j in (6.2); hence, we may concentrate only on ground state crystal properties.

Retaining only the nongradient terms in the expansion of $E_{xc}[\rho(\underline{r})]$, the exchange and correlation potential becomes

$$\frac{\delta E_{xc}[\rho(\underline{r})]}{\delta\rho(\underline{r})} \cong F_{ex}[\rho(\underline{r})] + F_{corr}[\rho(\underline{r})] \quad , \qquad (6.4)$$

where the exchange potential has the well-known form

$$F_{ex}[\rho(\underline{r})] = \frac{4}{3}\,\varepsilon_x[\rho(\underline{r})] \equiv -\left(\frac{3}{\pi}\,\rho(\underline{r})\right)^{1/3} \qquad (6.5)$$

which is 2/3 of the value given by SLATER [6.20]. The correlation energy of a uniform electron gas with local density $\rho(\underline{r})$ has been calculated from many-body theory by many authors using different techniques [6.21-23]. The agreement between the most recent results lies within 5-8 mRyd in the metallic density range. A particularly convenient form is to use the results of SINGWI et al. [6.23] fitted to an analytical expression by HEDIN et al. [6.24]. The total exchange and correlation energy is given at this level of approximation by

$$E_{xc}[\rho(\underline{r})] = \int \rho(\underline{r})\{\varepsilon_x[\rho(\underline{r})] + \varepsilon_c[\rho(\underline{r})]\}d\underline{r} \quad . \tag{6.6}$$

In the local spin-density functional (LSDF) theory, which is of great interest and utility for treating properties of magnetic systems, the Kohn-Sham exchange potential (6.5) is replaced by a spin polarized description such as that of VON BARTH and HEDIN [6.25]. This description has been found to be more appropriate because comparison of calculations with the Kohn-Sham potential and experiment by WANG and CALLAWAY [6.26] indicates that the calculated magnetic moment and exchange splitting in ferromagnetic Ni are too large. The Kohn-Sham potential evidently overestimates the tendency toward ferromagnetism, and this overestimate needs to be reduced through the use of a potential which incorporates additional correlation effects [6.26]. For example, the von Barth-Hedin potential for electrons of spin $\sigma(=\uparrow \text{or} \downarrow)$

$$V_{ex,\sigma} = A(\rho)(2\rho_\sigma)^{1/3} + B(\rho) \tag{6.7}$$

in which $\rho = \rho\uparrow + \rho\downarrow$ is the total charge density, apparently accomplishes this end (parametrized forms for A and B are given in their paper).

6.1.2 Linearized Muffin-Tin Orbital Method

Since the traditional energy band methods have a number of limitations when treating complex materials, several simplified energy band schemes have been proposed and shown to yield very good results. These methods, known as the linear muffin-tin orbital (LMTO) method [6.27] and the linearized augmented plane wave (LAPW) method [6.27,28] are 'linearized' versions of the KKR and APW schemes, respectively. Their major virtues are that they avoid some of the computational complexities and high costs of treating complex (many atoms per unit cell) systems inherent in the regular plane-wave based methods. These linearized methods retain relatively high accuracy and suffer little, if any, loss in computational speed compared to pseudopotential methods.

For the ternary compounds, the band structure calculations were performed using the LMTO band method [6.29]. This method is very efficient for computing band structures of complicated compounds such as the ternary borides and the Chevrels for the following reasons: (i) It utilizes the continous logarithmic derivative at the MT (or overlapping Wigner-Seitz, WS) radius to generate an *energy independent* basis so that the eigenvalue problem becomes linear in energy. (ii) Further, in the basis-set it is sufficient to include only 9 basis functions (s, p and d) per transition metal atom and to include higher ℓ-states in the three center terms to improve the convergence without increasing the dimensions of the eigenvalue matrices

[6.29]. In the case of Chevrel compounds having 15 atoms per unit cell, the matrices thus are of rank 135. (It would have been possible to obtain reasonably good convergence by using only s and p-basis for the eight sulfur sites per cell and so reduce the matrices to rank 95). (iii) The method leads to a separation of the band problem into a structure-dependent and a potential-dependent part. The structure part consists of calculating some quite complicated structure factors, but as these are independent of the size (i.e., lattice constant) of the system, it is not so difficult to study several compounds of the same structure.

Of course, the calculational method involves approximations which make it tractable. The potential is taken to be spherically symmetric within an overlapping WS-sphere geometry and, together with the low packing ratio of touching MT-spheres (e.g., $\sim$42% for the Chevrel structure), this is probably the most serious approximation. However, the Mo-S cluster in the Chevrels is almost fcc-like (packing ratio of 74%) so there the potential is quite well described within the model. Instead, it is states having a large amplitude around the edges of the unit cell which are affected by nonspherical potential contributions.

For the RE compounds, the f-states were treated like core states, i.e., the f-orbital is numerically determined and normalized over the given WS potential. In this process it turned out to be necessary to "embed" the WS potential in a "step-like" repulsive potential outside, in order to localize the f-state. The energy position of the f-level relative to the other states is therefore rather uncertain. In the spin-polarized calculations, where the majority f-band is completely filled, it would have been possible to extend the basis for the RE site only, leading to a modest increase of the matrix sizes, and let the f-state hybridize like the other valence states. But f-levels are very sensitive and may cause convergence difficulties in the self-consistent iterations, so we did not choose to include the f in the basis for those complicated Chevrel calculations. Also, an f-basis would not have been useful in the paramagnetic RE calculations.

In the calculation of electron-phonon coupling it is important to have realistic f-DOS values to correctly account for the d-f scattering, even if the f-DOS is very small on Mo or S sites. Here, it is a good approximation to determine the f-DOS from tail-overlap without using f in the basis. Tests on the C15 structure using this method and comparing them with a full f-basis calculation showed that the f-DOS was well described for non f-elements.

All calculations were performed self-consistently giving a convergence of 3 mRy or less for the eigenvalues. Core states are recalculated in each iteration which allows for studies of core level shifts and hyperfine fields in the spin-polarized calculations. The core states are fully relativistic while for the valence states, semi-relativistic j-averaged radial wave functions were used [6.30] so that all relativistic effects, except the spin-orbit coupling, are included.

For the exchange and correlation contribution to the potential, the local electron density (ρ) expression of HEDIN et al. [6.24] (HL) was used, to which the local spin-density expression of GUNNARSSON and LUNDQVIST [6.31] (GL) was added in the spin polarized calculation. Thus, in addition to the Coulomb potential, the total potential has an exchange and correlation contribution of the following form:

$$\mu_{KS}(\rho)\left[\beta_{HL}(\rho) \pm \frac{1}{3}\,\delta_{GL}(\rho)\xi/(1 \pm 0.297\xi)\right] \quad , \tag{6.8}$$

where $\mu_{KS}(\rho)$ is the Kohn-Sham exchange potential and ξ is the fractional magnetization density (the second term is absent in the paramagnetic calculations).

Further details of the calculations, specific to either the Chevrel or boride compounds, are given in their respective sections.

6.2 Chevrel Phase Compounds

6.2.1 Energy Band Calculations

As discussed in Chap.2 of Vol.I and as seen in Fig.6.1, the Chevrel phase structure consists predominantly of a cluster of Mo_6S_8 (or Mo_6Se_8) atoms in which the S atoms occupy the corners of a cube. On the surfaces, or rather slightly outside the surfaces, sit Mo atoms so that they form a Mo-octahedron. However, the X ray diffraction data by MAREZIO et al. [6.32] showed that the Mo-octahedron in $PbMo_6S_8$ was somewhat distorted giving different Mo-Mo distances which vary about 1% around the average value. While these variations probably differ among different compounds, the distortion is not large enough to be expected to affect most of the unique properties of the Chevrel phase compounds. Thus, in order to simplify the computations, we assume the Mo-octahedron to be undistorted in the band calculations in which case all the Mo atoms have the same potential. This assumption simplifies symmetry properties of the unit cell and the Brillouin-zone and thus also very much simplifies the band calculation. (BULLET [6.7] studied the effect of rhombohedral distortion of the molecular levels of a free Mo_6S_8 cluster; his results indicated that there was mostly a separation of previously degenerate levels due to the distortion. This is discussed in [Ref.6.9, Chap.6]).

The Mo_6S_8 cube is placed in a large cube of M atoms and rotated around the (111)-axis by about 15 degrees. This makes the surroundings of the two S atoms on the (111)-axis different from that of the other six S atoms so that there are two types of (spherically symmetric) sulfur potentials in the band calculation. In the following, we will use the notations S(2) for the two S atoms on the (111)-axis, and S(6) for the other six sulfur atoms.

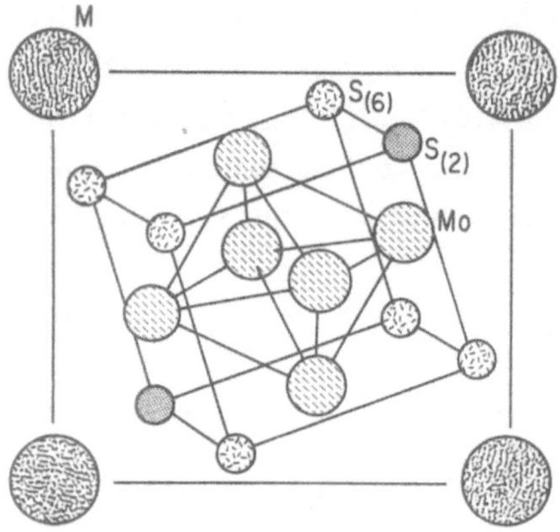

Fig.6.1. The unit cell of the Chevrel structure. Note the "filled" inner cube consisting of the Mo-S cluster. The two S atoms on the (111) diagonal are nonequivalent to the other six S atoms off the diagonal

There are two types of Chevrel phase structures, depending on what third element is added to the Mo_6S_8 cluster [6.32]. Smaller elements, like Cu, tend to occupy sites on the edges of the large cube, while larger elements, a rare-earth, Sn or Pb atom, occupy its corners. In the systems studied here, this latter type of structure is appropriate and, as a final simplification, we assume no rhombohedral distortion to the cubes. In reality, the rhombohedral angles α vary typically from 88.9 to 89.8 degrees for systems like those studied here, whereas we use 90.0 in the calculations as a typical average Chevrel structure. It has been observed, however, that scaled with the actual lattice constant, T_c is very sensitive to small variations in α (when varied by less than one degree around 89°). The physics behind such variations or the effects of nonstoichiometry cannot be interpreted easily from this study.

Our simplified structure is almost identical to that of Fig.2 in the paper of MAREZIO et al. [6.32]. Among the 15 atoms per cell there are 4 inequivalent potentials, one for the M site, one for Mo and two for the S sites. The cube dimension of the small cube in units of the lattice parameter $\underline{a}$ (equal to the dimension of the large cube) is 0.515. As stated earlier, the structure as a whole is not closely packed, with only about 42 percent of its total volume filled by touching spheres. However, the Mo_6S_8 cluster itself, in which its interesting physical properties are believed to originate, is very closely packed ($\sim$74% packing ratio) since the arrangement of atoms is similar to that of a fcc lattice. This justifies to a large extent the use of the LMTO method, with the overlapping sphere (ASA) concept, in studies of the Chevrel structure.

The potential is taken to be spherically symmetric within an overlapping WS sphere geometry. It is, instead, states having a large amplitude around the edges of the unit cell which are affected by the nonspherical potential contributions. The radii of the overlapping WS spheres were determined to give reasonably small potential discontinuities at the sphere boundaries. [For example, a preliminary

$SnMo_6S_8$ calculation resulted in the choice of 0.28 for Sn and 0.249 for atoms in the Mo_6S_8 cluster, in units of the lattice parameter]. The structure coefficients included the "combined correction terms" [6.27] which correct for the overlapping spheres and also improve the convergence properties of the LMTO method.

The first calculations were performed for $SnMo_6S_8$ with the final self-consistent band structure determined at 42 k-points in 1/12 of the irreducible Brillouin zone (IBZ) with the eigenstates self-consistently converged to about 3 mRy. For the calculations on the other compounds, a canonical band calculation scheme [6.27,29] was used in the initial stage of the self-consistent procedure with the $SnMo_6S_8$ results as input in order to minimize the computational costs.

For the magnetic state calculations, self-consistent results were obtained with spin-polarized calculations using 14 k-points in the IBZ. The DOS data given here are determined from a k-point weighted histogram technique averaged over an energy resolution of 5-10 mRy. However, the band structure plot shown, as well as values given for the Fermi velocities v_F, are determined by using a Fourier-fitting of the first-principle eigenvalues in one part of the BZ, namely, in the 1/48 of the BZ along the [111] direction. This, together with usual uncertainties connected with a Fourier-fitting procedure such as band crossings and high Fourier components, introduces quite large error bars to the resulting v_F values.

6.2.2 Results for the Paramagnetic States

a) *Band Structure, Density of States and Resulting Properties*

The electronic energy band structure of $SnMo_6S_8$, as determined from a Fourier series fit to the eigenvalues calculated at 35 k points in the IBZ, is shown in Fig.6.2 along the high symmetry lines. The results for the other compounds are very similar and so are not given here. The bands in Fig.6.2 are grouped according to their dominant atomic (and projected orbital) character. We see a large separation of the Sn-s band at about one Ryd below E_F and a clear gap between the S-p and Mo-d bands. The Mo-d bands are split in two with a substantial gap between them. E_F is seen to lie at the top of the lower half of the Mo-d band. The striking feature of the bands in Fig.6.2 is their flat (and hence dull) nature—perhaps the worst 'spaghetti' seen for E(k) plots.

The total DOS functions determined in a 2.5 mRy energy mesh and containing a 5 mRy broadening function, are shown in Figs.6.3-5. Note that in $SnMo_6S_8$ there are 44 occupied bands (including the low lying Sn 6s band), while in $EuMo_6S_8$ there are 43, but no such low lying occupied band. Hence, the E_F is in both cases one full band below the gap at 0.4-0.5 Ry. From the partial DOS functions shown in Figs. 6.6-8, it is seen that the large gap falls in the middle of the Mo-d band region and a second smaller gap is found between the Mo-d states and the S-p states. These main features of the band structure, which were seen in the early band results

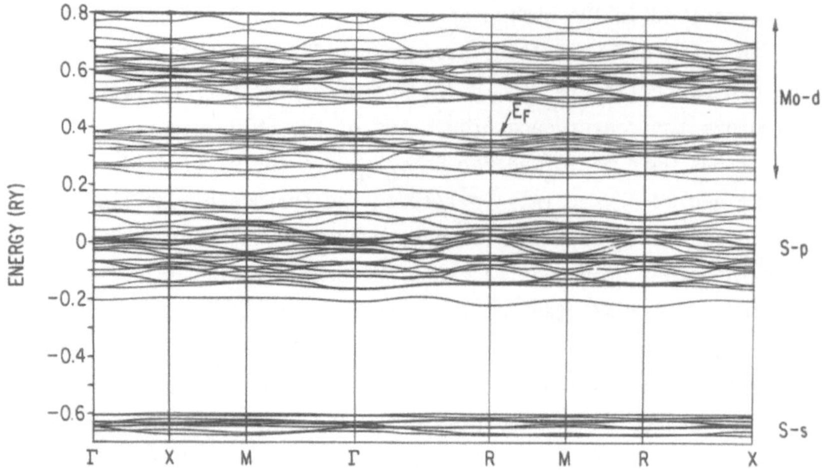

Fig.6.2. The bandstructure for $SnMo_6S_8$ as determined from a Fourier-fitting procedure in part of the IBZ. The horizontal line indicates the Fermi level

[6.7-9], are confirmed by experimental evidence for a gap in the d-band region because the Chevrel phase compound $Mo_2Re_4S_8$ is a semiconductor [6.33]. This can be understood from the fact that $Mo_2Re_4S_8$ has 2 more electrons than the Mo_6S_8 cluster in $SnMo_6S_8$. Thus in $Mo_2Re_4S_8$, one additional cluster band is occupied and the Fermi level falls in the gap.

The Sn-s band in $SnMo_6S_8$ is seen clearly below the S-p bands in Figs.6.2,3. In comparison with $EuMo_6S_8$ and $GdMo_6S_8$, there is an additional band separated from the S-p states in the smaller gap. This band emerges from the s-p band complex and hybridizes strongly with the Sn-s band. However, no bands of Sn-p character are found below the Fermi level so that a large charge transfer from Sn to the cluster takes place. Evidence of this is seen in Table 6.1 where the charge content is presented. Various experiments confirm that there is a large charge transfer from the single metal site to the cluster [6.12,33,34].

In the case of $EuMo_6S_8$, Eu has no occupied conduction bands with the Eu-s band appearing high above E_F. Therefore, in this case and in the Gd compound, a large charge transfer is found to proceed from the Eu-site to the cluster. The band structure here is probably very much like that of a Mo_6S_8 cluster itself, with the Eu atom providing additional charge to the cluster.

It is seen in Fig.6.2 that the bands have small dispersion which is responsible for the very "peaky" DOS character seen in Figs.6.3-5. Some of the bands in the lower part of the Mo-d complex have a width of only 3 mRy, while the few bands that cross the Fermi level have widths of 8-9 mRy. This very flat band structure gives, of course, possibilities for very high DOS peaks and in a rigid band model one can expect large variations of the DOS at E_F as a function of moderate changes in the charge transfer to the cluster.

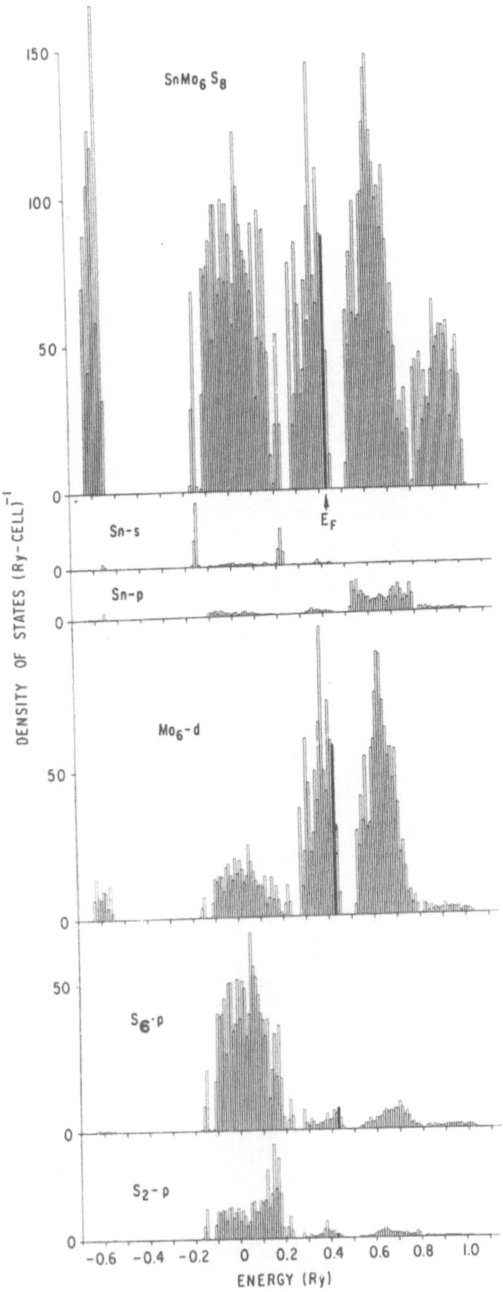

Fig.6.3. Total and selected partial (site and ℓ-decomposed) density of states for $SnMo_6S_8$ as obtained from sampling 42 k-points in the IBZ

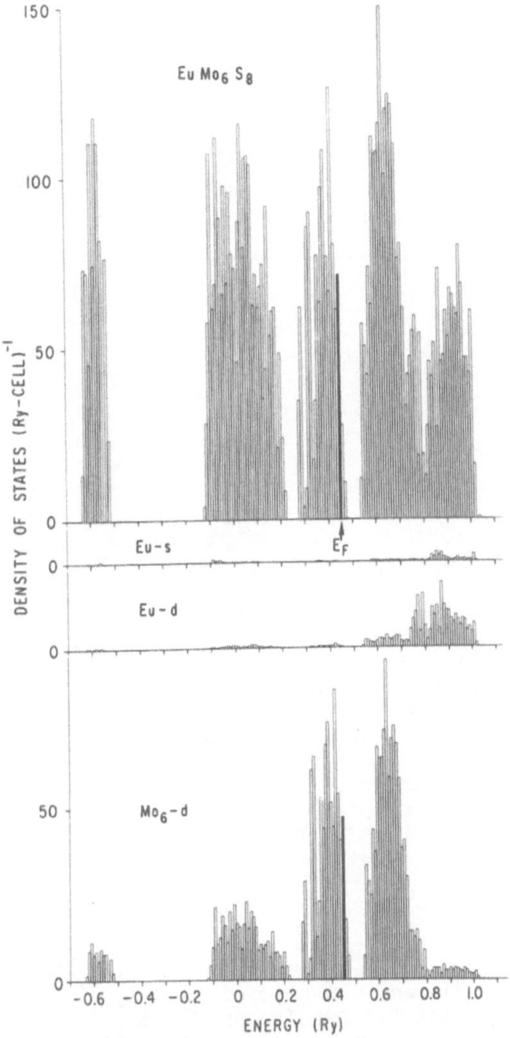

Fig.6.4. Total and selected partial (site and ℓ-decomposed) density of states for $EuMo_6S_8$ as obtained from sampling 42 k-points in the IBZ

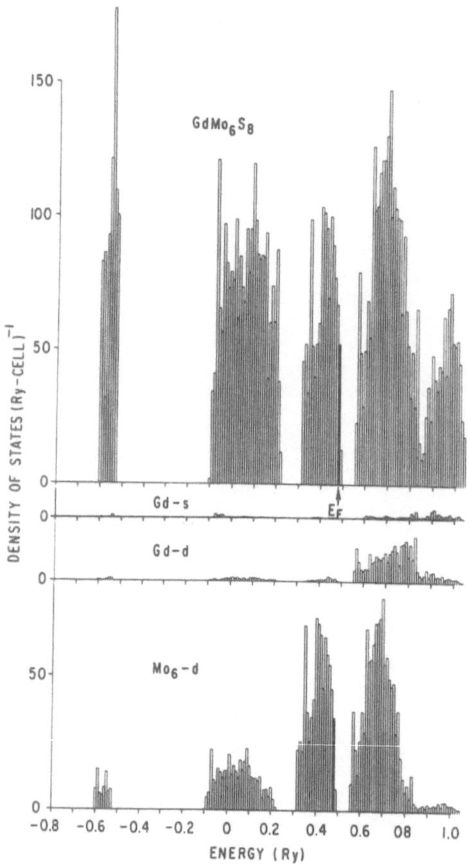

Fig.6.5. Total and selected partial (site and ℓ-decomposed) density of states for $GdMo_6S_8$ as obtained from sampling 42 k-points in the IBZ

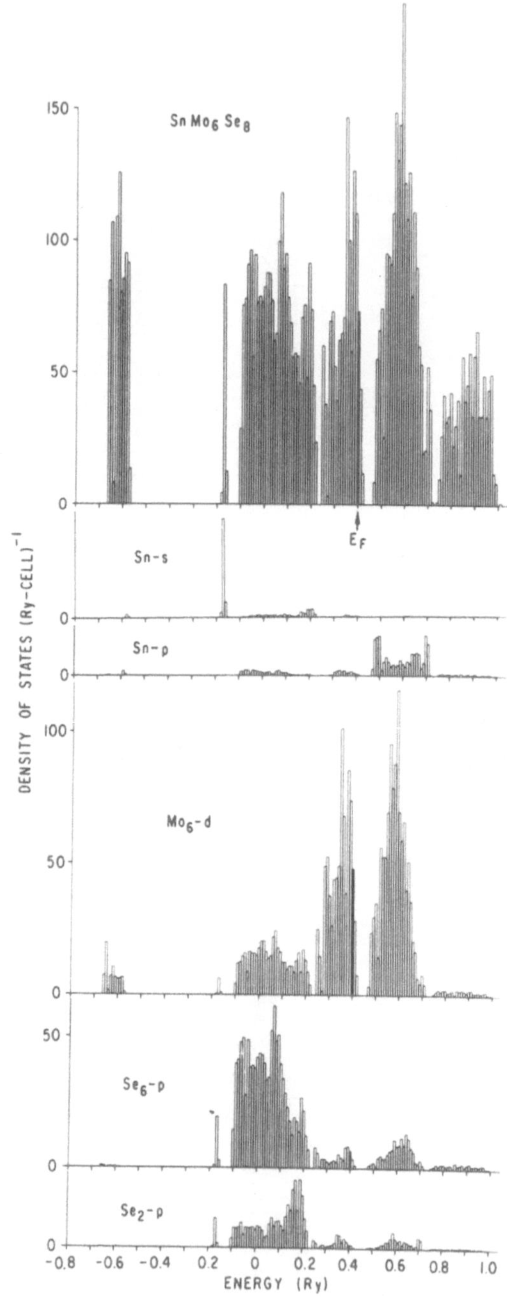

Fig.6.6. Total and selected partial (site and ℓ-decomposed) density of states for $SnMo_6Se_8$ as obtained from sampling 42 k-points in the IBZ

As core-like states, many of the S-s bands are extremely narrow having widths of about 1 mRy. The states at the two S atoms on the (111)-axis [denoted S(2)] are not quite so tightly bound as are the other (six) S atoms [denoted S(6)]. This is seen from the partial DOS diagrams in Figs.6.3-6; the S-s and p states have somewhat higher energy for the S atoms on the (111)-axis [type S(2)]. While the charge content within the S atoms of the different types do not differ significantly from each other, as can be seen in Table 6.1, the ones in the S(2) position have slightly more charge. Note, however, that the charge content given in Table 6.1 is determined by taking hybridization effects and tail contributions into account so that all sulfur p bands, bands 9 to 32 in, say, Fig.6.2, are occupied. The charge analysis shows, however, that a large fraction of the S-p functions penetrate the Mo-spheres and give some charge contributions into those spheres. This is the reason why the charge content in Mo and S spheres of equal size do not show the large charge transfer directly.

It has recently been reported that stable Chevrel phases, such as $Mo_6S_6Br_2$, can be made into good superconductors [6.35]. Further, it was also concluded that the halogen atoms prefer the sites on the (111)-axis because there is an evident saturation of 2 halogen atoms per unit cell. Hence, it is worthwhile to carefully analyze the band data to determine any detailed differences between the two types of sulfur atoms. The metal atom sites (M) have the S(2) atoms on the (111)-axis as nearest neighbors. In the partial Sn DOS shown in Fig.6.6, the Sn-s band was easily identified at about -0.25 Ry but, as was mentioned earlier, one S-p band shows large hybridization and tailing effects with the Sn-s states and so the band which emerges into the gap at 0.25 Ry has a large Sn-s content. This band is apparently made up from an S(2) p band since an inspection of the partial S-p DOS for this peak shows the largest p-content (per atom) for the S(2) atoms. It is also seen that the Sn-s peak gives more p charge at an S(2) site than at an S(6) site.

Not surprisingly, S(2) atoms on the (111)-axis show larger interaction with the M sites than do the (other) S(6) atoms. As seen from Table 6.1, the total charge within the different S-spheres is slightly larger at sites on the (111)-axis, and from the discussion above we may conclude that it is charge transferred from the Sn sites that goes primarily to the S on the (111)-axis. In compounds like $Mo_6S_6Br_2$, no atoms occupy the Sn site, but since only one electron is needed to make a closed shell of Br, it is likely that they can be more easily accommodated on the type 2 sites since these sites usually have more electronic charge than the type 1 sites.

The partial DOS values in Table 6.1 are obtained from a weighted k-point summation from the 42 points but scaled to agree with the Fourier-fit values for the total DOS. The total DOS at E_F obtained for $SnMo_6S_8$, 182 states per Ry-cell (or 2.23 states per eV - Mo atom), compares well with the bare DOS at E_F extracted from

Table 6.1. Charge content decomposed in site- and ℓ-character within the given WS radii, total DOS (given within a 5 mRy wide energy mesh), calculated electron phonon coupling constants λ and calculated and experimental superconducting transition temperatures T_c

	Charge ($el.atom^{-1}$)				DOS	λ	T_c(K)	T_c(K)
	s	p	d	total	($cell.Ry^{-1}$)		(calc)	(exp)
$SnMo_6S_8$					192	0.77	11.5	14.2
Sn	1.6	0.87	0.32	2.8	3.3			
Mo	0.60	0.84	4.3	5.8	143			
S(6)	1.6	4.3	0.42	6.3	33			
S(2)	1.6	4.4	0.45	6.4	11			
$EuMo_6S_8$					180	0.62	6.0	-
Eu	0.19	0.27	0.46	0.92	4.5			
Mo	0.60	0.85	4.4	5.8	130			
S(6)	1.6	4.3	0.42	6.3	33			
S(2)	1.6	4.2	0.46	6.3	11			
$GdMo_6S_8$					142	0.56	4.0	1.4
Gd	0.25	0.35	0.81	1.41	2.0			
Mo	0.58	0.84	4.3	5.8	105			
S(6)	1.6	4.3	0.44	6.4	29			
S(2)	1.7	4.4	0.49	6.5	6.1			
$SnMo_6Se_8$					188	0.69	7.0	6.8
Sn	1.7	1.1	0.3	3.2	3.9			
Mo	0.63	0.91	4.5	6.0	137			
Se(6)	1.6	4.1	0.4	6.1	39			
Se(2)	1.7	4.1	0.42	6.2	8			
$EuMo_6Se_8$					148	0.59	4.1	-
Eu	0.22	0.34	0.54	1.1	1.8			
Mo	0.64	0.91	4.5	6.0	110			
Se(6)	1.6	4.1	0.39	6.1	30			
Se(2)	1.7	4.1	0.40	6.2	5.4			
$GdMo_6Se_8$					128	0.46	1.3	5.6
Gd	0.32	0.42	0.98	1.7	2.6			
Mo	0.63	0.90	4.5	6.0	94			
Se(6)	1.6	4.1	0.42	6.1	26			
Se(2)	1.7	4.1	0.44	6.3	6.2			

Table 6.2. Fermi velocities, v_F (10^7 cm/s), obtained from the Fourier fitted bands and compared to results on some A15 compounds [Ref.6.29]

Chevrels		A15	
$SnMo_6S_8$	1.7	V_3Si	2.0
$EuMo_6S_8$	2.1	V_3Ga	2.1
$GdMo_6S_8$	1.8	V_3Ge	2.1
$SnMo_6Se_8$	1.8	Nb_3Al	2.5
$EuMo_6Se_8$	1.9	Nb_3Ga	2.6
$GdMo_6Se_8$	1.8	Nb_3Ge	2.2

specific heat measurements [6.36], namely, 2.6 ± 0.8 states per eV - Mo atom.

Returning to the very flat bands shown in Fig.6.2, it is clear that the electron velocities at E_F, v_F, will be very low. Values of v_F (in units of 10^7 cm/s), as obtained from the Fourier series fit to the bands, are compared in Table 6.2 with earlier calculated values for some A15 compounds [6.29]. Now, since the orbital contribution to the upper critical field $H^*_{c2}(0)$ is proportional to v_F^{-2}, it is readily seen why upper critical fields in the Chevrels can be high if some other mechanism like spin-orbit scattering is sufficiently strong to offset the Pauli paramagnetic limitation (the Clogston limit) — as has been demonstrated [6.9,33].

b) *Intra-Cluster Versus Inter-Cluster Effects and Superconductivity*

In order to study the role of the M atoms and to determine qualitatively the role of intra-cluster versus inter-cluster interactions, we have also carried out self-consistent LMTO energy band studies of the Mo_6S_8 compounds. Here there is no M atom and in the calculations both the "normal" Mo-S cube distances (used in full compounds) and expanded (by ~4.5%) distances relative to the unit cell cube were investigated. No major changes in the charge transfer were found to arise from the expansion of the Mo-S cluster. Somewhat surprisingly, the DOS at E_F was found to be smaller by ~30% for the expanded cluster compared to the normal cluster, accompanied by a slight broadening of the band width. This result could account for the low T_c value observed in Mo_6S_8. If one considers the effect of the cube expansion or the intra-versus-inter cluster distances, the large decrease in DOS at E_F shows that inter-cluster interactions are large in the Chevrel compounds: The intra-cluster S-S distance is smaller than the inter-cluster S-S distance for our "normal" structure, while in the "expanded" structure the situation is reversed; for the Mo-Mo in the "normal" structure, the intra-cluster distances are 15-20% smaller than the inter-cluster values but the difference is reduced to half that (~ 10%) for the "expanded" structure. The possible significance of these results to the observed superconductivity induced in $EuMo_6S_8$ [6.37,38] by application of pressure is being investigated.

c) *Charge Transfer and Magnetic Isolation*

As is clear from Table 6.1 and Figs.6.3-6, there is a significant charge transfer from the M site to the cluster. The values for the total charge at the M sites in Table 6.1 reflect to a great extent the "tail" contributions coming from the overlap of Mo and S (or Se) wave functions into the Wigner-Seitz (WS) spheres surrounding the M site atoms. Comparing for Mo in the Chevrels with an Mo sphere of the same size in A15 compounds [6.29], one sees that there is ~ 1 electron less in the Chevrels than in the A15's where no large charge transfer effects are seen. This yields the important result that there are essentially no occupied Eu or Gd conduction electron bands and gives for Eu a typically divalent isomer shift — in agreement with Mössbauer effect measurements [6.11,12]. It is also responsible, as we shall see, for the weak Ruderman-Kittel-Kasuya-Yosida (RKKY) magnetic interaction and the observed 'magnetic isolation' of the RE ions.

In RKKY, the magnetic ordering temperature T_m is given by

$$3k\,T_m = S(S+1)\mathscr{J}^2 N(E_F) \quad , \tag{6.9}$$

where S is the spin on the RE site and $\mathscr{J}$ is the 4f-conduction electron exchange interaction. We have calculated the direct exchange contribution to $\mathscr{J}$ given by

$$\mathscr{J}(o) = \mathscr{J}(\underline{k},k) = \sum_m \int d^3r \int d^3r' \psi_k^*(r)\phi_{f,m}^*(r') \frac{2}{|r - r'|} \phi_{f,m}(r)\psi_k(r') \tag{6.10}$$

using our LMTO results for the localized 4f and conduction electron ψ_k wave functions. The results for Eu and $GdMo_6S_8$ given by the orbital angular momentum component of the ψ_k are given in Table 6.3 and compared with the results of HARMON and FREEMAN for hcp/Gd metal [6.43]. We see very small variations between the Chevrel phase result and that of the pure RE metal indicating that $\mathscr{J}$ behaves in a 'normal' fashion. Hence it is the large reduction in the local (at Eu site) $N(E_F)$ values in the Chevrel compounds — as per Table 6.1 — which are responsible for their weak magnetic behavior. Note that our calculated $\mathscr{J}$ differs from experimentally derived values obtained from measurements of the product $\mathscr{J}N(E_F)$, since $N(E_F)$ is usually taken to be the total DOS of the system. This choice gives very small values for $\mathscr{J}$, although the product $\mathscr{J}N(E_F)$ agrees in both cases — as does the conclusion of rare-earth magnetic isolation.

Table 6.3. Calculated direct exchange interaction $\mathscr{J}$ in (eV) between the 4f and valence orbitals (normalized over the sphere) for Eu and Gd in $EuMo_6S_8$ and $GdMo_6S_8$, compared with the results for hcp Gd of Harmon and Freeman [Ref.6.43]

	Eu in $EuMo_6S_8$	Gd in $GdMo_6S_8$	Pure Gd (hcp)
$\mathscr{J}_s$	0.31	0.26	0.20
$\mathscr{J}_p$	0.76	0.49	0.30
$\mathscr{J}_d$	0.51	0.53	0.50

A further indication of the validity of our theoretical results is given by the Mössbauer effect determination of the spin-lattice relaxation rate W in the $Sn_{1.2(1-x)}Eu_xMo_{6.25}S_8$ compounds [6.11]. The results show that W, which is proportional to $|\mathscr{J}N(E_F)^2|$, is roughly one order of magnitude smaller than that measured in binary superconductors like Eu in $LaAl_2$. As seen from Table 6.1, the calculated values of $N(E_F)$ at the Eu or Gd sites in the sulfides are 1.3 and 1.5 states/Ry-atom whereas at the La site in $LaAl_2$ [6.44], the result is 17 states/Ry-atom and for Eu metal it is ~16 states/Ry-atom. Thus, it is clear that the reduction $|\mathscr{J}N(E_F)|^2$ observed by DUNLAP et al. [6.11] is brought about by an order of magnitude reduction of $N(E_F)$. Similar small values of $|\mathscr{J}N(E_F)|$ have been obtained from analyses of the systematic variation of T_c with RE for the two compound series $RE_xMo_6S_8$ and $RE_xMo_6Se_8$ [6.39,40], and from electron paramagnetic resonance of Gd in $Gd_xMo_6S_8$ [6.41] and $Gd_xMo_6Se_8$ [6.42].

Our $N(E_F)$ results of Table 6.1 and Figs.6.3-6 also permit a qualitative understanding of the weak suppression of the superconducting transition temperature T_c upon addition of rare-earth impurities in rather high concentrations to $SnMo_6S_8$. This is contrary to all other observations that the exchange interaction between the paramagnetic impurity spins and the conduction electron spins in a superconductor causes spin-flip scattering which acts as a pairbreaking mechanism and rapidly diminishes T_c according to the Abrikosov-Gor'kov relation for low concentrations:

$$\Delta T_c = -\frac{\pi^2}{2k}\mathcal{J}^2 N(E_F)S(S+1)\Delta x \quad , \tag{6.11}$$

where Δx is the impurity concentration. Apparently $N(E_F)$ at the RE site is so low and the RKKY interaction so weak that T_c is hardly suppressed even for high concentrations. Indeed, even the concentrated $REMo_6S_8$ (and Se_8) compounds are superconducting with $1 < T_c < 2$ K, despite their full and large magnetic moments.

d) *Magnetic Response of Conduction Electrons: Stoner Factor*

From the paramagnetic results for the Eu and Gd compounds, it is possible to study the relative magnetic response of the spin-degenerate Mo bands in terms of a Stoner-like criterion for magnetism. Here we follow the same procedure to calculate a Stoner criterion for compounds as was derived earlier for the C15 compounds [6.45]. The condition for a spontaneous spin-splitting of a paramagnetic band structure is fulfilled if the gain in exchange-correlation energy $\Delta\xi$ is larger than the loss of kinetic energy ΔT due to the splitting. This leads to the following condition:

$$\bar{S} = \sum_t \bar{S}_t = \frac{1}{N(E_F)}\sum_t \frac{1}{12}\int_0^{R_{WS}} r^2 \mu_{KS}(\rho)\delta_{GL}(\rho)\,\frac{\left[\sum_\ell N_{t\ell}(E_F)R^2_{t\ell}(E_F,r)\right]^2}{\rho(r)}\,dr \geq 1 \quad , \tag{6.12}$$

where (apart from earlier notation) t,ℓ are site and ordinary quantum number indices, $N(E_F)$ is DOS at the Fermi energy and $R_{t\ell}(E_F,r)$ is the t,ℓ radial wave function at energy E_F. Usually, the t summation covers all sites in the unit cell because we assume that a band state is split uniformly over the whole structure. However, since in this study we will look for the possibilities of a spin splitting of the Mo bands only (which are the superconducting bands), we restrict the t-summation to be over the Mo sites only and $N(E_F)$ is then the total Mo-DOS. This gives an upper limit for $\bar{S}$, because it costs a lot of kinetic energy to split the other bands whose local DOS is small, thereby reducing $\bar{S}$.

This procedure yields only the Stoner-like response of the Mo conduction electrons without regard to the dominant RKKY interaction so vital for magnetic ordering in rare-earth compounds. Clearly, to calculate a realistic Stoner-like criterion for local moment systems, we need to develop an approach which goes beyond the local density formalism with its emphasis on such quantities as $N(E_F)$, since in most cases the rare-earth 4f levels are located below E_F. Thus, the parameter $\bar{S}$ gives only a relative indication of the likelihood for magnetism to occur in the different compounds. The Mo contribution to $\bar{S}$ is determined to be ~ 0.62 for $EuMo_6S_8$ and ~ 0.46 for $GdMo_6S_8$, indicating that all else being equal (a rather poor assumption), Mo is a more effective carrier of magnetism in the Eu compound. We shall return to this assessment later when we present the results of our spin-polarized (ferromagnetic) studies.

e) *Superconductivity: Calculation of Electron-Phonon Coupling Parameter* λ

We have seen from Figs.6.3-6 that the considerable structure in the total DOS, particularly around E_F, arises from the Mo 4d electrons. Usually for Mo compounds, the d bands are occupied up to the "gap" region where the DOS is low, but in the Chevrel compounds a large charge transfer from Mo to S (about 1 electron per Mo atom) was found to occur and E_F falls in a high DOS region below the "gap". A partial DOS calculation, c.f. Table 6.1, shows that there is a high 4d DOS at E_F which is favorable for superconductivity. Indeed, the DOS per transition metal atom at E_F in the divalent systems is about 75% of that for the best superconducting A15 compounds [6.29]. FRADIN et al. [6.36] showed that the electron-phonon coupling parameter λ is proportional to $N(E_F)$ in a number of Chevrel compounds.

Estimates of the electron-phonon coupling parameter λ and superconducting transition temperature T_c may be made using our band results, the crude rigid muffin-tin approximation and strong-coupling theory. The McMillan equation [6.18] for strongly coupled superconductors expresses the superconducting transition temperature as

$$T_c = \frac{\langle\omega^2\rangle^{1/2}}{1.20} \exp\left(-\frac{1.04(1+\lambda+\mu_{sp})}{\lambda - (\mu^* + \mu_{sp})(1 + 0.62\lambda)}\right) , \tag{6.13}$$

where $\langle\omega^2\rangle^{1/2}$ is the averaged phonon frequency and λ, μ, and μ_{sp} are coupling constants for electron-phonon, electron-electron, and electron-spin interactions, respectively. The electron-phonon coupling parameter can be separated approximately into purely electronic (numerator) and purely phononic contributions (denominator):

$$\lambda = N(E_F)\langle I^2\rangle/M\langle\omega^2\rangle = \eta/M\langle\omega^2\rangle \tag{6.14}$$

where M is the atomic mass. In our work, η is calculated by the rigid-ion formula of GASPARI and GYORFFY [6.17]:

$$N(E_F)\langle I^2\rangle = \frac{E_F}{N(E_F)\pi^2}\left[\sum_\ell 2(\ell + 1)\sin^2(\eta_{\ell+1} - \eta_\ell)\,\frac{N_\ell(E_F)N_{\ell+1}(E_F)}{N_\ell^0(E_F)N_\ell^0(E_F)}\right] , \tag{6.15}$$

where η_ℓ is the phase shift of the ℓth wave and $N_\ell^0(E_F)$ is the DOS for a single WS sphere. [Here (6.15) is used for quantities in WS spheres rather than in nonoverlapping MT spheres as in the original Gaspari-Gyorffy theory]. For μ^* we have used the empirical formula given by BENNEMANN and GARLAND [6.46] which relates μ^* to the calculated or "bare" total DOS $N(E_F)$ in units of $(\text{eV atom})^{-1}$:

$$\mu^* = 0.26N(E_F)/[1 + N(E_F)] \quad . \tag{6.16}$$

In the absence of accurate estimates of μ_p, we have assumed this term to be zero.

As expected from the above, the dominant contribution to η comes from the Mo-d band with only small contributions from the M and chalcogen sites. The trend in η also follows the observed trend in T_c as expected if the phonon spectra in the different compounds are similar [6.4]. Our calculated η values are considerably larger than those of ANDERSEN et al. [6.9]. Estimates of λ given in Table 6.1, using the site decomposed $\langle\omega^2\rangle$ phonon data of BADER and SINHA [6.47] for the Sn compounds, give the correct magnitude for the T_c's but somewhat too low values for the electronic specific heat. About 90% of the total contribution to λ comes from the Mo sites indicating the important role played by the Mo 4d electrons in producing superconductivity. The much reduced T_c values for the trivalent rare-earths compared with those for the Sn (and Pb) and the divalent Yb compounds (T_c in $YbMo_6S_8$ = 9.1 K) is seen as a natural consequence of the greatly reduced $N(E_F)$ both total and Mo-4d, seen for $GdMo_6S_8$ in Table 6.1.

Thus there is satisfactory agreement for the T_c values, but still too low values for the specific heat; this indicates that the calculated λ on DOS values at E_F are too low for $SnMo_6S_8$ and $LaMo_6Se_8$. Clearly, a first principles calculation of the electron-phonon coupling associated with the various types of modes is an urgent necessity—as emphasized in [Ref.6.10, Chaps.7,8]. If, however, one assumes the validity of the theoretical estimates for η, the numerator in (6.14), then good agreement with experiment will be brought about by any mechanism which softens the frequencies and hence decreases the denominator, since this will increase λ and γ. Accurate theoretical determinations of the phonon spectra of these compounds is sorely lacking. As described in [Ref.6.10, Chap.8], recent isotope effect measurements and tunneling spectroscopy experiments have given important information about the strength of the electron-phonon coupling—and about the question as to wheter or not there are phonon modes which couple particularly strongly to the electrons in these compounds. As emphasized there, "these data hint at a special coupling to low frequency phonons" and lead to a rather stringent characterization of the important phonon modes, namely, they "should be soft, be coupled strongly to deformations of the Mo octahedra and should be equally strongly affected by the Mo and Se masses". Proof of the existence of these 'soft internal modes' is not available at present.

We have mentioned earlier the strong dependence of T_c on the rhombohedral angle α. This is another clear indication that distortions of the cluster may play an important role in determining the superconducting properties of the system and that the use of $\alpha = 90^\circ$ may severely affect the relevant band structure qualities we have determined.

6.2.3 Results for Ferromagnetic Structures

a) *Spin-Polarized Band Structure and Magnetization*

In the next stage of the calculations, the Eu and Gd-Mo_6S_8 sytems were treated in the local spin-density approximation. The self-consistent spin-polarized procedure can be viewed as a dynamic process in which the spin moments initially assumed on the RE site gradually penetrate the whole unit cell. In the paramagnetic calculations, the input potential is taken to be equal for both spin states so the resulting band structures for the two spins are identical. In the ferromagnetic calculations, the 7 core-like 4f electrons on the Eu or Gd were all assigned the same spin direction (here called majority or positive direction) and allowed to polarize the conduction electron states on that site. In the next iteration this polarization interacts with the Mo and S electrons and in turn polarizes them. We performed identical starting procedures for the two compounds and, at first, used very large potential mixing (26% of the new potential was mixed into the old potential but proved too large to allow stability for the continuation of the SC

Table 6.4. Magnetic moments after the first iteration and after self-consistency (in μ_B/atom)

	$\mu_B(1)$	$\mu_B(SC)$
$EuMo_6S_8$		
Eu	+0.005	+0.03
Mo	-0.10	-0.05
S(6)	-0.03	-0.02
S(2)	-0.05	-0.04
$GdMo_6S_8$		
Gd	+0.013	+0.08
Mo	-0.005	-0.01
S(6)	+0.005	+0.01
S(2)	+0.006	+0.03

procedure) in order to see the effect of the RE spin interaction with the Mo-S cluster. In the second iteration, a negative (minority) spin moment shows up at the MoS cluster in the Eu compound. In fact, it is somewhat larger than the final self-consistent results and amounts to $-0.1\mu_B$ per Mo atom, $-0.0.3\mu_B$ on each of the S(6) atoms and $-0.05\mu_B$ on each of the S(2) atoms on the (111)-axis closest to the Eu atoms. The Eu atom itself has a small positive conduction electron polarization of $+0.005\mu_B$. As shown in Table 6.4, the corresponding values for the Gd compound are $-0.005\mu_B$ (Mo), $+0.005\mu_B$ for S(6), $+0.006\mu_B$ for S(2) and $+0.013\mu_B$ for the Gd conduction bands. The differences between the two compounds show that Eu is more active in giving spin density (note negative or antiferromagnetic spin density!) to the MoS cluster than is Gd. In this context two things need to be understood: first, why is it dominantly negative spin which is carried to the MoS cluster and second, how to explain the differences between Eu and Gd. Since the charge transfer away from the RE site is very strong, the remaining occupied conduction electron wave functions are delocalized so that the radial RE charge density is highest at the WS boundary. The 4f spin density is only peaked far inside at about 0.8 a.u. and through the exchange interaction, acts to pull the valence majority spin density in towards that region, thus leaving negative valence spin at the outer part of the WS cell. Therefore it can be expected that tail functions tied onto the RE spheres will be negatively polarized. In the case of Gd with its extra valence electron, the presence of the occupied d-band makes the charge density a bit more localized than in Eu, and the negative tailing arguments are less effective in this case.

So far in the SC procedure the MoS system has not given any response to the polarized RE tails. Because of the weak polarization after the first iteration in the Gd compound, a magnetic field parallel to the minority spin direction was applied to the cluster in the next iteration in order to create a spin-polarization environment as in the Eu compound. (The "field" consists of making a rigid shift of the two spin potentials by 1 mRy, corresponding to a magnetic field of the order of several MGauss. This field is removed in the subsequent iterations). The Eu and Gd compounds are carried to self-consistency by continuing from the first iteration with a potential mixing of about 15%.

The result of the final self-consistent magnetization (in μ_B/atom), as given in Table 6.4, shows clearly that an antiferromagnetic magnetic moment resides on the MoS cluster in the Eu compound, while in the Gd compound the "forced" antiferromagnetic starting configuration fades away. The convergence is not as complete for the Gd as it is for the Eu compound so the small magnetization numbers for the Gd compound may be in the noise of the calculation. (Our experience is that it requires very many iterations to completely get rid of a *forced* magnetic initial state; for instance in Cu, starting from $\sim 0.5\mu_B$, it is possible to have $0.005\mu_B$ remaining after several iterations). The total conduction electron magnetization is very small compared with that in ordinary magnetic systems and the exchange splitting for the Mo-d band in the Eu compound is only 1-2 mRy.

Consider the separate contributions to the magnetization in μ_B per atom for $EuMo_6S_8$ and $GdMo_6S_8$ shown in Table 6.4. In both cases we see a small positive conduction-electron contribution to the moment on the rare-earth site, with this contribution being somewhat larger in the case of the Gd component. The table also clearly shows that there is an antiferromagnetically aligned magnetic moment which resides on the MoS cluster in the Eu compound and essentially a zero contribution in the case of the Gd compound. This greater (negative) polarization of the Mo conduction electrons in Eu than in Gd is consistent with our Stoner criteria calculations mentioned above for the paramagnetic compounds. Note also the small (but significant) negative moments which reside on the sulfur sites with the larger negative contribution being on the two sulfur atoms that lie closest to the Eu atoms along the (111)-axis.

b) *Increase of H_{c2} with Doping of Magnetic Impurities: Jaccarino-Peter Mechanism*

As stated earlier, one of the most dramatic and unusual effects observed in the Chevrel compounds is the large *increase* in H_{c2} found when magnetic RE impurities (Eu) are doped into a superconductor ($SnMo_6S_8$). For Eu concentrations up to $\sim$50%, H_{c2} is observed to increase rapidly to $\sim$400 kG at 50% Eu from its value of $\sim$275 kG for 0% Eu (both determined at 2 K in these $T_c \simeq$ 10 K superconductors). These results are in sharp conflict with usual theoretical ideas and all other experimental observations on the pairbreaking effects of magnetic impurities on H_{c2}. FISCHER et al. [6.13] took these observations as evidence for the validity of the Jaccarino-Peter mechanism whereby the conduction electrons are negatively polarized by exchange interactions with the local moments and this results in a negative exchange field which partially compensates the externally applied field (for a further discussion, also see Chap.3 of this volume).

The first direct determination of the conduction electron spin polarization were made by FRADIN et al. [6.12] in $Sn_{0.5}Eu_{0.5}Mo_6S_8$ by means of bulk magnetization, ^{95}Mo NMR and ^{151}Eu Mössbauer effect measurements. Among other results they found that (i) the enhancement of H_{c2} by approximately 100 kG arises from a negative (assumed to be s band) polarization at the Mo site and (ii) a small positive polarization at the Eu site and (as mentioned above) a weak Korringa relaxation which is consistent with the weak depression of T_c.

The theoretical magnetization results for $EuMo_6S_8$ given in Table 6.4 clearly demonstrate the negative magnetization on the Mo (and S) sites and so firmly establish the validity of the JACCARINO-PETER [6.15] mechanism invoked by FISCHER et al. [6.13] and FRADIN et al. [6.12] as being responsible for the high upper critical fields in this system. This negative magnetization agrees with the sign of the results of FRADIN et al. [6.11,12] and their conclusion that this magnetic contribution can destroy the superconductivity by its pair-breaking effect on the superconducting Mo d electrons. This result is also consistent with $\rho(T)$ anomalies and H_{c2} enhancement in $La_{1.2-x}Eu_xMo_6S_8$ [6.48].

c) *Hyperfine Fields*

In their studies of the spin polarization in the high H_{c2} compound $Sn_{0.5}Eu_{0.5}Mo_6S_8$, FRADIN et al. [6.11,12] used Mössbauer and NMR methods to study the hyperfine field and Knight shift at the Eu and Mo nuclei, respectively. The Eu hyperfine field was measured to be -285 ± 15 kGauss. FRADIN et al. took as the core polarization contribution the measured free ion Eu^{2+} field (-340 kGauss) and so estimated the remaining total valence contribution to be +55(±15) kGauss. These authors also found an isomer shift for Eu which indicated an extremely low s conduction electron density. Their measured value is that obtained typically for Eu^{2+} in ionic compounds – a result, as they emphasized, which is anomalous for conducting systems. From our spin-polarized ferromagnetic calculations we may determine the Fermi contact contribution to the hyperfine field at the Eu nucleus in $EuMo_6S_8$. The shell by shell contribution of the core s electrons to the spin density (in a.u.) at the nucleus is given in Table 6.5. (The field is proportional to the spin density from the s electrons only because the $p_{1/2}$ core electrons make only a negligible contribution at the nucleus). While the 1s, 2s and 3s densities are negative, the 4s and 5s are positive and of almost equal magnitude so that the resultant field is -345 kGauss. It is remarkable that this first solid state energy band calculation of a rare-earth hyperfine field gives excellent agreement for the core polarization Fermi contact term with that measured for an ionic compound [this result also confirms indirectly the experimental and theoretical result which shows that Eu is so magnetically isolated that it has a divalent ion (Eu^{2+}) isomer shift]. Our calculated conduction electron spin density at the nucleus (also given in Table 6.5) yields a contact contribution of +25 kGauss. Subtracting the total contact contribution of core and valence electrons from the total measured value gives +35 kGauss for the remaining contributions from dipolar and unquenched angular momentum hyperfine terms. This value appears to be reasonable in view of results obtained on other systems.

Table 6.5. Theoretical Fermi contact hyperfine fields (in kG) obtained for Eu and Mo in $EuMo_6S_8$. The experimental total value for Eu is -285 ± 15 kG and for Mo the Knight shift is negative

Eu	Mo
1s - 120	
2s - 555	
3s - 6250	
4s + 270	
53 + 6305	
Total core = -345	Total core = -11
Valence 6s = + 25	Valence = + 3
Total = -320 kG	Total = - 8

For Mo, FRADIN et al. [6.11,12] found a negative Knight shift, much of which is not quenched at the superconducting transition temperature T_c. From the proportionality of the magnetization and the Knight shift, FRADIN et al. concluded that the Knight shift arises from a negative (5s) conduction electron polarization at the Mo site due to a rare-earth conduction-electron coupling of the form $-2J_{sf}\bar{S}_{4f}\bar{s}_{5s}$. From our spin-polarized results we find that the core electrons give a Fermi contact field of -11 kGauss (predominantly from the 4s shell, as in spin-polarized free atom results) with the conduction electrons giving a smaller contact field of +3 kGauss. Thus the dominant contribution to the negative Knight shift (hyperfine field in our case) arises from exchange polarization of the core electrons by the 4d conduction electrons (polarized in turn by the 4f magnetic moment).

6.3 Ternary Rhodium Borides

In the ternary boride MRh_4B_4 compounds (with M: a RE metal), MATTHIAS et al. [6.49] found that either ferromagnetism (M = Gd, Tb, Dy, and Ho) or superconductivity (M = Y, Nd, Sm, Er, Tm, and Lu) existed at temperatures $T < 12$ K. This behavior is in sharp contrast to that observed in the ternary Chevrel phase compounds $M_xMo_6S_8$ and $M_xMo_6Se_8$ where each RE compound was found to be superconducting except for Ce and Eu. The sharp break in properties between the M = Ho and Er boride compounds is of especially great interest—with magnetism for Ho ($T_m = 6.6$ K) and superconductivity for Er ($T_c = 8.7$ K), even though the effective magnetic moments of both RE ions ($10.6\mu_B$ for Ho and $9.6\mu_B$ for Er) differ by only a small amount (~ 10%). Surprisingly, $ErRh_4B_4$, which becomes superconducting at 8.7 K, was found to become magnetic at T = 0.9 K with the return of the system to a normal conducting state ("re-entrant magnetism" in a superconductor) [6.50].

This section discusses some of our studies [6.1-3] of the origin of these phenomena using results of *ab initio* self-consistent band structure calculations carried out on three of these ternary borides, notably M = Y (which has no 4f electrons), Ho and Er. We are particularly concerned with the competing mechanisms for producing magnetism and superconductivity in the RE systems. However, only a qualitative understanding is at present possible because our band calculations are for the paramagnetic states.

6.3.1 Energy Band Calculations

As described in Sect.6.2, the band calculations [6.1,2] for the full 18-atom/unit-cell structures were performed self-consistently using the LMTO method and other related methods on the structure shown in Fig.6.7. The potential contained the Hedin-Lundqvist treatment for the exchange and correlation and for the heavier elements (Ho and Er), the relativistic Dirac equation without spin-orbit splitting was used. The potential was defined to be spherically symmetric around each site out to the overlapping-sphere radii. In the derivation of the Madelung contribution to the potential, however, a nonoverlapping-sphere model was used. The energies were calculated at 18 $\vec{k}$ points in the irreducible wedge of the BZ and ℓ_{max} was 1 for boron atoms and 2 for the other atoms, resulting in 122×122 eigenvalue matrices. The ℓ convergence was improved by setting the maximum ℓ in the internal summations equal to $\ell_{max} + 1$. The self-consistency convergence is estimated to be better than 3 mRy for energy states below E_F. The matrix elements included the corrections to the overlapping spheres. At first the relative sizes of the atoms were deduced entirely from given atomic-radii data, but preliminary canonical calculations showed a poor potential at the M sites; the most repulsive potential was reached about 15% inside the spheres. Therefore, the size of the M volumes was decreased and that of the Rh and B volumes was increased.

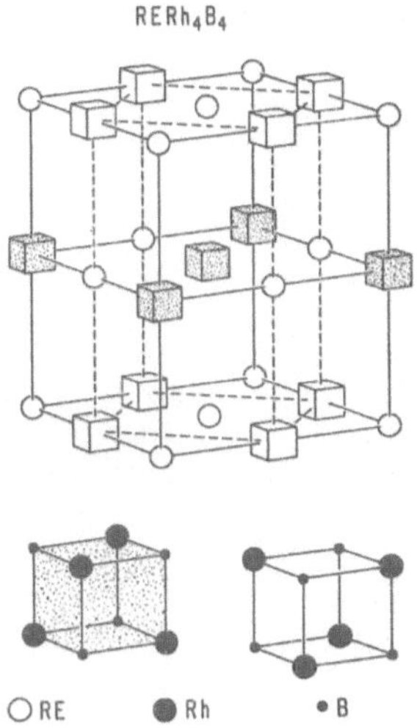

Fig.6.7. The unit cell for ternary rhodium borides (from [6.53])

6.3.2 Energy Band Results

Since the basic energy band structures are complex (51 bands are occupied up to a Fermi energy) and are not particularly informative for the phenomena considered, we do not give these results here. Instead, we focus on the more physical quantities such as total and separate ℓ-decomposed DOS by atom species and the wave function character involved in the basic interactions leading to long-range ordering.

The total DOS for one spin for $HoRh_4B_4$ and the ℓ-decomposed contributions to the DOS for the two Ho and eight Rh atoms in a unit cell are shown in Fig.6.8. The other compounds show very similar behavior as is seen from Table 6.6 which gives for each of these compounds a partial (by atom type and ℓ-value) and total DOS at E_F in states/Ry-cell. The DOS for the B atoms shows a not insubstantial p contribution of ~20 states/Ryd-cell-spin at E_F but a negligible s contribution. The relevance of these results to the ^{11}B NMR studies of FRADIN et al. is discussed in Chap.7. This figure provides us with a qualitative understanding of the properties of the alloys. We focus on the fact that E_F falls at a peak in the DOS and

Table 6.6. Total and ℓ-decomposed by atom density of states (in states/Ryd-cell)

	M			Rh			B		Total
	s	p	d	s	p	d	s	p	
Er	0.57	3.19	15.40	3.35	16.8	92.3	2.38	21.1	155
Ho	0.61	2.88	14.10	3.40	16.1	72.7	2.01	20.0	131
Y	0.5	3.36	17.45	3.91	18.2	79.8	2.36	21.7	147

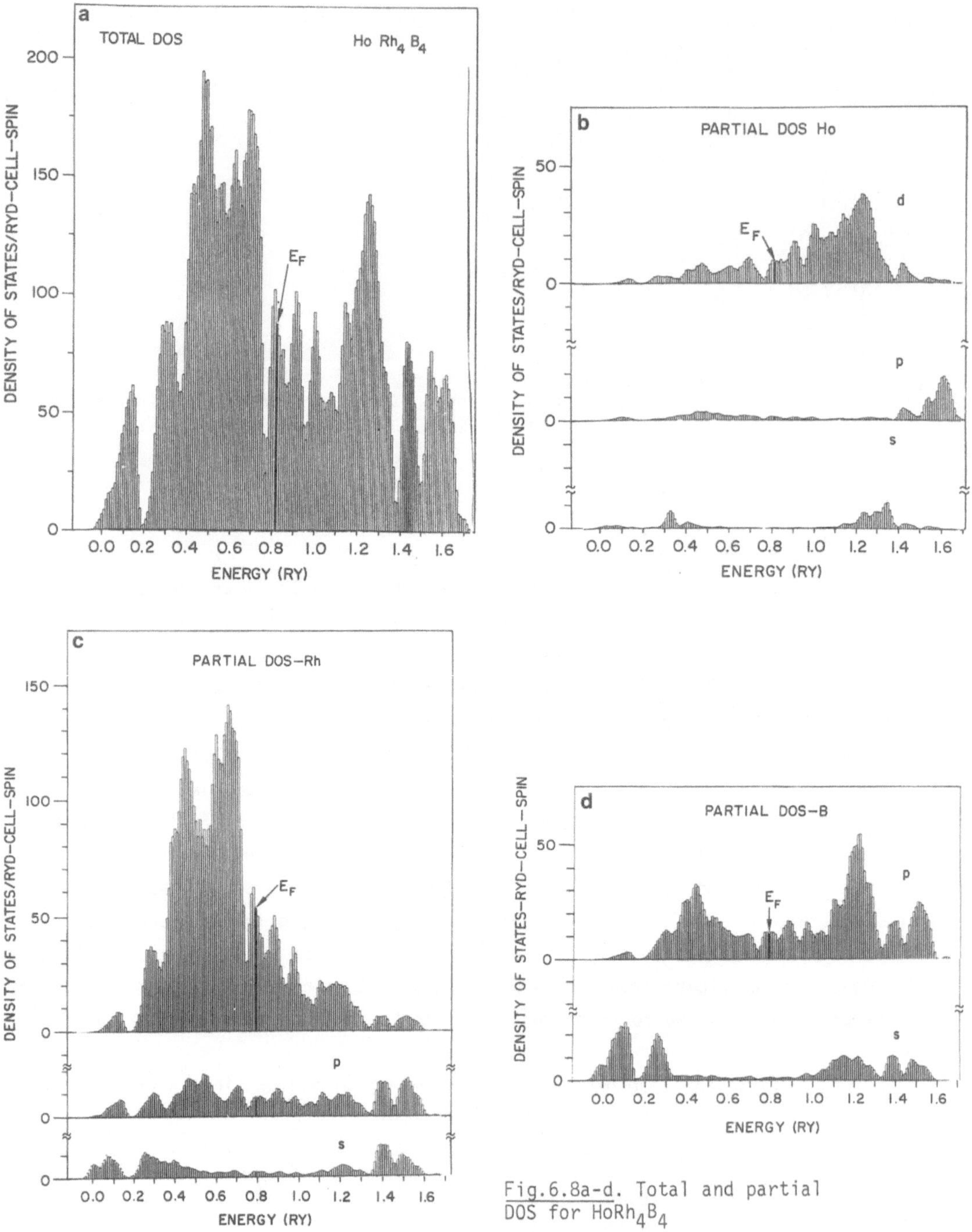

Fig.6.8a-d. Total and partial DOS for $HoRh_4B_4$

that this peak structure arises from the structure in the Rh 4d contribution. (Since E_F did not occur at a peak in the DOS in the non-self-consistent calculations, this indicates the importance of the charge transfer effects which are taken into account in the self-consistent calculations).

6.3.3 Magnetic Ordering

A crude estimate of the possible occurence of magnetism in these compounds may be obtained by considering the RKKY interaction [cf. (6.9)] for T_m. For the heavy RE metals, the exchange interaction $\mathcal{J}_{f-d}$ is between the localized 4f electrons and the dominant 5d conduction band electrons (and not the s-p electrons). In these borides, the lack of any sizeable s or p DOS makes the f-d interaction even more dominant. Since the Ho 5d partial DOS at E_F in $HoRh_4B_4$, ~12.9 states/Ry-cell spin, is ~1/3 the value of $N(E_F)$ in Ho metal, T_m in the ternary borides for those RE with a sizeable spin S (i.e., Gd, Tb, and Dy) will certainly order magnetically, provided that the exchange interaction $\mathcal{J}_{f-d}$ is not greatly different in the ternary than it is in the metal [cf. (6.9)]. Figure 6.9 shows the 4f radial charge density in $ErRh_4B_4$ and $HoRh_4B_4$ and the $\ell = 2$ (5d) part of the conduction band radial density evaluated at $E = E_F$. Comparing the results of Fig.6.9 with those obtained by HARMON and FREEMAN [6.43] for Gd metal, one sees a strong similarity between the respective 4f densities indicating that the 4f electrons are highly localized and that the 5d wave functions at E_F show a good deal of bonding and hence differ substantially from the free atom results. Through the exchange interaction between the localized 4f and RE 5d electrons (hybridized with the Rh 4d's), the magnetic

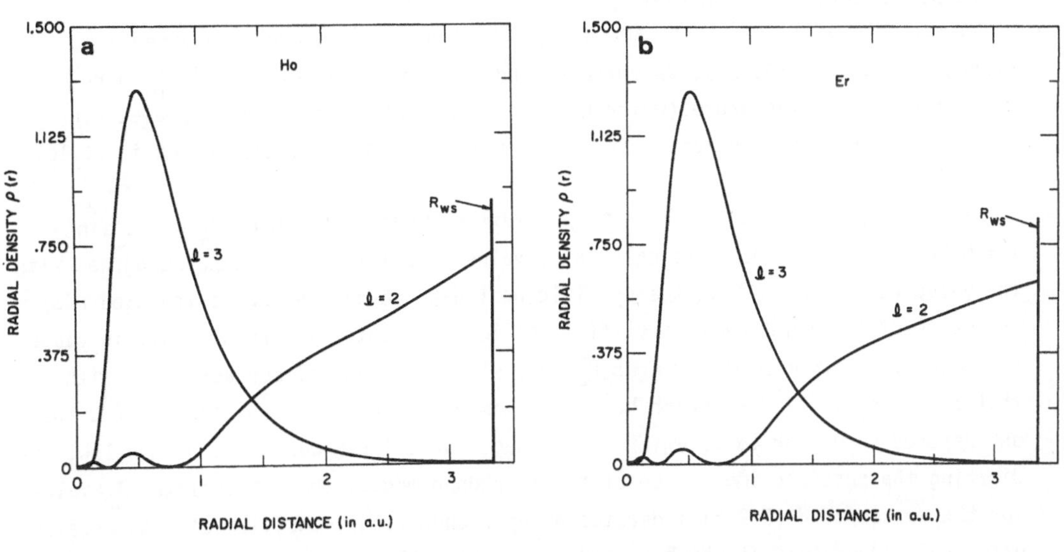

Fig.9a,b. Radial d and f charge densities for Er and Ho in the corresponding Rh_4B_4 compound

(RKKY)-type coupling between the 4f local moments may lead to a still sizeable magnetic ordering at a temperature T_m. A reduction in $\mathcal{J}_{f-d}$ in the ternary to approximately 1/3 of the value in the metals would, therefore, according to (6.9) and the large reduction seen for $N(E_F)$, lead to T_m values of the order of the values actually observed.

6.3.4 Superconductivity

If one examines the DOS shown in Fig.6.7, one observes that the large transition metal DOS from the Rh contribution indicates that a crucial requirement for the occurrence of superconductivity is satisfied. Further, a crude estimate of the electron-phonon coupling parameter λ, obtained using our calculated bare total DOS and the measured electronic specific heat in $ErRh_4B_4$, yields $\lambda = 0.96$ and indicates that in strong-coupling theory a large T_c value would result. As in Sect. 6.3 for the Chevrels, a crude estimate of λ may be made using the rigid muffin-tin approximation of Gyorffy-Gaspari [cf. (6.14,15)] and the quantities derived from our band structure calculations. From this estimate of λ one may then use McMillan's strong coupling relation to obtain an estimate of T_c [cf. (6.13)].

It needs to be re-emphasized that no rigorous derivation of an equation like (6.13) has been given for a ternary compound such as MRh_4B_4. Thus, any estimate must of necessity be crude—especially since phonon-frequency data are not available.

We have calculated λ for the Er and Ho compounds using the Gyorffy-Gaspari rigid muffin-tin formula (6.15) for $\eta = N(E_F)\langle I^2\rangle$ evaluated at the WS radii instead of at the muffin-tin radii.

An estimate of T_c was then made using McMillan's strong-coupling extension of BCS theory [cf. (6.13)]. We further crudely approximate $\langle\omega^2\rangle$ as $\frac{1}{2}\,\theta_D^2$. Here θ_D is the Debye temperature (which we take to be 200 K) and μ^* is the electron-electron interaction constant which we evaluate from the relation given in (6.16) above.

For $ErRh_4B_4$ we find $\lambda = 0.89$ and $T_c = 10.1$ K, whereas for $HoRh_4B_4$ we obtain $\lambda = 0.58$ and $T_c = 3.9$ K. This calculated value of λ for the Er compound agrees with the value we estimated from the specific heat data [6.53] and our calculated DOS; the value of T_c, which omits any effect of local moments, is, remarkably, in close agreement with experiment for $ErRh_4B_4$ (8.7 K) and in even closer agreement with that for $LuRh_4B_4$ (11.8 K) which has no 4f moment. However, it should be noted that our assumed value for $\langle\omega^2\rangle$, which is the same for all compounds, is rather low, indicating the possible importance of "soft" phonon modes. The extrapolated T_c value for the Ho compound (5.9 K) indicates a superconducting transition at a temperature below the temperature at which it orders magnetically, and hence superconductivity will not occur. For YRh_4B_4, we calculated λ to be 0.69 and $T_c = 5.6$ K which is considerably lower than the measured value of 11.3 K. The discrepancy may be due to the use of atomic parameters [6.49], which later were shown to be incorrect by YVON and GRÜTTNER [6.54].

Note however, that these estimates take no account of the magnetic or spin-fluctuation depression on T_c. Our λ values are thus a result derived only from electron-phonon coupling for a typical MRh_4B_4 compund.

6.3.5 Effect of Local Moments on T_c

Let us now include the 4f local moments in our estimates of T_c. Since the 4f electrons are highly localized, they may be considered as being well separated from the Rh 4d "superconducting" carriers and their local moments to lower T_c in the way expected of dilute impurities. In the view of ABRIKOSOV-GOR'KOV [6.13], the change in superconducting temperature expected in the Er and Tm alloys, ΔT_c, due to dilute impurities [cf. (6.11)] should be proportional to the de Gennes factor $(g - 1)^2 J(J + 1)$ where g is the gyromagnetic ratio and J is the total angular momentum of the 4f shell of the trivalent RE ion. Referring to the data of MATTHIAS et al. [6.49] and using the Lu compound's experimental T_c as a base, one sees indeed that the measured ΔT_c for Er and Tm relative to Lu, plotted in Fig.6.10, are proportional to $(g - 1)^2 J(J + 1)$. Further, this simple estimate—assuming all other factors such as $N(E_F)$ and $\mathcal{J}$ being equal—shows that in the absence of magnetic ordering, the 4f local moments in the Ho alloy would result in a ΔT_c of ~5.5 K and a T_c of ~6 K (compared with an observed [6.49] magnetic-ordering temperature, T_m ~6.6 K). Thus, $HoRh_4B_4$, unlike $ErRh_4B_4$, is not expected to be a superconductor—in agreement with the estimate of T_c based on the calculation for λ given above. Strong confirmation to this view of weak interaction between the Er 4f and Rh 4d electrons and hence the model of dilute impurities is given by the Mössbauer experiments of SHENOY et al. [6.51]. They find from relaxation measurements that the spin-flip scattering in $ErRh_4B_4$ is weaker than in the superconducting ternaries $ErMo_6(S,Se)_8$ (which exhibit similar properties and are similarly describable). The importance of crystal field effects is discussed in Chaps.4,7,8.

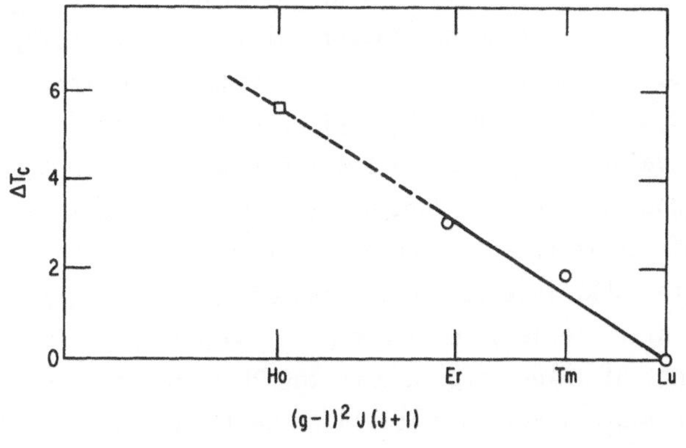

Fig.6.10. Change in T_c due to magnetic impurities in rhodium borides (from [6.49]) extrapolated to $HoRh_4B_4$

6.3.6 Re-Entrant Magnetism in $ErRh_4B_4$

The interplay of superconductivity and magnetism, which is the dominant characteristic feature of the ternary compounds, results in a number of unusual magnetic and superconducting properties. As illucidated by the band calculations, the unique electronic structure of the ternaries (but more so in the Chevrels than in the rhodium borides) which makes possible the existence of superconductivity in compounds made up of RE constituents containing large magnetic moments, is their relative magnetic isolation, i.e., a weak coupling of the conduction electrons to these localized magnetic moments. As we have seen, this makes for weak exchange interactions, lowered magnetic transition temperatures and small superconducting pairbreaking. One of the most exotic phenomena displayed in the ternaries by this interaction of magnetism and superconductivity is that of re-entrant magnetism in the superconductors $ErRh_4B_4$ [6.16] and $HoMo_6S_8$ [6.50]. Early on, we have used the results of our self-consistent band studies [6.1-3] to provide a *qualitative* discussion of possible consequences of this interaction of superconductivity and magnetism including the question of re-entrant magnetism. Recently, sophisticated theoretical studies, which account for both the superconducting and magnetic interactions, have provided a detailed description of these phenomena [6.55-58]. These developments are described in Chap.9. Here we discuss qualitatively some of the implications of these early band results.

We have seen that for the compounds from Gd to Ho, the large effective spin moment leads to a magnetic interaction which dominates the superconducting interaction and $T_m > T_c$. At the other end, in the absence of 4f local moments, Lu, Tm, and Er are predicted to be superconductors; the effect of the local moment is to reduce this T_c value. More specifically, in the Er and Tm compounds—with their smaller spin moments, reduced 4f-5d exchange integrals (caused by the lanthanide contraction) and somewhat lower 5d DOS than in Ho—the "effective" magnetic ordering temperature T_m^* these systems would have in the absence of the onset of the superconducting state is reduced and so $T_c > T_m$. For $ErRh_4B_4$, the actual magnetic-ordering temperature $T_m = 0.9$ K is below the superconducting transition temperature $T_c = 8.9$ K. It is also below reasonable values (3-4 K) estimated from the T_m values for the Dy and Ho compounds using a de Gennes factor relation as in (6.10). As we have indicated previously [6.1-3], this may be taken as evidence for the fact that the occurrence of superconductivity in $ErRh_4B_4$ affects its magnetic properties. Briefly, the *observed* value of T_m in Er is smaller than this effective ordering temperature T_m^*, because once the superconducting state has been achieved the conduction electron susceptibility is greatly reduced—at least for the Rh electrons. Hence, for this compound, the RKKY interaction is less effective compared to that in Ho (and earlier elements in the series); dipole-dipole coupling may also contribute to the magnetic ordering. At lower temperatures the RKKY interaction is still sufficiently strong to order magnetically the localized Er 4f moments to produce a sufficiently large exchange field to destroy the superconducting state.

This results in the low observed temperature, 0.9 K, at which the re-entrant magnetic state orders.

Some evidence for the depression of T_m by the onset of superconductivity is seen in the data of JOHNSTON et al. [6.52] who studied the boundaries between the normal paramagnetic, superconducting and normal magnetically-ordered phases in the $(Er_{1-x}Ho_x)Rh_4B_4$ pseudoternary system. Re-entrant behavior was found to persist to $x_{cr} \approx 0.89$ and only magnetic ordering was found for $x > x_{cr}$. Extrapolating the T_c vs x data for $0.8 \leq x \leq x_{cr}$, they observed a downward curvature of the T_c vs x phase boundary. This sort of behavior is expected from the arguments given above concerning the reduction of the conduction electron susceptibility in the superconductor. Also of interest is that their extrapolation of the T_c vs x data for $x \lesssim 0.8$ indicates that if it were not for the occurrence of magnetic order near x = 1 at 6-7 K, $HoRh_4B_4$ would become superconducting at a somewhat lower temperature of 5-6 K.

6.3.7 Mixed State of Magnetism and Superconductivity in $ErRh_4B_4$

We have seen some evidence that the collective interactions leading to magnetism (RKKY) and superconductivity (BCS) are competing and not totally independent. This is confirmed by recent theoretical treatments [6.55-58]. As we have emphasized [6.1-3], the suppression of the effective magnetic-ordering temperature T_m^* a system like $ErRh_4B_4$ would have in the absence of superconductivity in a magnetically ordered state, suggests some possible new experiments to explore these effects. Very simply, any suppression of the superconducting order parameter would raise the measured magnetic ordering temperature T_m (since $T_m^* > T_m$). In particular, based on some physical arguments, we have proposed that the application of an external magnetic field H_{ext}, greater than the upper critical field H_{c2}, at a temperature $T_m < T < T_m^*$ in $ErRh_4B_4$, would produce a magnetically ordered state—with an effect on T_m well beyond that expected for the application of an external field on a ferromagnet [6.3]. Further, if one then reduces H_{ext} to zero, we predicted for the first time that it may be possible to either induce a metastable ferromagnetic state or to form a *mixed state* in which separate (normal *and*) ferromagnetic regions of the compound co-exist with regions of superconductivity. Such a state is possible because energy is required to break up the induced ferromagnetic state and to set up superconductivity over a coherence length.

The existence of a mixed state is now a subject of great experimental and theoretical interest [6.55-58]. The effects of competing ferromagnetic and superconducting order interactions for the case of weak exchange interactions are found, in mean field theory, to result in at least five possible phases [6.55-58]. In addition, interesting precursor effects are predicted in the superconducting state above the magnetic transition. Neutron scattering and resistivity measurements [6.59], most recently, on *single crystals* by SINHA et al. [6.60], clearly show that the transition from superconductivity to ferromagnetism proceeds through an intermediate mixed state which displays both superconductivity and long-range magnetic order. As seen in Fig.6.11, the mixed state first appears upon cooling at about 1.2 K, where both a ferromagnetic moment and a modulated moment with a period of 92 Å are seen in neutron scattering experiments. At 0.7 K, the crystal enters a purely ferromagnetic phase, with the modulated moment and superconductivity disappearing simultaneously. Hysteresis in the lower transition to pure ferromagnetism indicates a first-order phase transition, while the upper transition to pure superconductivity is second order. These experiments show conclusively that a mixed state similar to those which have been predicted theoretically actually exists, though some of its features differ significantly from the early theories.

One of the most striking features of the mixed state can be seen in magnetization measurements [6.60]. If the crystal is cooled to a temperature slightly above 0.7 K, it exhibits superconductivity in the presence of a small external field. Application of a large enough field quenches superconductivity and induces a ferromagnetic, rather than paramagnetic, state. Now when the external field is removed, the sample does not return to a superconducting state *but remains ferromagnetic* in agreement with our earlier prediction [6.3]. Such unusual behavior clearly shows the presence of the mixed state and demonstrates the kind of new and unexpected behavior that occurs when the tendency to superconductivity and ferromagnetism are so delicately balanced in a single material.

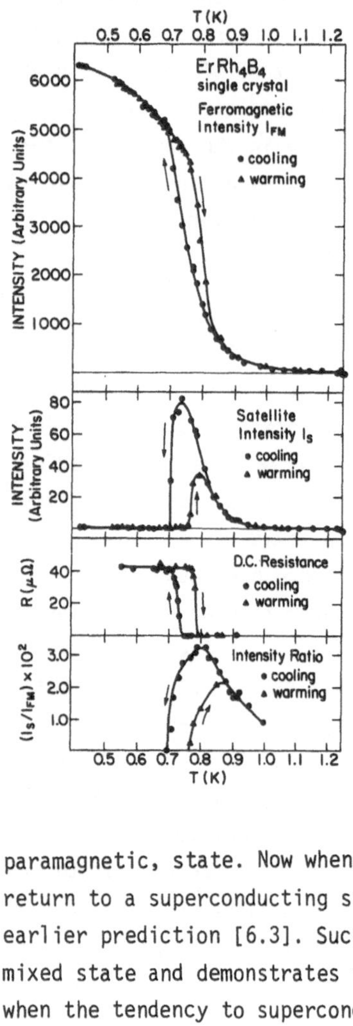

Fig.6.11. Temperature dependence of the ferromagnetic intensity from the (101) Bragg peak, the satellite intensity, the dc resistance and the ratio of the satellite to the ferromagnetic intensity for the (101) reciprocal lattice point (after [6.60])

References

6.1 T. Jarlborg, A.J. Freeman, T.J. Watson-Yang: Phys. Rev. Lett. *39*, 1032 (1977)
6.2 A.J. Freeman, T. Jarlborg, T.J. Watson-Yang: J. Mag. Magn. Matls. *7*, 296 (1978)
6.3 A.J. Freeman, T. Jarlborg: J. Appl. Phys. *50*, 1876 (1979)
6.4 T. Jarlborg, A.J. Freeman: Phys. Rev. Letts. *44*, 178 (1980); in *Superconductivity in d- and f-Band Metals*, ed. by H. Suhl, M.B. Maple (Academic, New York 1980) p.521; J. Mag. Magn. Matls. *15-18*, 1579 (1980)
6.5 A.J. Freeman, T. Jarlborg: In *Proc. Intern. Conf. on Ternary Superconductors*, ed. by G.K. Shenoy, B.D. Dunlap, F.Y. Fradin (North-Holland, Amsterdam 1981) p.59
6.6 T. Jarlborg, A.J. Freeman: J. Mag. Magn. Matls. *27*, 135 (1982)
6.7 D.W. Bullett: Phys. Rev. Lett. *39*, 664 (1977)
6.8 L.F. Matthiess, C.Y. Fong: Phys. Rev. B*15*, 1760 (1977)
6.9 O.K. Andersen, W. Klose, H. Nohl: Phys. Rev. B*17*, 1209 (1978) and Chap.6 of Vol.1
6.10 Ø. Fischer, M.B. Maple (eds.): *Superconductivity in Ternary Compounds I*, Topics in Current Physics, Vol.32 (Springer, Berlin, Heidelberg, New York 1982)
6.11 B.D. Dunlap, G.K. Shenoy, F.Y. Fradin, C.D. Barnet, C.W. Kimball: J. Mag. Magn. Matls. *13*, 319 (1979)
6.12 F.Y. Fradin, G.K. Shenoy, B.D. Dunlap, A.T. Aldred, C.W. Kimball: Phys. Rev. Lett. *38*, 719 (1977) and [Ref.6.10, Chap.7]
6.13 Ø. Fischer, M. Decroux, S. Roth, R. Chevrel, M. Sergent: J. Phys. C*8*, L474 (1975)
6.14 A.A. Abrikosov, L.P. Gor'kov: Sov. Phys.-JETP *12*, 1243 (1961)
6.15 V. Jaccarino, M. Peter: Phys. Rev. Lett. *9*, 290 (1962)
6.16 W.A. Fertig, D.C. Johnston, L.E. DeLong, R.W. McCallum, M.B. Maple, B.T. Matthias: Phys. Rev. Lett. *38*, 987 (1977)
6.17 G.D. Gaspari, B.L. Gyorffy: Phys. Rev. Lett. *28*, 801 (1972)
6.18 W.L. McMillan: Phys. Rev. *167*, 331 (1968)
6.19 P. Hohenberg, W. Kohn: Phys. Rev. *136*, 864 (1964); W. Kohn, L.J. Sham: Phys. Rev. A*140*, 1133 (1965)
6.20 J.C. Slater: Phys. Rev. *81*, 385 (1951)
6.21 P. Nozières, D. Pines: Phys. Rev. *14*, 442 (1958)
6.22 D. Pines: *Elementary Excitations in Solids* (Benjamin, New York 1963)
6.23 K. Singwi, A. Sjölander, D.M. Tosi, R.H. Land: Phys. Rev. B*1*, 1044 (1970)
6.24 L. Hedin, B.I. Lundqvist, S. Lundqvist: Solid State Commun. *9*, 537 (1971)
6.25 U. von Barth,L. Hedin: J. Phys. C*5*, 1629 (1972)
6.26 C.S. Wang, J. Callaway: Phys. Rev. B*9*, 4897 (1972); Phys. Rev. B*11*, 2417 (1975); Phys. Rev. B*15*, 298 (1977)
6.27 O.K. Andersen: Phys. Rev. B*15*, 3060 (1975)
6.28 D.D. Koelling, G. Arbman: J. Phys. F*5*, 2041 (1975)
6.29 T. Jarlborg, G. Arbman: J. Phys. F (Metal Phys.) *7*, 1635 (1977); J. Phys. F (Metal Phys.) *6*, 189 (1976); T. Jarlborg: J. Phys. F*9*, 283 (1979)
6.30 D.D. Koelling, B.N. Harmon: J. Phys. C (Sol. St. Phys.) *10*, 3107 (1975)
6.31 O. Gunnarson, B.I. Lundqvist: Phys. Rev. B*13*, 4274 (1976)
6.32 M. Marezio, P.D. Dernier, J.P. Remeika, E. Corenzwit, B.T. Matthias: Mat. Res. Bull. *8*, 657 (1973)
6.33 Ø. Fischer: Appl. Phys. *16*, 1 (1978)
6.34 Ø. Fischer, R. Odermatt, G. Bongi, H. Jones, R. Chevrel, M. Sergent: Phys. Lett. *45*A, 87 (1973)
6.35 M. Sergent, Ø. Fischer, M. Decroux, C. Perrin, R. Chevrel: J. Sol. State Chem. *22*, 87 (1977)
6.36 F.Y. Fradin, G.S. Knapp, S.D. Bader, G. Cinader, C.W. Kimball: In *Proc. of Second Rochester Conf. on Superconductivity in d- and f-Band Metals*, ed. by D.H. Douglass (Plenum, New York 1976) p.297
6.37 C.W. Chu, S.Z. Huang, C.H. Lin, R.L. Meng, M.K. Wu: Phys. Rev. Lett. *46*, 276 (1981)
6.38 D.W. Harrison, K.C. Lin, J.D. Thompson, C.Y. Huang, P.D. Hambourger, H.L. Luo: Phys. Rev. Lett. *46*, 280 (1981)

6.39 M.B. Maple, L.E. DeLong, W.A. Fertig, D.C. Johnston, R.W. McCallum, R.N. Shelton: In *Valence Instabilities and Related Narrow Band Phenomena,* ed. by R.D. Parks (Plenum, New York 1977) pp.17-29
6.40 Ø. Fischer, M. Ishikawa, M. Pellizzone, A. Treyvad: J. de Physique (Colloque C5) *40*, C5-89 (1979)
6.41 R. Odermatt, M. Hardeman, J. van Meijel: Solid State Commun. *32*, 1227 (1979)
6.42 S. Oseroff, R. Calvo, D.C. Johnston, M.B. Maple, R.W. McCallum, R.N. Shelton: Solid State Commun. *27*, 201 (1978)
6.43 B.N. Harmon, A.J. Freeman: Phys. Rev. B*10*, 4849 (1974)
6.44 T. Jarlborg, A.J. Freeman, D.D. Koelling: J. Magn. Magn. Matls. *23*, 291 (1981), and to be published
6.45 T. Jarlborg, A.J. Freeman: Phys. Rev. B*22*, 2332 (1980); Phys. Lett. *74*A, 349 (1979); Phys. Rev. Lett. *45*, 653 (1980)
6.46 K. Bennemann, J. Garland: A.I.P. Conf. Proc. *4*, 103 (1972)
6.47 S.D. Bader, S.K. Sinha: Phys. Rev. B*18*, 3082 (1978) and [Ref.6.10, Chap.7]
6.48 M.S. Torikachvili, M.B. Maple: Solid State Commun. *40*, 1 (1981)
6.49 B.T. Matthias, E. Corenzwit, J.M. Vandenberg, H.E. Barz: Proc. Nat. Acad. Sci. USA *74*, 1334 (1977)
6.50 M. Ishikawa, Ø. Fischer: Solid State Commun. *24*, 747 (1977)
6.51 G.K. Shenoy, B.D. Dunlap, F.Y. Fradin, C.W. Kimball, W. Potzel, F. Pröbst, G.M. Kalvius: J. Appl. Physics *50*, 1872 (1979)
6.52 D.C. Johnston: Solid State Commun. *26*, 141 (1978)
6.53 L.D. Woolf, D.C. Johnston, H.B. MacKay, R.W. McCallum, M.B. Maple: J. Low Temp. Phys. *35*, 651 (1979)
6.54 K. Yvon, A. Grüttner: In *Superconductivity in d- and f-Band Metals,* ed. by H. Suhl, M.B. Maple (Academic, New York 1980) p.515
6.55 E.I. Blount, C.M. Varma: Phys. Rev. Lett. *42*, 1079 (1979)
6.56 M. Tachiki, H. Matsumoto, T. Koyama, H. Umezawa: Solid State Commun. *34*, 19 (1980)
6.57 C.G. Kuper, M. Revzen, A. Ron: Phys. Rev. Lett. *44*, 1545 (1980)
6.58 See the review by M. Tachiki: In *Proc. Intern. Conf. on Ternary Superconductors,* ed. by G.K. Shenoy, B.D. Dunlap, F.Y. Fradin (North Holland, Amsterdam 1981) p.267
6.59 D.E. Moncton, D.B. McWhan, P.H. Schmidt, W. Thomlinson, M.B. Maple, H.B. MacKay, L.D. Woolf, Z. Fisk, D.C. Johnston: Phys. Rev. Lett. *45*, 2060 (1980)
6.60 S.K. Sinha, H.A. Mook, D.G. Hinks, G.W. Crabtree: Bull. A.P.S. *16*, 277 (1981): Phys. Rev. Lett. *48*, 950 (1982)

7. NMR and Mössbauer Studies in Ternary Superconductors[1]

F. Y. Fradin, B. D. Dunlap, G. K. Shenoy, and C. W. Kimball

With 11 Figures

Nuclear magnetic resonance (NMR) played an important historical role [7.1] in the early microscopic confirmation of the BARDEEN, COOPER, SCHRIEFFER (BCS) [7.2] theory of superconductivity. Hyperfine techniques, particularly the Mössbauer effect and NMR, are making important contributions to the microscopic picture developing for the ternary superconductors. Shortly after the discovery of superconductivity in the Chevrel phases [7.3] by MATTHIAS and coworkers [7.4], Mössbauer studies of the localized vibrations of ^{119}Sn in $SnMo_6S_8$ led to a description of the soft, anharmonic, resonance modes of the loosely bound tin atom in the network of Mo_6S_8 units [7.5]. This work led to other efforts using heat capacity and inelastic neutron scattering to characterize the lattice modes of the Chevrel phases. In Sect.7.1 we present a brief survey of the structural and localized vibrational properties of a number of Chevrel phase compounds studied by the Mössbauer effect.

Following the discovery by FISCHER and coworkers [7.6] that paramagnetic substitutions for Sn or Pb in the Chevrel phases resulted in only a weak depression of T_c and an enhancement of H_{c2}, NMR and Mössbauer studies [7.7] were used to establish the microscopic aspects of the compensation of the external field by the negative spin polarization of the conduction electrons and the weak local moment-conduction electron exchange coupling [7.7]. In Sect.7.2 we present a brief summary of NMR, the Mössbauer effect and electron paramagnetic resonance studies of conduction electron-local moment interactions in paramagnetic Chevrel phases and rare-earth Rh_4B_4 phases.

The discovery by FERTIG and coworkers [7.8] of the return to the normal state at the ferromagnetic transition in $ErRh_4B_4$, led to more general investigations of the competition and coexistence between superconductivity and magnetism in ternary compounds. Neutron scattering [7.9] and Mössbauer effect [7.10] studies have described the anomalous state of the ordered moment on Er in these unusual compounds. In Sect.7.3, we review the experimental situation with regard to the crystal field ground states and the ordered rare-earth moments in ternary superconductors studied by the Mössbauer effect.

---

1 Work supported by the U.S. Department of Energy, the National Science Foundation, and the Air Force Office of Scientific Research.

In Sect.7.4 we review Mössbauer effect studies of crystal chemistry and charge transfer in a number of classes of ternary superconductors. Concluding remarks are given in Sect.7.5. It is hoped that this short review of the application of hyperfine techniques to the complex and novel ternary superconductors will serve to demonstrate the utility of these techniques for a microscopic understanding of the localized phonon, electron and magnetic behavior.

7.1 Mössbauer Effect Studies of Local Vibrational Modes

The relevance to superconductivity of vibrational properties obtained from Mössbauer data received considerable attention following McMILLAN's [7.11] extension of the BCS theory to strong-coupled superconductors. Within the approximations of current theoretical treatments of strong-coupled superconductors, the Mössbauer results yield measures of the phonon distribution important in determining the electron-phonon coupling parameter λ, and hence, T_c [7.12]. In the harmonic approximation, the recoil-free fraction f is related to the mean-squared displacement $\langle x^2\rangle$ and the thermal shift $\delta_{th}(T)$ to the mean-squared velocity $\langle v^2\rangle$; these quantities, in turn, can be related by model calculations to moments of the phonon distribution. For example, in an isotropic, harmonic system, the recoil-free fraction is given by

$$f = \exp(-K^2\langle x^2\rangle) \quad , \tag{7.1}$$

where K is the wave vector of the Mössbauer gamma ray and the thermal shift is given by

$$\delta_{th}(T) = -\langle v^2\rangle/2c \quad , \tag{7.2}$$

where $\langle x^2\rangle$ and $\langle v^2\rangle$ are thermally averaged over the phonon spectrum of the system. A phonon spectral density, or some appropriate model, may then be used to interpret the experimental results. If, for example, a Debye model is used, then

$$\langle v^2\rangle = \langle v^2\rangle_o + \frac{9kT^4}{\theta^3 M}\int_0^{\theta/T}\frac{x^3dx}{e^x-1} \tag{7.3}$$

and

$$\langle x^2\rangle = \frac{h^2}{Mk\theta}\left[\frac{1}{4} + \left(\frac{T}{\theta}\right)^2\int_0^{\theta/T}\frac{x\,dx}{e^x-1}\right] \quad , \tag{7.4}$$

where M is the mass of the atom being investigated and θ is the Debye temperature. This technique has been applied to studies of the A15 compounds Nb_3Sn [7.12] and $V_3Ga_{1-x}Sn_x$ [7.13] and the Chevrel phase $SnMo_6S_8$ [7.5,14].

High T_c materials frequently show low temperature mode softening and anharmonicity, as well as a tendency to undergo lattice distortions [7.15,16]. If the lattice vibrations are anisotropic, the recoil-free fraction will be dependent on crystal direction; the lattice anisotropy for a noncubic point symmetry will be reflected in temperature dependent changes in the intensities of the quadrupole split spectral lines. Analysis of such data delineates the anisotropy of atomic vibrations. The effect of aharmonicity and phonon-mode softening are reflected in the details of the f and δ temperature dependencies.

Measurements of f(T) and δ(T) by KIMBALL et al. [7.5] and BOLZ et al. [7.14], using the ^{119}Sn Mössbauer resonance, showed very unusual temperature dependencies of both the recoil-free fraction and the thermal shift in the Chevrel phase superconductor $SnMo_6S_8$. While many of these materials show lattice instabilities with a first-order phase transition, a transition for the Sn-based compounds is not observed with either crystallographic and resistivity, or heat capacity and susceptibility measurements [7.15,16]. Mössbauer spectra obtained in an external magnetic field (30 kG) show that the quadrupole interaction parameter e^2qQ is positive. This means that the σ line (transitions $\pm 1/2 \rightarrow \pm 1/2$) is lower in energy than the π line (transitions $\pm 1/2 \rightarrow \pm 3/2$) [7.5]. As the temperature is increased, the π line becomes increasingly more intense than the σ line. This behavior is due to an anisotropic Debye-Waller factor (and is not characteristic of preferred particle orientation).

Both high T_c members of the Mo_6Se_8 structural family, $SnMo_6S_8$ and $PbMo_6S_8$, have been found to retain the rhombohedral $R\overline{3}$ structure to 4.2 K. The $SnMo_6S_8$ structure consists of a distorted-cubic network of S atoms with every fourth cube occupied by a Sn atom or an Mo_6 octahedron [7.16]. Due to the axial site symmetry at the Sn site, the asymmetry parameter in the electric field gradient is taken to be zero ($\langle x^2\rangle \approx \langle y^2\rangle$). The anisotropy in the Debye-Waller factor can then be determined from the ratio of intensities $R = I_\pi/I_\sigma$ (the Goldanskii-Karyagin effect) [7.17]. For a random orientation of particles, these intensities are

$$I_\pi = \exp(-K^2\langle x^2\rangle) \int_0^1 (1 + u^2)\exp(-\varepsilon u^2)du \tag{7.5}$$

and

$$I_\sigma = \exp(-K^2\langle x^2\rangle) \int_0^1 (\tfrac{5}{3} - u^2)\exp(-\varepsilon u^2)du \quad , \tag{7.6}$$

where $u = \cos\theta$, θ is the angle between the principal axis of the electric field gradient tensor and the direction of observation, K^2 is the square of the γ-ray wave vector, and $\varepsilon = K^2(\langle z^2\rangle - \langle x^2\rangle)$. In the present case, the data indicates $R > 1$ so that $\varepsilon < 0$ [7.18]. From the temperature dependence of R, $\varepsilon(T)$ is obtained; using these values in (7.5,6) the temperature dependence of $\langle z^2\rangle$ and $\langle x^2\rangle$ shown in Fig.7.1 is found [7.14]. A dramatic deviation from the harmonic expectation for the slope occurs in the vicinity of 80 K. The decreasing value of the slope at high temperatures for both $\langle x^2\rangle$ and $\langle z^2\rangle$ results from anharmonic contributions to the Sn vibrational motion. BOLZ et al. [7.14] found that $\langle x^2\rangle$ decreases with increasing temperature; i.e., that there is a contraction of the Sn surroundings in

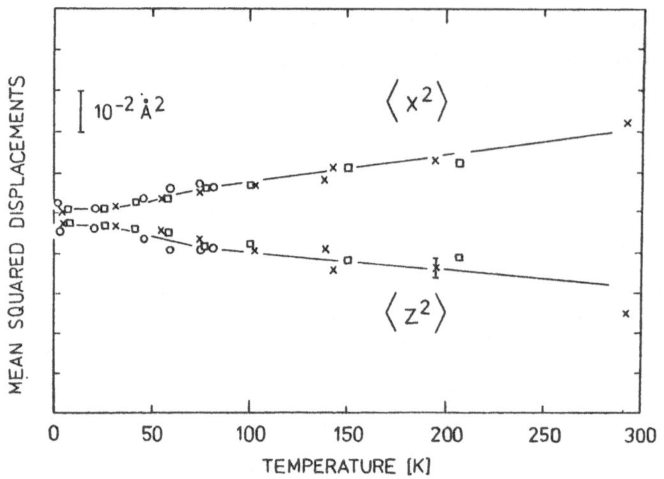

Fig.7.1. Temperature dependence of $<x^2>$ and $<z^2>$ for ^{119}Sn in $SnMo_6S_8$ [7.14]

the direction normal to the <111> axis. This effect is observed to be especially strong in the temperature region between 50 and 110 K. Recent X ray studies by CHEVREL [7.19] confirm the contraction of the lateral motion ($<x^2>$) with increasing temperature. The Sn vibrational behavior does not appear to be affected by substitution of Eu for up to 67% of the Sn concentration.

The thermal shift ($\sim <v^2>$) also indicates strong anharmonic behavior; that is, it cannot be accounted for on the basis of a temperature independent phonon density of states [7.5,14]. Heat capacity results for $SnMo_6S_8$ show that the entropy Debye temperature varies from less than 200 K near T_c ($\sim$ 12 K) to $\sim$400 K at T_c = 300 K [7.20]. The entropy result is an average over all the lattice vibrations but indicates a large deviation from Debye behavior. Further, from the harmonic theory the moments of the phonon distribution $\omega(n) = (\sum_i \omega_i^n)^{1/n}$ must monotonically increase with n. The moment $\omega(1)$ obtained from the thermal shift is smaller than the moment $\omega(-1)$ obtained from the mean-square displacements, again indicating anharmonic behavior. A variational approach, due to HUI and ALLEN [7.21], has been used to calculate the temperature dependence of the recoil free fraction of Sn in $SnMo_6S_8$ [7.22]. Quasiharmonic frequencies were found for one-dimensional single particle potentials involving fourth-order or sixth-order anharmonicity. The temperature dependence of the recoil free fraction required large contributions from either a fourth or sixth-order term in the potential.

The difference between the superconducting transition temperature of $SnMo_6S_8$ ($T_c \sim$ 12 K) and the isomorphic binary compound Mo_6Se_8 ($T_c \sim$ 6 K) was attributed, in early studies, to the additional soft modes associated with Sn in $SnMo_6S_8$. In $SnMo_6S_8$ there is an additional contribution to $F(\omega)$ [and presumably to $\alpha^2(\omega)F(\omega)$] of weight equal to 3 times the number of tin atoms and energy ω, 40 K $\lesssim \omega \lesssim$ 100 K, which might be expected to make a significant contribution to the electron-phonon coupling parameter and, hence, T_c. As discussed below, it is not clear at this time how significant the role of the metal atom vibrations are, except that they influence the structural character of the unit cell and therefore the behavior of the compound.

In considering heat capacity, Mössbauer and neutron diffraction results, BADER et al. [7.23] suggested a molecular crystal model of the lattice vibrations and also proposed that the observed mode-softening is a result of hybridization between the resonance modes associated with the Sn atom and the acoustic modes related to the Mo_6S_8 complex. The phonon spectrum is accounted for by depicting the Chevrel phase as a diatomic crystal comprising Mo_6S_8 units and metal atoms; that is, the phonons arise from acoustic and optic modes of the metal and librational, acoustic and internal modes of Mo_6S_8.[2] Recent inelastic neutron scattering measurements of the phonon density of states by SCHWEISS et al. [7.25] confirmed their previous observation that the superconducting compounds (e.g., $SnMo_6S_8$) exhibit a pronounced phonon softening for low frequencies (below ~ 18 meV) at low temperature. These results indicate that the soft modes should be attributed to the external modes within the framework of the molecular crystal model. Behavior consistent with a proposed hybridization between a flat TO branch and the normal transverse acoustical mode is observed. POBELL [7.26] observed coupling of external modes to intramolecular modes of the Mo_6S_8 clusters by point contact electron tunneling measurements. Pobell's isotope effect measurements, however, do not confirm the concept of a molecular crystal model with respect to a dominant electron-phonon coupling of external modes. The isotope results for Sn are especially significant for the Mössbauer studies of the vibrational properties of the metal atom. It is inferred from the Sn isotope results that T_c is not dependent on the Sn mass and that there is weak overlap with the Mo d-orbitals.

In recent Mössbauer studies at Sn sites in $Sn_xMo_yS_8$ ($x \leq 1$) for various stoichiometries, WAGNER and FREYHARDT [7.27] obtained a temperature dependent shift in agreement with that found by KIMBALL et al. [7.5] and BOLZ et al. [7.14]; however, the temperature dependence of the mean square displacement was found to be much less anharmonic than that reported by KIMBALL et al. [7.5]. WAGNER and FREYHARDT [7.27] found two distinct superconducting transitions as the stoichiometry was varied, one on the Mo-rich and one on the Mo-poor side of the phase diagram; their Mössbauer results, however, do not show clear differences in the behavior of Sn.

Measurements by FRADIN et al. [7.28] show that T_c of $SnMo_6S_8$ is depressed by the substitution of Se for S, changing from 13 K for $SnMo_6S_8$ to 6.5 K for $SnMo_6S_6Se_2$, and then remaining roughly constant to $SnMo_6Se_8$. Concomitantly, there is a lattice expansion of 9.4% that does not appear to be correlated with T_c; however, a change in the rhombohedral angle (which measures the orientation of the Mo_6S_8 complex within

2 See chapters on neutron scattering, isotope, and tunneling measurements elsewhere in this volume and in [7.24] for detailed descriptions of the behavior of the Chevrel phases.

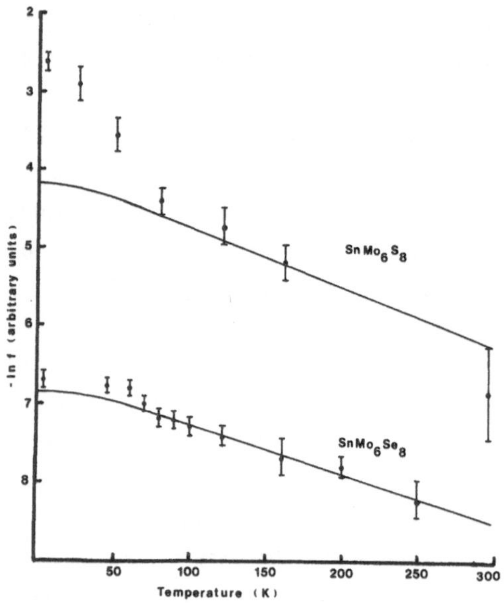

Fig.7.2. Average mean square displacement for ^{119}Sn in $SnMo_6S_8$ ($T_c \sim 13$ K) and $SnMo_6Se_8$ ($T_c \sim 6.5$ K) [7.29]

the unit cell) shows a weak correlation with T_c. In order to contrast the phonon anomalies in a low T_c and a high T_c compound, Mössbauer measurements were made at ^{119}Sn in $SnMo_6Se_8$ [7.29]. In the selenide the quadrupole interaction is about 40% smaller than in the sulfide. This is attributed primarily to an increase in the lattice constant. Because the two lines of the quadrupole doublet are unresolved, the decomposition of the recoil-free fraction into parallel and perpendicular components is difficult. However, both the recoil-free fraction and the thermal shift show mode-softening around 80 K. These anomalies are weaker than in the sulfide and are in keeping with the difference in their T_c values (6.5 K for $SnMo_6Se_8$ vs 13 K for $SnMo_6S_8$), as can be seen from the temperature dependence of the recoilless fraction shown in Fig.7.2. The solid lines indicate the behavior expected on the basis of a Debye model fitted to the high temperature data. Measurements have been made for $LaMo_6Se_8$ (T_c = 11 K) with 2% Sn subsituted for La [7.29]. Again, as for $SnMo_6Se_8$, the ^{119}Sn spectra in $La_{0.98}Sn_{0.02}Mo_6Se_8$ show a very small quadrupole splitting and the phonon anomalies are much less pronounced than in $SnMo_6S_8$.

The superconducting properties of nonstoichiometric Chevrel phase compounds, e.g., $M_{1.2}Mo_6S_8$, are often superior to those with stoichiometric composition. At present, location of the excess atoms in the Chevrel lattice (or in a second phase) is uncertain. In order to clarify the microscopic chemistry of the excess metal atoms, the temperature dependence of the ^{119}Sn effect in $Sn_{1.2}Mo_6S_8$ has been measured and clearly indicates that, in contrast to Sn in $Sn_{1.0}Mo_6S_8$, the Sn atom in $Sn_{1.2}Mo_6S_8$ is in two distinct electronic sites [7.29]. One site is similar to that of Sn in $Sn_{1.0}Mo_6S_8$, but the other has a larger Debye temperature and is much less anharmonic. Experimental results have not yet clearly determined whether the second electronic state of Sn is due to interstitial atoms in the Chevrel phase unit cell, or due to an interspersed second phase, such as SnS.

7.2 Conduction Electron-Local Moment Interactions in Paramagnetic Ternary Compounds

7.2.1 Chevrel Phases

A remarkable feature of the high critical field H_{c2} Chevrel phase superconductors is that alloying paramagnetic rare-earth ions into the compounds can enhance H_{c2} without substantially altering the superconducting transition temperature. For example, FISCHER and co-workers [7.30] report that in the compounds $Sn_{1.2(1-x)}Eu_xMo_{6.35}S_8$, H_{c2} (2 K) = 400 kG and T_c = 10.2 K for x = 0.5, while H_{c2} (2K) = 275 kG and T_c = 10.4 K for x = 0.0. They relate this result to the JACCARINO-PETER [7.31] mechanism in which exchange coupling of the conduction electrons to the localized moment yields a negative conduction-electron polarization (Chap.3). This in turn produces a negative exchange field which partially compensates the externally applied field. This phenomenon may often be present, but the enhancement of H_{c2} is not usually seen since a decrease in T_c (and hence H_{c2}) due to exchange scattering from the localized moment usually predominates. However, in the Chevrel-phase compounds, the dependence of T_c on the magnetic-impurity concentration shows that exchange scattering is weak and, therefore, a net enhancement of H_{c2} may occur.

The local moment-conduction electron interaction in $Sn_{1-x}Eu_xMo_6S_8$ has been extensively studied using ^{95}Mo nuclear magnetic resonance (NMR) by FRADIN et al. [7.7] and ^{151}Eu Mössbauer spectroscopy by DUNLAP et al. [7.32]. Recently, FREEMAN and JARLBORG [7.33] have carried out spin-polarized band structure calculations for a number of Chevrel phase compounds. The transition temperatures are T_c = 13.3 K for $SnMo_6S_8$ and T_c = 13.0 K for $Sn_{0.5}Eu_{0.5}Mo_6S_8$ [7.28]. Mössbauer spectra showed a prominent single line with an isomer shift corresponding to Eu^{2+}, and a minor second line indicating ~5% of the Eu content is trivalent.

The susceptibility (Faraday method) of $SnMo_6S_8$ increases only by 8×10^{-6} emu/Mo between 77 and 20 K and hence is essentially temperature independent. The isotropic Knight shift is K_{iso} = + 1.55%, also independent of temperature in the normal state. Below T_c there is a change, ΔK_{iso} = + 0.4%, due to a superconducting gap which opens at the Fermi level and leads to a quenching of the predominantly d-like Pauli paramagnetism. Since the core polarization field is negative [7.34], a negative term in the Knight shift is quenched. (It should be noted that SANO et al. [7.35] find no change in the Knight shift of ^{207}Pb and ^{77}Se in $PbMo_6Se_8$).

As shown in Fig.7.3, the results for $Sn_{0.5}Eu_{0.5}Mo_6S_8$ are very different. The magnetization σ follows a Brillouin function $B_{7/2}(x)$, appropriate for Eu^{2+}, with a saturation value slightly less than that of the free-ion value. This smaller

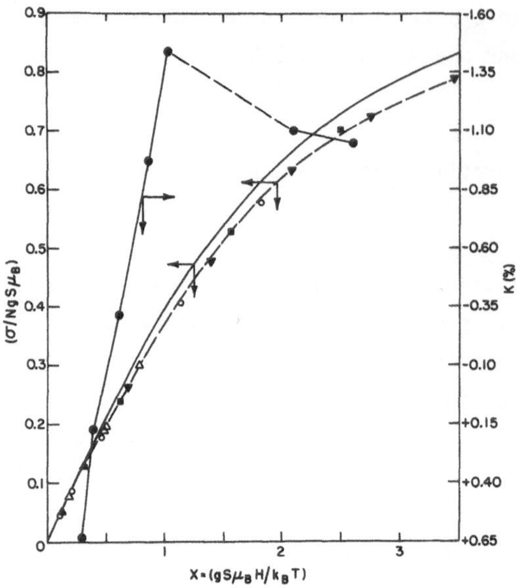

Fig.7.3

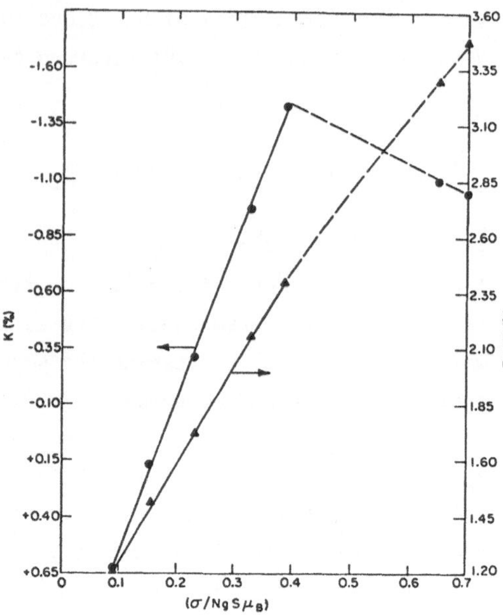

Fig.7.4

Fig.7.3. Reduced magnetization ($\sigma/NgS\mu_B$) and ^{95}Mo Knight shift K as a function of $x = gS\mu_B H/k_B T$ for $Sn_{0.5}Eu_{0.5}Mo_6S_8$. Normal-state magnetization (dashed curve) obtained at 13.6 K (▼), 15.1 K(■), 20.5 K (o), 48.6 K (Δ), and 75 K (Δ). Solid curve for $\sigma/NgS\mu_B$ is the Brillouin function $B_{7/2}(x)$. Knight-shift data for $x > 1.5$ was obtained in the superconducting state. The values of H are the applied field corrected for the demagnetizing field. Absolute shift can be obtained by multiplying 46 863 G by K(1 + K) [7.7]

Fig.7.4. Knight shift K and full width at half-maximum (FWHM) linewidth of ^{95}Mo as a function of reduced magnetization ($\sigma/NgS\mu_B$). NMR data in superconducting state indicated by dashed lines. Absolute shift and linewidth can be obtained by multiplying 46 863 G by K/(1 + K) and the FWHM by 1/(1 + K), respectively [7.7]

saturation value is partially due to the nonmagnetic Eu^{3+} in the material, but may also reflect a net negative conduction-electron polarization. The Knight shift shows a very large temperature dependence in the normal state. On passing through T_c (X = 1 in the figure), there is a change ΔK_{iso} = + 0.38% which is similar to that found in the pure $SnMo_6S_8$ and presumably arises from the same source. However, much of the large negative shift is not quenched at T_c. This behavior appears to indicate two-band superconductivity, i.e., at T_c a gap opens in the d band, but the s band is unaffected. Part of the Knight shift at ^{95}Mo may be due to spin transfer via molecular orbitals. However, this would not be an effective mechanism for enhancing H_{c2}. In Fig.7.4, we have plotted K_{iso} vs σ. For $T > T_c$, one finds $K_{iso} = [1.265 - 6.95\sigma(NgS\mu_B)^{-1}]\%$. The proportionality between K_{iso} and σ suggests that the Knight shift arises from a negative conduction-electron polarization at the Mo site due to a rare-earth-conduction electron coupling of the form $-2J_{sf}\bar{S}_{4f}\cdot\bar{s}_{5s}$. We assume, based on the ^{95}Mo and ^{151}Eu linewidth results, very weak coupling to the

Mo-derived 4$\underline{d}$ states. The linewidth of the ^{95}Mo resonance is shown as a function of reduced magnetization in Fig.7.4. The proportionality to σ indicates a distribution of Knight shifts probably due to different Eu^{2+} environments about the ^{95}Mo, i.e., random replacements of Sn by Eu^{2+}. Since there is only a small change in slope of the linewidth versus σ in the superconducting state, it is clear that the coupling of the conduction electrons to the Eu^{2+} in the superconducting state is essentially unchanged from that in the normal state, i.e., s-band coupling to Eu^{2+}.

The ^{151}Eu Mössbauer effect has been measured in $Sn_{0.5}Eu_{0.5}Mo_6S_8$. Values of the hyperfine field H_{hf}^{Eu} at 4.2 K have been obtained in external fields of 15 and 60 kG. When extrapolated to saturation using a $B_{7/2}$ Brillouin function, both measurements give a hyperfine field of - 285 ± 15 kG. Fully polarized Eu^{2+} ions in unsulators show a saturation hyperfine field due to core polarization of -340 kG [7.36]. Therefore, we have a contribution due to conduction-electron polarization of ~ + 55 kG. If this contribution is due to a contact interaction from polarized $\underline{s}$ conduction electrons,[3] then the conduction-electron polarization at the Eu site is positive. In other Eu intermetallics, the conduction-electron contribution to the hyperfine field varies from + 40 to + 200 kG. The value obtained here, at the lower extreme of these values, indicates a rather small value of the $\underline{s}$-state conduction-electron spin polarization at the Eu site. A low total $\underline{s}$-electron density at the Eu is also indicated by the observed isomer shift of - 14.0 mm/s. Such a shift is typically obtained for Eu^{2+} in ionic compounds and is anomalous for conducting systems. In Eu^{2+} intermetallics, isomer shifts of about half this value are more common [7.37].

The other aspect of this problem, the weak depression of T_c, can be studied through the Korringa relaxation of the Eu^{2+} ions. Superconducting pairbreaking occurs in the presence of paramagnetic impurities due to the exchange interaction between the localized spin $\underline{S}$ and conduction electron spin σ:

$$\underline{H}_{ex} = - 2\underline{J}(g_J - 1)\underline{J} \cdot \sigma = - 2\underline{J}\ \underline{S} \cdot \sigma \quad , \tag{7.7}$$

where $\underline{J}$ is the exchange integral and J is the rare-earth angular momentum. Within the framework of the ABRIKOSOV-GORKOV theory [7.38], this interaction will produce a decrease in T_c which is linear in the concentration of paramagnetic ions x with a slope for small x given by

$$\Delta T_c = \frac{\pi^2}{2k} \underline{J}^2 N(E_F) S(S + 1) \Delta x \quad . \tag{7.8}$$

3 Contributions to the hyperfine field from transferred effects due to the other Eu ions will not be significant here since the Eu-Eu distance is very large (~ 6.5 Å). In addition, there may also be a contribution through core polarization from polarized $\underline{d}$ electrons, but this is generally much smaller than that arising from the contact term.

Here $N(E_F)$ is the density of states at the Fermi energy projected on the local spin and $\underline{J}^2$ now represents $\langle \underline{J}(q)\rangle^2$, where the average is over the Fermi surface. Another manifestation of the exchange coupling is the electron-spin relaxation rates of the impurity spins. The interaction yields an electronic-spin relaxation rate given by the Korringa expression

$$\underline{W}_K = \frac{1}{\underline{t}_K} = \frac{2\pi}{\hbar} [(g_J - 1)\underline{J}\, N(E_F)]^2\, kT \quad . \tag{7.9}$$

Measurement of paramagnetic ion relaxation rates thus provides values for $|\underline{J}\, N(E_F)|$. In the case when the relaxation rate is large compared with nuclear precession rates, an excess linewidth $\Delta\Gamma$ is obtained which is proportional to the relaxation time, $T_1 = 1/\underline{W}$.

The detailed relationship between $\Delta\Gamma$ and T_1 has been discussed for the limit of fast relaxation by BRADFORD and MARSHALL [7.39]. The problem is simplified substantially here due to the fact that Eu^{2+} is an S-state ion and therefore the magnetic hyperfine interactions that occur in the absence of external magnetic fields are isotropic, i.e., independent of the crystal symmetry (zero-field splittings are small compared to our measuring temperatures). Following the procedures of BRADFORD and MARSHALL[4] one obtains for the 21.6 keV transition of ^{151}Eu

$$W(s^{-1}) = 2.75 \times 10^8/\Delta\Gamma\ [\mathrm{mm/s}] \quad . \tag{7.10}$$

Taking $\Delta\Gamma = \Gamma_{exp} - \Gamma_0$, where $\Gamma_0 = 4.4$ mm/s is the linewidth in the limit of very fast relaxation, (7.10) yields the relaxation rates shown in Fig.7.5.

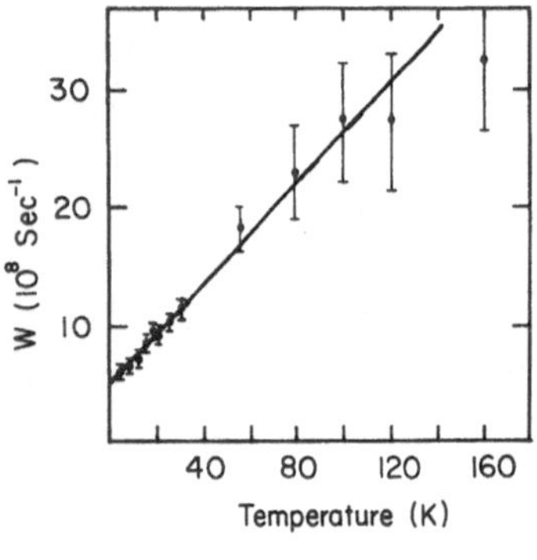

Fig.7.5. Paramagnetic relaxation rate vs temperature for Eu in $Sn_{0.75}Eu_{0.25}Mo_6S_8$ [7.32]

From the figure one sees that the relaxation process consists of a temperature-independent part, presumably due to spin-spin interactions (W_{ss}) which causes the nonzero intercept for T = 0, and a Korringa part (W_K) which is linear in T. No effects on W_K are seen below T_c due to, for example, superconducting gap effects

4 The expression of (7.10) has been generalized by F. HARTMAN-BOUTRON [7.40] in a form applicable to other Mössbauer transitions as well.

Table 7.1. Value of $|N(E_F)\underline{J}|$ in Ternary Superconductors

Compound	$\lvert N(E_F)\underline{J}\rvert$ per atom-spin	References
$Eu_{0.25}Sn_{0.75}Mo_6S_8$	0.0033	[7.32]
$GdMo_6Se_8$	0.0047	[7.42]
1% Gd in $SnMo_6S_8$	0.0061	[7.43]
0.91% Gd in $PbMo_6S_8$	0.0050	[7.43]
$ErRh_{1.1}Sn_{3.6}$	0.019	[7.70]

because the relaxation is dominated in this region by W_{ss}. From the slope of W vs T and (7.9), one finds

$$|\underline{J}N(E_F)| = 0.0033/\text{Eu atom-spin} \quad . \tag{7.11}$$

For comparison we note that EPR measurements on Eu in the binary superconductor $LaAl_2$ gave a value [7.41] of $|\underline{J}N(E_F)| = 0.03$ eV, an order of magnitude larger. This directly demonstrates the weak interaction of the Eu ions with the conduction electrons in the Chevrel phase systems, in agreement with the weak dependence of T_c on concentration. Recent band structure calculations by FREEMAN and JARLBORG [7.33] indicate that the very small coupling is due to a very small density of states at the Fermi level at the Eu^{2+}.

OSEROFF et al. [7.42], using EPR measurements on Gd^{3+} in $GdMo_6Se_8$, have found the value of $|\underline{J}N(E_F)| = 0.0047(4)$/Gd atom-spin from the Korringa relaxation contribution to the EPR linewidth. ODERMATT et al. [7.43] have determined $|\underline{J}N(E_F)|$ for 1% Gd substituted $SnMo_6S_8$ and $PbMo_6S_8$ from the Korringa relaxation of the Gd EPR signal; they found values of 0.0061(6) and 0.0050(6), respectively. In Table 7.1 we list values of $|\underline{J}N(E_F)|$ determined for a number of ternary superconductors from EPR and Mössbauer effect measurements.

We may summarize all the results of this section within the context of a simplified view of these complex materials. We consider the system to contain two bands which we label the "$\underline{s}$ band" and the "$\underline{d}$ band." The superconducting properties are dependent on the $\underline{d}$ band, derived largely from the Mo ions. Below the superconducting transition temperature, a gap develops in the $\underline{d}$ band, causing the observed change in the Knight shift. When paramagnetic ions are added to the material, a rather weak exchange interaction occurs, largely with the $\underline{s}$ band. Also, a shielding of the Mo $\underline{d}$ electrons from the Eu site, which is surrounded by sulfur atoms, inhibits exchange scattering of the $\underline{d}$ electrons off the Eu^{2+} moments and so T_c is weakly dependent on the magnetic-ion concentration. Therefore, the primary effect of the magnetic ions is to cause a polarization of the $\underline{s}$-band electrons when the Eu^{2+} moments are aligned in an external field. This polarization is parallel to the field at the Eu site but shows a spatial dependence such that it is negative at the Mo site. Below T_c, no gap opens in the $\underline{s}$ band so this polarization remains.

As a result, there is a partial compensation of the applied field at the Mo sites due to weak s-d exchange coupling, inhibiting pairbreaking of the d-band electrons and causing an enhancement of the critical field over that in the pure $SnMo_6S_8$.

7.2.2 Rhodium Borides

Compared to the unusual but reasonably well understood properties of the magnetically doped Chevrel phases, the experimental results on the rhodium borides ($RERh_4B_4$, where RE is yttrium or a lanthanide) are not well understood at present. There are differences reported in the ordered moments in the ferromagnetic state of $ErRh_4B_4$ (Sect.7.3), departures from ABRIKOSOV-GORKOV (A-G) behavior [7.38] in the depression of T_c (and H_{c2}) with paramagnetic ion addition in $Y_{1-x}Er_xRh_4B_4$, and unusual magnetic correlations observed for $T_M < T < T_c$ in a number of rhodium borides.

Early work on concentrated $RERh_4B_4$ superconducting compounds appeared consistent with the expectations from A-G behavior, assuming weak exchange coupling $|\underline{J}N(E_F)|$ similar to that of the Chevrel phases. However, detailed experiments on dilute paramagnetic additions, e.g., $Y_{1-x}RE_xRh_4B_4$, indicate important local moment-conduction electron coupling. Band structure calculations also indicate that the rare-earth ions in the $RERh_4B_4$ compounds have large values of the projected $N(E_F)$ in contrast to the Chevrel phases.

In Table 7.2 we list the paramagnetic moment μ_{eff} in $Y_{1-x}Er_xRh_4B_4$ obtained from the temperature dependence of the susceptibility (Faraday method) [7.44]. For $ErRh_4B_4$, μ_{eff} is essentially equal to the free ion value $\mu_{FI} = 9.58\ \mu_B/Er$. Since the crystal field splittings are relatively small, μ_{eff} is approximately independent of temperature for $T \gtrsim 40$ K. The enhanced value of μ_{eff} for small Er concentrations can be ascribed to conduction electron spin polarization due to the effective exchange interaction. The susceptibility of $Y_{1-x}Er_xRh_4B_4$ can be written for small x, ignoring both crystal field interactions of the local moment and the exchange enhancement of the host susceptibility, as

$$\chi = \chi_{YRh_4B_4} + xgJ\mu_B B_J(X)[1 + 2(g - 1)\underline{J}\ N(E_F)/g] \quad , \qquad (7.12)$$

where $X = gJ\mu_B H/k_B T$. Using the value 6/5 for the Landé g-factor for Russell-Saunders coupling of Er^{3+}, $^4I_{15/2}$, we derive the values of $\underline{J}N(E_F)$ from the paramagnetic moment μ_{eff} listed in Table 7.3. A large value of $\underline{J}N(E_F)$ is also found from the enhancement of the saturation moment for the 1% Er compound (Table 7.4). This is a very large effective exchange-coupling strength. A corresponding value for Eu^{2+} $^8S_{7/2}$ in $SnMo_6S_8$ is $|\underline{J}N(E_F)| = 0.003$. Clearly the values of $|\underline{J}N(E_F)|$ for $Y_{1-x}Er_xRh_4B_4$ at low x are unphysically large and may indicate the importance of exchange enhancement of the host. Additional evidence [7.44] for strong local moment-conduction electron coupling is found in the rather large line broadening

Table 7.2. Properties of $Y_{1-x}Er_xRh_4B_4$

x	0	0.003	0.01	0.03	0.1	0.5	1.0
$\mu_{eff}(\mu_B/Er)$	-	11.0	10.8	10.1	9.8	-	9.6
$\mu_{sat}(\mu_B/Er)$	-	-	9.85	-	-	-	8.5
$\lvert \underline{J}N(E_F)\rvert$	-	0.96	0.80	0.33	0.14	-	-
T_c[K]	10.28	10.31	10.30	10.36	10.27	9.51	8.61

Note: For $Y_{0.98}Gd_{0.02}Rh_4B_4$ $\mu_{eff} = 8.35\ \mu_B$/Gd, compared to the free ion value of 7.94 μ_B/Gd, yielding a value of $J_{sf}N(E_F) = 0.11$; for $Y_{0.90}Gd_{0.10}Rh_4B_4$ μ_{eff} is equal to the free ion value.

Table 7.3. Magnetic moments (μ) on the rare-earth (RE) ion in $RERh_4B_4$ compounds, and approximate position (Δ) of first excited CEF level

RE	$\mu[\mu_B]$	Δ[K]
Dy	9.2 ± 0.2	~ 25
Er	8.3 ± 0.2	~ 15
Tm	6.4 ± 0.2	~ 15

Table 7.4. Valence Electron Concentrations (VEC) in Chevrel Phase Compounds

Compound	T_c[K]	VEC/Mo_6 Cluster	References
Mo_6S_8	-	20	[7.79]
Mo_6Se_6	6.3	22	[7.79]
Mo_6Te_8	-	28	[7.80]
$Mo_6S_6I_2$	14.0	22.3	[7.81]
$Mo_6Se_7I_1$	7.6	22.75	[7.79]
$Mo_6Te_6I_2$	2.6	28.3	[7.81]
$Pb^{2+}Mo_6S_8$	15.2	22	[7.81]
$Sn^{2+}Mo_6S_8$	14.0	22	[7.81]
$RE^{3+}Mo_6S_8$	<2.0	23	[7.81]

of the ^{11}B NMR signal in the dilute Er compounds; the broadening scales as H/T at high temperature due to the inhomogeneous Knight shift resulting from the random substitution of Er for Y. Another striking feature of the data in Table 7.2 is the negligible depression of T_c at low Er concentration.

In order for paramagnetic ions to be effective pair breakers, they must be free to fluctuate. The fluctuations of the local moments can be sensitively measured by the enhancement of the ^{11}B nuclear spin-lattice relaxation rate relative to the Korringa value T_1T = 45 s-K for ^{11}B in YRh_4B_4. Although the Korringa relaxation rate in YRh_4B_4 is consistent with the density of states at the boron site calculated

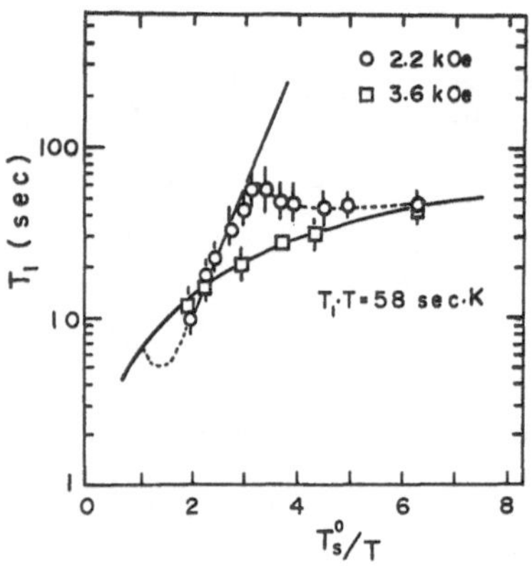

Fig.7.6. Temperature dependence of ^{11}B spin-lattice relaxation time T_1. T_c is the superconducting transition temperature of 8.15 K in $ErRh_4B_4$. Lower field data obtained at 3 MHz; higher field data obtained at 5 MHz [7.48]. For the lower field data, the solid line is the relaxation expected for a BCS superconductor. For the higher field data, the solid line is the Korringa relaxation T_1T = 58 s-K [7.48]

by JARLBORG et al [7.45], large enhancements of T_1^{-1} due to local moment-nuclear moment dipole-dipole coupling [7.44] are found above T_c for the dilute rare-earth substituted $Y_{1-x}RE_xRh_4B_4$. KUMAGAI et al. [7.46] report very rapid relaxation (T_1) values at 300 K for ^{11}B of 120, 128 and 200 μs in $TbRh_4B_4$, $DyRh_4B_4$, and $ErRh_4B_4$, respectively. (Note: High temperature Knight shift and linewidth analyses have been reported for $LuRh_4B_4$ and $GdRh_4B_4$ by JOHNSTON and SILBERNAGEL [7.47]). By comparison, $SmRh_4B_4$, which orders antiferromagnetically below T_c, has a longer T_1 for ^{11}B and a complex temperature dependence. Clearly, one expects Sm ions to behave in a complex manner because of the close lying Hund's rule states of different J.

Even more surprising is the fact that in $ErRh_4B_4$ the local moment fluctuations appear to be frozen out at low temperatures $T < T_c$. As can be seen in Fig.7.6, KUMAGAI et al. [7.48] find that at low temperatures $T \lesssim 4.2$ K, T_1 of ^{11}B in $ErRh_4B_4$ is long. In fact, at an applied field of 3.6 kOe, which is sufficient to quench superconductivity, T_1 obeys the Korringa relation T_1T = 58 s-K. At a field of 2.2 kOe, where $ErRh_4B_4$ is superconducting, T_1 first grows exponentially as $\exp[\Delta(o)/k_BT]$ with decreasing temperature as expected of a BCS superconductor with gap $\Delta(o)$. However, for temperatures less than 2.4 K the superconducting gap appears to close and T_1 again approaches the Korringa value. The open question is the mechanism by which the local moment fluctuations become frozen out above the magnetic transition.

Typically in all the concentrated $RERh_4B_4$ compounds studied at low temperatures [7.46,48], the ^{11}B linewidths are several kOe broad. The broadening may be due to a combination of particle shape demagnetization broadening and dipole or RKKY coupling of the local moment to the ^{11}B nuclear spins. It has been suggested [7.48] that the closing of the gap (Fig.7.6) for $T_c/T \gtrsim 3$ in $ErRh_4B_4$ is due to field induced ferromagnetism, since below this temperature range the ^{11}B spin-spin relaxation time T_2 also begins to lengthen, as is typical in a ferromagnetic transition.

Returning to YRh_4B_4 and the dilute substitution of rare earths, a striking feature of the ^{11}B NMR line shapes is the sudden large, field-independent, inhomogeneous magnetic broadening that sets in below T_c. (Note: the magnetic nature of

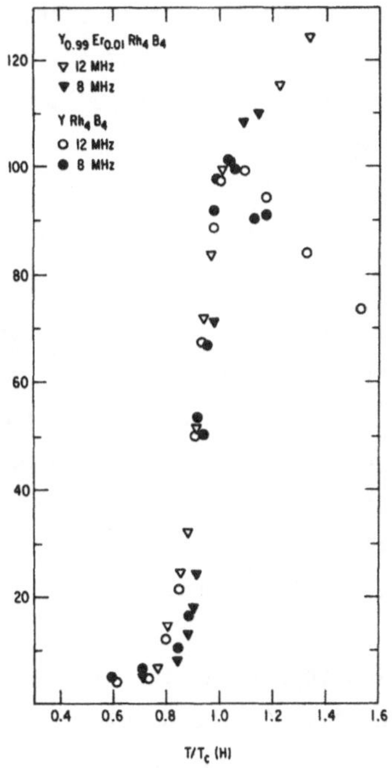

Fig.7.7. The reduced temperature $T/T_c(H)$ dependence of the normalized integrated resonance intensity of ^{11}B obtained from the spin echo in YRh_4B_4 and $Y_{0.99}Er_{0.01}Rh_4B_4$ at 8 and 12 MHz. From susceptibility measurements T_c was determined to be 8.2 K and 7.2 K at $H \simeq 6$ kOe and $H \simeq 8.8$ kOe, respectively [7.44]

the inhomogeneous broadening is unambiguously determined from the rf pulse turn angle conditions necessary to maximize the NMR spin-echo signal). This effect is found in YRh_4B_4, $LuRh_4B_4$ and dilute alloys with either Er or Gd. Because of the field (0.2 to 20 kOe) independent character and the large magnitude (> 400 Oe) of the broadening, both particle shape demagnetization and vortex line broadening can be discarded as possible causes of the inhomogeneity in these strong type II superconductors [7.44]. The sharp loss of the integrated (±100 Oe) signal at $T_c(H)$ due to the broadening is displayed in Fig.7.7. Note that for $T > T_c$, the ^{11}B signal follows the nuclear Curie law for pure YRh_4B_4, but due to inhomogeneous Knight shift and particle-shape demagnetization broadening there is already some loss of signal above T_c in the Er-doped compound. [Signals have been normalized to unity at $T = T_c(H)$ in Fig.7.7].

The loss of NMR signal is quite reminiscent of the behavior observed by BARNES and CREEL [7.49] for the ^{11}B resonance in the itinerant antiferromagnet CrB_2. As the magnetic-order parameter (staggered magnetization) increases below the Néel temperature, a large inhomogeneous distribution of magnetic hyperfine fields occurs which broadens the resonance so that the signal is unobservable. The inhomogeneity is due to the incommensurate nature of the conduction electron spin magnetization. In order that itinerant antiferromagnetic transitions occur in YRh_4B_4 and $LuRh_4B_4$, the magnetic-order parameter would have to be strongly coupled to the superconducting order parameter since the loss of NMR signal follows the magnetic field dependence of the superconducting transition temperature. Although the coexistence of antiferromagnetism and superconductivity has been found in ternary compounds, the order parameter of the antiferromagnetic state is associated with localized 4f electron moments. Clearly, in YRh_4B_4 and $LuRh_4B_4$ there are no localized 4f electron moments. As in the discussion of the paramagnetic moment enhancement, exchange enhancement of YRh_4B_4 would have to play an important role.

In summary, we find an anomalous state of affairs in the $RERh_4B_4$ compounds, namely:

(i) Dilute Er substitutions in YRh_4B_4 appear to give rise to a large induced conduction electron polarization but very weak depression of T_c.

(ii) Local moment fluctuations strongly enhance the ^{11}B spin-lattice relaxation rate in the $RERh_4B_4$ compounds; however, in $ErRh_4B_4$, the fluctuations appear to be frozen out below T_c. Very recent results on $Y_{1-x}RE_xRh_4B_4$ also indicate very slow ^{11}B relaxation rates below T_c for small x.

(iii) A correlated magnetic state appears to set in below T_c but above $T_M(0)$ in $Y_{1-x}Er_xRh_4B_4$ $(0 \leq x \leq 1)$.

(iv) Local moment-conduction electron exchange coupling appears to be dramatically reduced in the concentrated $RERh_4B_4$ compounds relative to the dilute $Y_{1-x}RE_xRh_4B_4$ compounds.

7.3 Local Moments in Magnetically Ordered Rare-Earth Compounds

Information on local magnetic moments is essential to any discussion of the magnetic interactions in ternary superconductors. This is closely related to the crystalline electric field (CEF) interactions which, in the case of non-S state rare-earth atoms, will modify both the ground state magnetic moment and its anisotropy. This interaction will in turn modify coupling of the local moment to the superconducting electrons and to the other magnetic atoms in the lattice. In addition, T_c will be affected by exchange induced transitions (Van Vleck type) to the excited CEF states. No single technique can provide this information uniquely. Amongst the resonance methods, the Mössbauer effect has made many experimental contributions to this area which will be discussed in this section. In the next section we will introduce the magnetic parameters measured in a Mössbauer experiment and indicate their relationship to the magnetic properties. Thereafter, various examples are discussed.

7.3.1 Hyperfine Fields and Magnetic Moments

The interaction of the dipole moment of the nucleus with the hyperfine magnetic field generated by surrounding electrons causes a Zeeman splitting of the nuclear ground and excited levels. When the electronic spin on the atom (3d or 4f) is ordered, as in a ferro or antiferromagnet, this results in a splitting of the hyper-

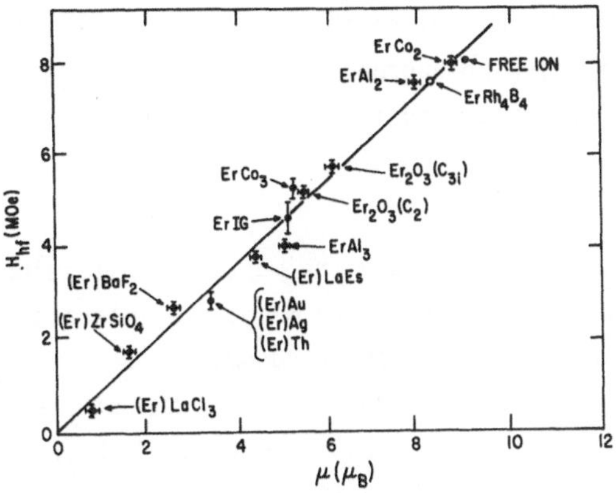

Fig.7.8. Linear dependence of Er magnetic hyperfine field on Er magnetic moment in various compounds [7.52]

fine (hf) spectra. However, this splitting will also be observed in paramagnets if the spin is fluctuating at a frequency ν_s, small compared to the hyperfine frequency ν_{hf} (the "slow relaxation" limit).

For a paramagnetic rare-earth ion, the interaction Hamiltonian is given by

$$\underline{H}_m = \sum_{i=x,y,z} A_i I_i J_i \quad , \tag{7.13}$$

where I_i and J_i are components of the nuclear and electronic angular momenta, respectively, and the A_i are proportional to electronic g factor components. In the case of a magnetically-ordered material or a highly anisotropic paramagnet ($A_x = A_y = 0$), this is equivalent to a simple vector interaction between the nuclear dipole moment and a hyperfine magnetic field H_n (the "effective field" case).

In general, contributions to the hyperfine field arise from the polarization of core electrons by the unpaired 3d or 4f spin, from the unquenched orbital currents (primarily in 4f systems), from spin dipolar interactions and from conduction electron polarization. For rare-earth atoms, the core polarization is less than a percent of the orbital contribution, except in S-state atoms (Eu^{2+}, Gd^{3+}) where the orbital contribution is zero. The dipolar fields, arising from other spins in the lattice, and conduction electron contributions are important only in the case of S-state atoms. The orbital field for a rare-earth case is given by [7.50]

$$\vec{H}_{orb} = -2\mu_B \langle r^{-3}\rangle\langle J||N||J\rangle\vec{J} \quad , \tag{7.14}$$

where $\langle r^{-3}\rangle$ is the expectation value for the 4f electrons and $\langle N\rangle$ represents a reduced matrix element. If there is magnetic ordering, the hyperfine field $H_n \approx H_{orb}$ is proportional to $\langle J_z\rangle$ where the expectation defines the quantum mechanical and thermal averages.

The above discussion implies a linear proportionality between the measured hf magnetic field and the local magnetic moment in rare-earth ions (except for S-state cases). This has been demonstrated for Dy, Er, Tm, and Yb systems [7.51-53] and is illustrated in Fig.7.8 for Er systems. Measurement of the hyperfine field therefore provides a direct measure of the local magnetic moment on the rare-earth atom. In the case of 3d conducting systems, there is also an appropriate proportionality between the 3d moment and H_n. This is well demonstrated in many Fe systems [7.54].

7.3.2 Rare-Earth Rhodium Borides

Compounds having the general formula $RERh_4B_4$ have provided the most common testing ground for studies of the ways in which superconductivity and magnetism interact. In these systems the strength of various interactions such as conduction electron coupling of the rare-earth moments (RKKY-type), superconducting, CEF, and dipolar coupling, are of comparable magnitude. This results in a wide spectrum of observed phenomena. In deducing the magnetic information, bulk magnetization measurements are limited since severe magnetic anisotropies due to CEF make the interpretation of data on powder samples difficult, and no data on single crystals are available. The Mössbauer effect measurements even on powders do not have this difficulty since they measure only local magnetic properties, often without the need of external magnetic fields.

Mössbauer spectra have been obtained for compounds with RE = Dy, Er and Tm and for some pseudoternaries containing Ho and Gd [7.55-57]. Well-resolved magnetic hf spectra are observed in all cases even above the magnetic transition, indicating $\nu_{hf} \gg \nu_s$. If ν_s is completely described by the conduction electron coupling discussed earlier (Sect.7.2.1), this implies a small value for $\underline{J}N(E_F)$. However, it is also likely (for non-S state ions experiencing CEF) that large anisotropies are present, which will reduce any fluctuation of the spin because $g_x = g_y \approx 0$. As discussed earlier, under these circumstances the splitting directly yields the value of H_n which is proportional to the paramagnetic moment. Mössbauer spectra of Dy and Er below the magnetic transition temperature of $DyRh_4B_4$, $ErRh_4B_4$ and $(Er,Gd)Rh_4B_4$ show little change in the Mössbauer spectra except for a sharpening of some of the hf lines [7.55]. One then concludes that the magnetic moments are unchanged and that the magnetic interaction is relatively weak compared to the CEF interactions.

In Table 7.3 results obtained for the magnetic moments on the rare-earth atoms in various $RERh_4B_4$ compounds are given [7.56,57]. In all cases, the measured moment is near the free-ion value. This shows the presence of a state predominantly determined by the angular momentum component $|J_z| = J$ due to the CEF interaction.

The CEF Hamiltonian relevant to the $\overline{4}2m$ symmetry of the RE atom is

$$\underline{H}_{CEF} = B_2^0O_2^0 + B_4^0O_4^0 + B_6^0O_6^0 + B_4^4O_4^4 + B_6^4O_6^4 \quad . \tag{7.15}$$

A simple point charge model predicts that B_2^0 will be the dominant term, being positive for Er and Tm and negative for Dy and Ho. If the exchange interaction is isotropic so the CEF determines the anisotropy, then we expect the moment to be along the basal plane for Er and Tm compounds and along the c-axis for Ho and Dy compounds.

This has been verified by neutron diffraction studies on $ErRh_4B_4$ and $HoRh_4B_4$ [7.9,8]. On the other hand, the B_2^0 term alone would predict different ground state moments from those given in Table 7.3. Thus, other CEF parameters play an important role in these compounds.

Although a complete analysis is not available at the present time, the ground state seems to be made up of two close lying doublets in the case of $ErRh_4B_4$ and $TmRh_4B_4$. A rapid thermal averaging between these states then produces the observed moment. This conclusion is consistant with the measured magnetic entropy of these materials which shows the ground state to be composed of four levels lying close to one another [7.59]. At higher temperatures, the Mössbauer spectra show severe line-broadening due to spin relaxation through the next excited CEF level [7.57]. This allows one to obtain an estimate of the locations of the first excited CEF states, and these are given in Table 7.3. These results are qualitatively in agreement with those found in specific heat studies [7.59].

It must be pointed out that the Mössbauer measurements in the ordered state do not give the direction of the moment with regard to the crystal axis. Also, since one is in the so-called slow spin-relaxation regime ($\nu_{hf} \gg \nu_s$), the magnetic transition in $ErRh_4B_4$ is not seen from the spectra. The spectrum measured at 100 mK (T_m = 0.93 K) permits us to obtain the magnetic moment on Er to be 8.3 ± 0.2 μ_B. This is 30% higher than that measured from neutron diffraction (Fig.7.9). Such a difference in the magnetic moment deduced by the two techniques is quite unusual, having previously been seen only in some disordered magnetic solids. The discrepancy is much too large to be due to experimental errors in the two experiments and must represent real differences in the two types of measurements. Two possible explanations have been suggested to explain this anomaly:

(i) Due to the nature of neutron diffraction, only the long-range coherently-ordered moment is observable. Hyperfine techniques, on the other hand, measure a single ion property and when $\nu_{hf} \gg \nu_s$, as in this case, the total moment on the atom is obtained independent of the magnetic state of the material. One could therefore explain the two measurements with a complex magnetic state which has a total moment of 8.3 μ_B, and a ferromagnetically ordered component of 5.6 μ_B, with the remaining component being disordered even at very low temperatures.

(ii) The material may not be homogeneous but may contain domains of ferromagnetism embedded in a spin-disordered matrix. In this case, the conversion of neutron intensities into a magnetization would be in error by assuming all the material to be ordered. However, in the slow relaxation regime, the Mössbauer spectra are the same for the ordered and disordered regions and so are insensitive to the different phases. One possible suggestion for the disordered regions is unusually wide domain walls.

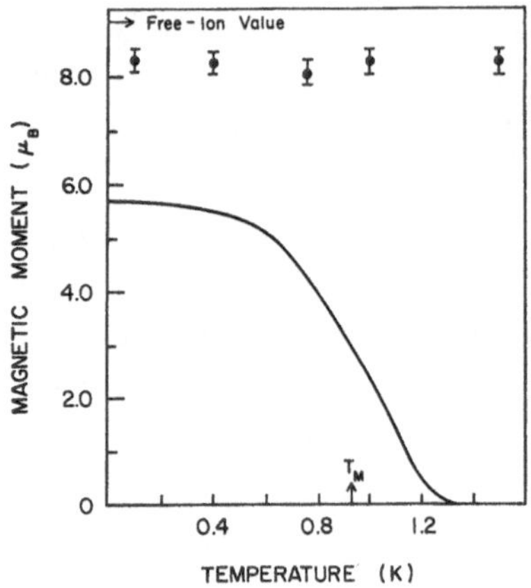

Fig.7.9. Temperature dependence of total moment (solid circles) as determined by Mössbauer spectroscopy and coherent moment (solid line) as determined by neutron diffraction on the Er atom in $ErRh_4B_4$

At present, insufficient evidence is available to allow a proper selection among these possibilities. However, it is clear from the above suggestions that the magnetic state of $ErRh_4B_4$ is not that of a simple ferromagnet. The origin of such behavior is generated by various competing effects. The closeness of two CEF doublets may generate temperature dependent anisotropies in the Er moment. The dipolar interaction between Er moments will tend to drive the system into the antiferromagnetic state [7.60]. The RKKY coupling will perhaps favor either ferromagnetism or helical magnetism [7.61,62] depending on the q vector dependence of the susceptibility of the conduction electrons. The observed behavior is probably a complex resultant of these different effects.

KUMAGAI et al. [7.46] have carried out Gd zero-field NMR measurements on $(Y_{1-x}Gd_x)Rh_4B_4$ in the ferromagnetic state. For x = 1.0, sharp peaks are seen at 38.5 and 50.5 MHz corresponding to resonances in ^{155}Gd and ^{157}Gd, respectively. At x = 0.8, these peaks shift upward in frequency by a factor of approximately two and become very broad. At higher concentrations up to x = 0.26, little additional shift is seen, but the peaks again become very sharp. The reasons for these changes are not well understood, although two suggestions are made related to a change in the Fermi surface of the materials, or a fixing of the magnetization to certain crystalline axes such that the signal would come from magnetic domains rather than from Bloch walls.

Attempts were also made to see if remanent superconductivity exists in the Bloch walls. Measurements of the temperature dependence of the nuclear relaxation rate T_1 failed to show the exponential dependence expected for superconductivity. However, one cannot rule out the possibility that the observed signal comes from domains where no superconductivity will be present, or that the material exhibits gapless superconductivity in the Bloch walls.

7.3.3 Ternary Silicides

The class of compounds having the general formula $M_2Fe_3Si_5$ (M = Sc, Y, RE) have been shown to exhibit either superconductivity or magnetic behavior or both, depending on M (see also Chap.2). The values of T_c in these compounds are as high as 6.0 K for M = Lu [7.63] while the M = Tm compound shows re-entrant behavior [7.64]. These have the highest values of T_c known for compounds containing large concentrations of iron, and the behavior may be contrasted with the Chevrel phase materials where small concentrations of Fe destroy the superconductivity. For most cases where M is a rare-earth, the compounds show antiferromagnetic order and are not superconducting. The superconducting compounds are of particular interest since none of the constituents are those classically associated with superconductivity. Therefore, questions arise concerning the origin of the superconducting properties and the role played by the various components.

Mössbauer spectroscopy has been used to study compounds with M = Sc, Y, Tb, Dy, Er, Tm, and Lu [7.65-68]. Spectra taken using the ^{57}Fe resonance in these compounds can be decomposed into two electric quadrupole doublets, each representing one of the two nonequivalent crystallographic sites occupied by the Fe atoms. In the absence of external magnetic fields, no magnetic hyperfine splitting is seen in compounds with M = Sc, Y, Lu, implying either the absence of a magnetic moment on Fe or a fast spin relaxation such that $\nu_s \gg \nu_{hf}$. One can distinguish between these two possibilities by the application of an external magnetic field H_{ext} to the material. In that case, the measured hyperfine field is

$$H_n = H_{ext} + H_i + H_t \quad , \tag{7.16}$$

where H_i is the field produced by the local magnetic moment when aligned by the external field, and H_t is a transferred hf field arising from the polarization of conduction electrons due to other magnetic atoms (M) in the lattice. Analysis of the data in large external fields on compounds with M = Sc and Lu gives $H_n = H_{ext}$, showing that there is no magnetic moment on the Fe atom. This nonmagnetic character of Fe therefore makes it possible for these compounds to be superconducting. In fact, the large values of T_c suggest that d-electrons participate in the superconductivity. If we accept the criterion that the clustering of metal atoms is essential for superconductivity [7.69], then the crystal structure suggests that the 3d electrons of the Fe atoms may play an essential role [7.67].

Although many cases are now known where antiferromagnetism can coexist with superconductivity in ternary compounds, such is not the case here. For example, $Dy_2Fe_3Si_5$ is antiferromagnetic below 4 K but shows no superconducting behavior. Mössbauer spectra for ^{161}Dy at 1.5 K yield a magnetic moment of 7.0 ± 0.2 μ_B on Dy in this compound, while the field at the Fe nucleus indicates less than 0.005 μ_B moment on the atom in the antiferromagnetic state. Data have also been obtained for compounds with M = Dy, Tb in external magnetic fields at 4.2 K. While one

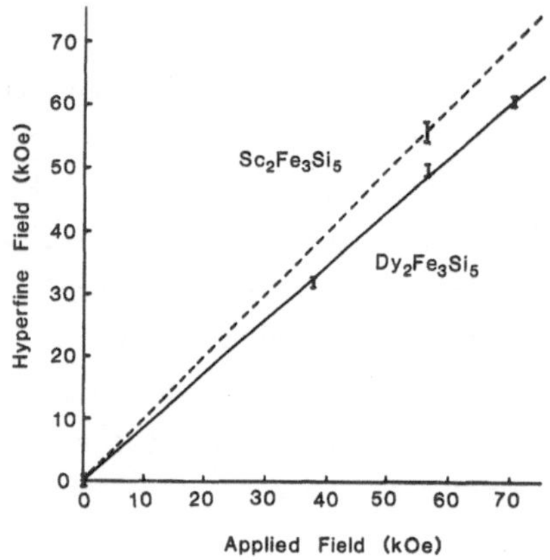

Fig.7.10. Dependence of hyperfine field at the Fe site on external field at 4.2 K in $Sc_2Fe_3Si_5$ and $Dy_2Fe_3Si_5$

finds $H_n = H_{ext}$ in $Sc_2Fe_3Si_5$, one always finds $H_n < H_{ext}$ in compounds with M = Dy and Tb. This implies a negative spin polarization at the Fe nucleus induced by the moments on the rare-earth atom as depicted in Fig.7.10. The resulting exchange interaction caused by the proximity of the M and Fe atoms in this particular structure therefore may preclude superconductivity in the antiferromagnetic state.

7.3.4 Ternary Stannides and Chevrel Phase Compounds

Ternary compounds with the general formula MRh_xSn_y form a group having a complex crystal chemistry in which three phases are known. The materials exhibit a variety of interesting magnetic and superconducting phenomena [7.70,71] including the re-entrant superconductivity in $ErRh_{1.1}Sn_{3.6}$. The re-entrant superconductor $ErRh_{1.1}Sn_{3.6}$ also behaves similarly to $ErRh_4B_4$. The Mössbauer studies [7.70] yield 7.5 μ_B for the moment on Er well below T_m (= 0.46 K), while the neutron measurements [7.71] show no long-range correlations at all in its ferromagnetic state. The discussion presented for the behavior of $ErRh_4B_4$ is pertinent also to $ErRh_{1.1}Sn_{3.6}$.

The contribution of Mössbauer spectroscopy in understanding the crystal chemistry will be discussed in the next section. Here, we mention the most important results on magnetic studies [7.72]. Mössbauer spectra taken for ^{119}Sn in MRh_xSn_y (M = Dy, Er) show a small magnetic hyperfine splitting, indicating a nonzero spin density at the Sn site. Since the dipolar field at the Sn site is negligible compared to the measured hyperfine field, this must arise from polarization of the conduction electrons by the spins on the M atoms. The Dy compound is ferromagnetic and such a transferred field at the Sn site is expected. In the Er compound, however, the transferred field is also observed in the paramagnetic state of the material. This will occur if the spin-relaxation frequency for the Er magnetic moment generating the field is small compared to the hyperfine frequency of the ^{119}Sn due to the field. Analysis of the Sn and Er hyperfine patterns shows this to be the case. It will be of interest to study this conduction electron polarization below T_c, since it is presumed that the superconductivity is due to electrons from Sn metal clusters in this material.

Magnetic investigations of Chevrel compounds using the Mössbauer effect are rather limited. In the pseudoternary compounds $Eu_xSn_{1-x}Mo_6S_8$, the magnetic order is of spin-glass type [7.73]. Similar behavior is found for $Fe_xSn_{1-x}Mo_6S_8$ while $Fe_xMo_6S_8$ appears to be antiferromagnetic [7.74]. The iron compounds show no superconductivity for iron concentration in excess of about 5 atomic percent. Measurements on $YbMo_6S_8$ show that it contains approximately equal amounts of Yb^{2+} and Yb^{3+} ions. The magnetic Yb^{3+} ions show antiferromagnetic order below 2.75 K [7.75], with a magnetic moment of 1.7 μ_B. This is smaller than the free-ion value for Yb^{3+} and again shows the importance of the CEF in determining the magnetic properties.

7.4 Crystal Chemistry and Charge Transfer

The measurement of the Mössbauer isomer shifts and quadrupole interactions provides a great deal of insight into the crystal chemistry of materials and charge transfer between various atoms [7.76]. Oxidation states can also be identified with such measurements. In addition, information on the occupation of particular atoms in the available crystallographic sites can be obtained. Here we present some crystal chemistry results concerning ternary superconductors.

The chemical and magnetic state of Fe atoms in the superconducting $R_2Fe_3Si_5$ compounds can be evaluated from the isomer shifts of Fe. When Fe is coordinated by 12 or 13 atoms in this lattice, it goes into a low-spin state with a filled d-band or filled d-orbital. This results in a unique isomer shift for Fe. Fe also has no moment in the superconductor Th_7Fe_3 in which the Fe atom has a simple coordination. However, in the latter case, the isomer shift reflects a low density of d-states at the Fe site, so the electronic state is distinct from that in $M_2Fe_3Si_5$ compounds [7.77].

The complex nature of the crystal chemistry of MRh_xSn_y has been described by CHENAVAS et al. [7.78]. The Mössbauer spectrum in $ErRh_{1.1}Sn_{3.6}$ at 77 K is made up of two quadrupole split patterns suggesting two crystallographically distinct Sn atoms in the structure. The isomer shifts for these Sn atoms are also distinct, indicating a large chemical dissimilarity between these atoms. CHENEVAS et al. reason that the structure can be described by $Sn(1)Er_3Rh_4Sn(2)_{16}$ in which Sn(1) is cation-like and Sn(2) is anion-like. The isomer shifts relative to $CaSnO_3$ are 1.53 mm/s and 2.15 mm/s, which have been assigned to Sn(1) and Sn(2) sites, respectively. The systematics of the isomer shift have been used to suggest that Sn(2) behaves like α-Sn metal, while the Sn(1) atom is positively charged. The population of the sites deduced from the spectrum is approximately 1:12 which compares favorably with the ratio 1:16 based on structural considerations [7.72].

One of the important factors which determines the value of T_c in Chevrel phase compounds is the valence electron concentration (VEC) on the Mo_6-cluster [7.16]. The magic number of 22 seems to produce the highest value for T_c (Table 7.4). Mössbauer spectroscopy has been used to deduce the charge on Te atoms in the $Mo_6(Te_{1-x}Se_x)_8$ system [7.80] and on I atoms in $Mo_6X_6I_2$ (X = S, Te) compounds [7.81]. From the evaluation of the isomer shifts and quadrupole interactions, it is found that Te and I in these compounds have -1 and -0.85 electron charges, respectively. Assuming sulfur atoms to have a charge of -2, the VEC on the Mo_6-cluster can be deduced, as shown in Table 7.4. While the highest T_c is obtained for the VEC of about 22 in the sulfides, no such correlation can be drawn for the tellurides. This may be due to the formation of p-d hybridized bands from the nonlocal 5p electron on Te.

The Chevrel phase compound $Fe_xMo_6S_8$ ($1 < x < 4$) is a defect structure; there are twelve available sites per unit cell for the Fe atom with a maximum occupation of approximately four. The small metal atom leads to a contraction along the rhombohedral axis and a "delocalization" of the Fe atom from the center of symmetry. Two adjacent rings of six sites result, with the inner ring (I) occupied at low metal concentrations, but with some ring II sites also occupied at higher metal concentrations [7.16]. Mössbauer measurements on $Fe_1Mo_6S_8$ show a single quadrupole split pattern at ~ 50 K and two such patterns at ~ 100 K. Figure 7.11 is a plot of the ratio of the intensity of Fe in the two electronic states vs temperature. X ray measurements show that $FeMo_6S_8$ undergoes a symmetry lowering transformation at a temperature of ~ 100 K [7.82]. Previous X ray studies on $Fe_2Mo_6S_8$ [7.83] and $Fe_{1.32}Mo_6S_8$ [7.84] have shown that these compounds are rhombohedral at high temperature, but undergo a transformation to the triclinic phase at low temperature with the transition temperature lower for lower Fe concentration. YVON has deduced that at high temperature the metal atom is distributed among the twelve available sites, but as the temperature is lowered, the inner site tends to be preferentially occupied [7.83]. In $Fe_2Mo_6S_8$ the Fe atoms chemically order and give rise to a cell with triclinic symmetry and only two iron sites per cell. The Mössbauer results represent a microscopic observation of the order-disorder transformation leading to the symmetry lowering phase transition [7.74]. The curve in Fig.7.11 is proportional to the occupation of the two electronic states n_{II}/n_I assumed by Fe atoms through the phase change since the Mössbauer intensity is a product of the occupation number and the Debye-Waller factor. The total Mössbauer intensity abruptly increases in the low temperature phase, indicating that the low temperature phase has a higher Debye temperature. At temperatures above 150 K the Mössbauer lines begin to broaden. Between 210 and 220 K the broadening rapidly increases, giving rise to line shapes characteristic of atomic diffusion [7.74].

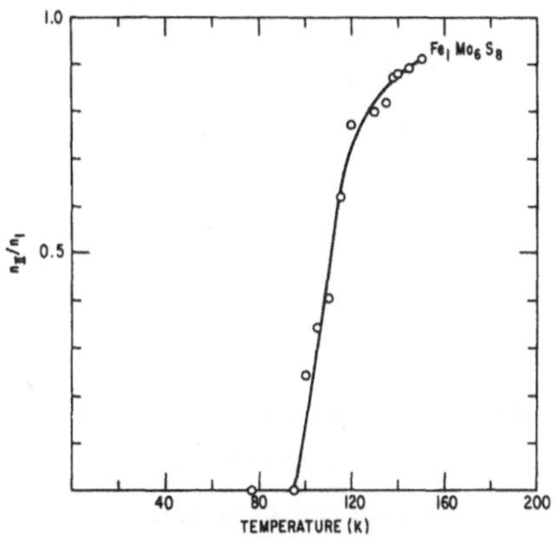

Fig.7.11. Order-disorder transition in $FeMo_6S_8$ near 95 K measured by temperature-dependent change in Mössbauer intensities

Mössbauer measurements have been made at ^{57}Fe sites in Chevrel phase compounds [$Fe_xSnMo_6S_8$ (x = 0.05-0.4)] between 4.4 and 1100 K to delineate characteristics of site occupation and electronic behavior of excess metal atoms [7.85]. In $SnMo_6S_8$ the Sn atom is near the center of inversion symmetry. The $SnMo_6S_8$ unit cell is larger than that of $FeMo_6S_8$ and the character of the central metal site is different. The addition of Fe to $SnMo_6S_8$ causes a further increase in the volume of the unit cell [7.28]. Neutron diffraction measurements [7.86] on $Fe_{0.4}SnMo_6S_8$ show that Fe occupies one crystallographic site which corresponds closely to the ring II site in $Fe_xMo_6S_8$.

At 300 K, all $Fe_xSnMo_6S_8$ (x = 0.05 to 0.4) compounds show similar, slightly broadened, quadrupole split-doublet spectra at ^{57}Fe nuclei. Over this concentration range, the Fe atoms are thus in similar crystallographic environments and behave alike electronically. In addition, the parameters derived from these spectra are close in value to those attributed to Fe in ring II in $FeMo_6S_8$. Two electronic states are observed by the Mössbauer effect for the Fe atom in $Fe_xSnMo_6S_8$, although Fe has been found to be in a single crystallographic site by neutron diffraction. The two states are inferred to arise from the presence of a fraction of Fe atoms having nearest-neighbor ring II sites occupied in adjacent cells. The two electronic states are then due to Fe monomers and dimers. Random placement of the Fe atoms leads to a small probability of pair formation ($\sim$ 0.01); the fraction of pairs inferred from the Mössbauer intensities is $\sim$0.3. A nonrandom, short-range ordering process is inferred to take place in which pairs of Fe atoms are formed between nearest-neighbor ring II sites in adjacent unit cells. The distance between Fe atoms in a pair is approximately the same distance as that in iron pairs in $Fe_2Mo_6S_8$ [7.83]. The shift for both states of Fe in $Fe_xSnMo_6S_8$ is similar, as would be expected from the site assignment. However, the quadrupolar coupling for the two states begins to change dramatically near 400 K and the rate of change of the splitting is different for the two states.

It is possible that some iron atoms may occupy central cell positions vacated by tin atoms in $Fe_xSnMo_6S_8$. Recent stoichiometry studies of $SnMo_6S_8$ by WAGNER and

FREYHARDT [7.27] show that lattice parameter and T_c changes are connected with the generation of defects, rather than precipitation of additional phases; for example, ^{119}Sn in Mo, MoS_2, and Mo_2S_3 were shown by Mössbauer measurements to comprise less than one atomic percent of the alloy.

7.5 Conclusions

In this review of NMR and Mössbauer effects in ternary superconductors, we have tried to demonstrate the utility of hyperfine techniques for measuring the localized vibrational, magnetic and charge characteristics of these cluster compounds. A number of critical issues still need to be addressed in magnetic superconductors, such as the nature of the ground state in ferromagnetic superconductors (i.e., the question of inhomogeneous magnetization and the question of the existence of superconducting regions in the domain walls) and the nature of the correlated magnetic state above re-entrant transitions in ferromagnetic superconductors. Hyperfine techniques including zero-field magnetic resonance and pure quadrupole resonance have the necessary sensitivity and appropriate local interaction tensors to attack these problems dealing with coupled superconducting and magnetic-order parameters.

References

7.1 L.C. Hebel, C.P. Slichter: Phys. Rev. *113*, 1504 (1957)
7.2 J. Bardeen, L.N. Cooper, J.R. Schrieffer: Phys. Rev. *108*, 1175 (1957)
7.3 R. Chevrel, M. Sergent, J. Prigent: J. Sol. State Chem. *3*, 515 (1971)
7.4 B.T. Matthias, M. Marezio, E. Corenzwit, A.S. Cooper, H.E. Barz: Science *175*, 1465 (1972)
7.5 C.W. Kimball, L. Weber, G. VanLanduyt, F.Y. Fradin, B.D. Dunlap, G.K. Shenoy: Phys. Rev. Lett. *36*, 412 (1976)
7.6 Ø. Fischer, M. Decroux, S. Roth, R. Chevrel, M. Sergent: J. Phys. C*8*, L474 (1975)
7.7 F.Y. Fradin, G.K. Shenoy, B.D. Dunlap, A.T. Aldred, C.W. Kimball: Phys. Rev. Lett. *38*, 719 (1977)
7.8 W.A. Fertig, D.C. Johnston, L.E. Delong, R.W. McCallum, M.B. Maple, B.T. Matthias: Phys. Rev. Lett. *38*, 987 (1977)
7.9 D.E. Moncton, D.B. McWhan, J. Eckert, G. Shirane, W. Thomlinson: Phys. Rev. Lett. *39*, 1164 (1977)
7.10 B.D. Dunlap, G.K. Shenoy, F.Y. Fradin, C.W. Kimball: J. de Physique *39*, 379 (1978)
7.11 W.L. McMillan: Phys. Rev. *167*, 331 (1968)
7.12 T.A. Kitchens, P.P. Craig, R.D. Taylor: In *Mössbauer Effect Methodology*, Vol.5, ed. by I.J. Gruverman (Plenum, New York 1970) p.123;
J.S. Shier, R.D. Taylor: Phys. Rev. *174*, 346 (1968)
7.13 C.W. Kimball, S.P. Taneja, L. Weber, F.Y. Fradin: *Mössbauer Effect Methodology*, Vol.9, ed. by I.J. Gruverman, C.W. Seidel, and D.K. Dieterly (Plenum, New York 1974) p.93;
C.W. Kimball, L.W. Weber, F.Y. Fradin: Phys. Rev. B*14*, 2769 (1976)
7.14 J. Bolz, H. Hauck, F. Pobell: Z. Physik B*25*, 35 (1976)
7.15 Ø. Fischer: Appl. Phys. *16*, 1 (1978)

7.16 K. Yvon: In *Current Topics in Materials Science*, Vol.3, ed. by E. Kaldis (North-Holland Publishing Co., NY 1979) p.53
7.17 V.I. Goldanskii, R.H. Herber: In *Chemical Applications of the Mössbauer Effect*, ed. by V.I. Goldanskii, R.H. Herber (Academic, New York 1968) p.1
7.18 L.H. Bowen, C.L. Heinbach, B.D. Dunlap: J. Chem. Phys. *59*, 1390 (1973); T.C. Gibb: J. Chem. Soc. A*1970*, 2503 (1970)
7.19 R. Chevrel: To be published
7.20 S.D. Bader, G.S. Knapp, A.T. Aldred: Ferroelectrics *17*, 321 (1977)
7.21 J.C.K. Hui, P.B. Allen: J. Phys. C: Solid State Phys. *8*, 2923 (1975)
7.22 C.W. Kimball, G. Van Landuyt, C. Barnet, G.K. Shenoy, B.D. Dunlap, F.Y. Fradin: J. Physique *39*, C6-367 (1978)
7.23 S.D. Bader, G.S. Knapp, S.K. Sinha, P. Schweiss, B. Renker: Phys. Rev. Lett. *37*, 344 (1976);
S.D. Bader, S.K. Sinha: Phys. Rev. B*18*, 3082 (1978)
7.24 Ø. Fischer, M.B. Maple (eds.): *Superconductivity in Ternary Compounds I*, Topics in Current Physics, Vol.32 (Springer, Berlin, Heidelberg, New York 1982)
7.25 B.P. Schweiss, B. Renker, R. Flükiger: In *Ternary Superconductors*, ed. by G.K. Shenoy, B.D. Dunlap, F.Y. Fradin (North-Holland, Amsterdam 1981) p.29
7.26 F. Pobell: In *Ternary Superconductors*, ed. by G.K. Shenoy, B.D. Dunlap, and F.Y. Fradin (North-Holland, Amsterdam 1981) p.35
7.27 H.A. Wagner, H.C. Freyhardt: Physica *107*B, 657 (1981)
7.28 F.Y. Fradin, J.W. Downey, T.E. Klippert: Mat. Res. Bull. *11*, 993 (1976)
7.29 B.L. Stafford, C.D. Barnet, C.W. Kimball, F.Y. Fradin: Bull. Am. Phys. Soc. *24*, 389 (1979)
7.30 Ø. Fischer, M. Decroux, R. Chevrel, M. Sergent: In *Superconductivity in d- and f-band Metals*, ed. by D.H. Douglass (Plenum, New York 1976) p.175
7.31 V. Jaccarino, M. Peter: Phys. Rev. Lett. *9*, 290 (1962)
7.32 B.D. Dunlap, G.K. Shenoy, F.Y. Fradin, C.D. Barnet, C.W. Kimball: J. Magn. Mag. Mat. *13*, 319-321 (1979)
7.33 A.J. Freeman, T. Jarlborg: In *Ternary Superconductors*, ed. by G.K. Shenoy, B.D. Dunlap, and F.Y. Fradin (North-Holland, Amsterdam 1981) p.59
7.34 A. Narath, D.W. Alderman: Phys. Rev. *143*, 328 (1966)
7.35 N. Sano, T. Taniguchi, K. Asayama: Solid State Commun. *33*, 419 (1980)
7.36 J.M. Baker, F.I.B. Williams: Proc. Roy. Soc. London, Ser. A*267* (1962); see also [7.33]
7.37 I. Nowik, B.D. Dunlap, J.H. Wernick: Phys. Rev. B*8*, 238 (1973)
7.38 A.A. Abrikosov, L.P. Gor'kov: Sov. Phys. JETP *12*, 1243 (1961)
7.39 E. Bradford, W. Marshall: Proc. Phys. Soc. (London) *87*, 731 (1966)
7.40 F. Hartman-Boutron: J. Physique *41*, C1-223 (1980)
7.41 G. Koopman, U. Engel, K. Baberschke, S. Hüfner: Solid State Commun. *11*, 1197 (1972)
7.42 S. Oseroff, R. Calvo, D.C. Johnston, M.B. Maple, R.W. McCallum, R.N. Shelton: Solid State Commun. *27*, 201 (1978)
7.43 R. Odermatt, M. Hardiman, J. VanMeijel: Solid State Commun. *32*, 1227 (1979)
7.44 F.Y. Fradin, P.K. Tse, A.T. Aldred: In *Ternary Superconductors*, ed. by G.K. Shenoy, B.D. Dunlap, F.Y. Fradin (North-Holland, Amsterdam 1981) p.141;
P.K. Tse, A.T. Aldred, F.Y. Fradin: Phys. Rev. Lett. *43*, 1825 (1979); *44*, 1094 (1980)
7.45 T. Jarlborg, A.J. Freeman, T.J. Watson-Yang: Phys. Rev. Lett. *39*, 1032 (1977)
7.46 K. Kumagai, Y. Inoue, Y. Kohori, K. Asayama: In *Ternary Superconductos*, ed. by G.K. Shenoy, B.D. Dunlap, F.Y. Fradin (North-Holland, Amsterdam 1981) p.185
7.47 D.C. Johnston, B.G. Silbernagel: Phys. Rev. B*21*, 4996 (1980)
7.48 K. Kumagai, Y. Inoue, K. Asayama: Solid State Commun. *35*, 531 (1980)
7.49 R.G. Barnes, R.B. Creel: Phys. Lett. *29*A, 203 (1969)
7.50 S. Ofer, I. Nowik, S.G. Cohen: In *Chemical Application of Mössbauer Spectroscopy*, ed. by V.I. Goldanskii, R.H. Herber (Academic Press, New York 1968) p.428
7.51 J.D. Cashion, G.K. Shenoy: To be published
7.52 G.K. Shenoy, B.D. Dunlap, F.Y. Fradin, C.W. Kimball, W. Potzel, F. Pröbst, G.M. Kalvius: J. Appl. Phys. *50*, 1872 (1979)
7.53 G.M. Kalvius, G.K. Shenoy, B.D. Dunlap: *Les Elements des Terres Rares*, Tome II (C.N.R.S., Paris 1969) p.477
7.54 F. Van der Woude, G.A. Sawatzky: Phys. Rep. *12*C, 335 (1974)

7.55 G.K. Shenoy, B.D. Dunlap, F.Y. Fradin, S.K. Sinha, C.W. Kimball, W. Potzel, F. Pröbst, G.M. Kalvius: Phys. Rev. *21*, 3886 (1980)
7.56 G.K. Shenoy, B.D. Dunlap, F.Y. Fradin, C.W. Kimball: Bull. Am. Phys. Soc. *23*, 206 (1978)
7.57 G.K. Shenoy, P.J. Viccaro, D. Niarchos, J.D. Cashion, B.D. Dunlap, F.Y. Fradin: In *Ternary Superconductors*, ed. by G.K. Shenoy, B.D. Dunlap, F.Y. Fradin (North-Holland, Amsterdam 1981) p.163
7.58 G.H. Lander, S.K. Sinha, F.Y. Fradin: J. Appl. Phys. *50*, 1990 (1979)
7.59 L.D. Wolf, D.C. Johnston, H.B. Mackay, R.W. McCallum, M.B. Maple: J. Low Temp. Phys. *35*, 651 (1979)
7.60 M. Redi, P.W. Anderson: Proc. Nat. Acad. Sci., USA *78*, 27 (1981)
7.61 B. Coqblin: *The Electronic Structure of Rare-Earth Metals and Alloys* (Academic, New York 1977)
7.62 P.W. Anderson, H. Suhl: Phys. Rev. *116*, 898 (1959)
7.63 H.F. Braun: Phys. Lett. *75*A, 386 (1980)
7.64 C.U. Segre, H.F. Braun: Phys. Lett. *85*A, 372 (1981)
7.65 J.D. Cashion, G.K. Shenoy, D. Niarchos, P.J. Viccaro, C.M. Falco: Phys. Lett. *79*A, 454 (1980)
7.66 J.D. Cashion, G.K. Shenoy, D. Niarchos, P.J. Viccaro, A.T. Aldred, C.M. Falco: J. Appl. Phys. *52*, 2180 (1981)
7.67 J.D. Cashion, G.K. Shenoy, D. Niarchos, P.J. Viccaro: Proc. Intern. Conf. on Appl. Mössbauer Effect, Jaipur, India, Dec. 14-18, 1981, to be published
7.68 G.K. Shenoy, D. Niarchos, D. Noakes, C.U. Segre, H.F. Braun: To be published
7.69 J.M. Vandenberg: In *Ternary Superconductors*, ed. by G.K. Shenoy, B.D. Dunlap, F.Y. Fradin (North-Holland, Amsterdam 1981) p.21
7.70 G.K. Shenoy, F. Pröbst, J.D. Cashion, P.J. Viccaro, D. Niarchos, B.D. Dunlap, J.P. Remeika: Solid State Commun. *37*, 53 (1980)
7.71 J.P. Remeika, G.P. Espinosa, A.S. Cooper, H. Barz, J.M. Rowell, D.B. McWhan, J.M. Vandenberg, D.E. Moncton, Z. Fisk, L.D. Woolf, H.C. Hamaker, M.B. Maple, G. Shirane, W. Thomlinson: Solid State Commun. *34*, 923 (1980)
7.72 G.K. Shenoy, P.J. Viccaro, J.D. Cashion, D. Niarchos, B.D. Dunlap, F. Pröbst, J.P. Remeika: In *Ternary Superconductors*, ed. by G.K. Shenoy, B.D. Dunlap, F.Y. Fradin (North-Holland, Amsterdam 1981) p.233
7.73 J. Bolz, G. Crecelius, H. Maletta, F. Pobell: J. Low Temp. Phys. *28*, 61 (1977)
7.74 J.M. Friedt, B.D. Dunlap, G.K. Shenoy, A.T. Aldred, F.Y. Fradin, C.W. Kimball: Physica *107*B, 61 (1982)
7.75 P. Bonville, J.A. Hodges, P. Imbert, G. Jehanno, R. Chevrel, M. Sergent: Rev. Physique Appl. *15*, 1139 (1980)
7.76 See, for example: *Mössbauer Isomer Shifts*, ed. by G.K. Shenoy, F.E. Wagner (North-Holland, Amsterdam 1978)
7.77 J.D. Cashion, G.K. Shenoy, D. Niarchos, P.J. Viccaro, C.M. Falco: In *Nuclear and Electron Resonance Spectroscopies as Applied to Materials Science*, ed. by E.N. Kaufmann, G.K. Shenoy (North-Holland, Amsterdam 1981) p.315
7.78 J. Chenavas, J.L. Hodeau, A. Collomb, M. Marezio, J.P. Remeika, J.M. Vandenberg: In *Ternary Superconductors*, ed. by G.K. Shenoy, B.D. Dunlap, F.Y. Fradin (North-Holland, Amsterdam 1981) p.219
7.79 M. Sergent, Ø. Fischer, M. Decroux, C. Perrin, R. Chevrel: J. Solid State Chem. *22*, 87 (1977)
7.80 R.W. McCallum, F. Pobell, R.N. Shelton: Poster Paper, Intern. Conf. Mössbauer Spectroscopy, Portoroz, Yugoslavia (1979)
7.81 G.V. Subba Rao, D. Niarchos, G.K. Shenoy, J.D. Cashion, D. Hinks, A.M. Umarji, S. Janaki: Proc. Intern. Conf. Appl. Mössbauer Effect, Jaipur, India, December 14-18, 1981, to be published
7.82 H. Knott, M. Mueller: Private communication
7.83 K. Yvon: Acta Crystal., to be published
7.84 J. Guilleon, O. Bars, D. Grandjean: Acta Crystal. B*32*, 1338 (1976)
7.85 B.L. Stafford, T.F. Karlov, C.W. Kimball, F.Y. Fradin, as well as J.M. Dugan, B.L. Stafford, C.W. Kimball, P.J. Viccaro, F.Y. Fradin, A.T. Aldred, B.D. Dunlap: In *Ternary Superconductors*, ed. by G.K. Shenoy, B.D. Dunlap, F.Y. Fradin (North-Holland, Amsterdam 1981)
7.86 J. Jorgenson, D.G. Hinks, F.J. Rotella: In *Ternary Superconductors*, ed. by G.K. Shenoy, B.D. Dunlap, and F.Y. Fradin (North-Holland, Amsterdam 1981)

8. Neutron Scattering Studies of Magnetic Ordering in Ternary Superconductors

W. Thomlinson, G. Shirane, J. W. Lynn and D. E. Moncton

With 12 Figures

The possibility of the microscopic coexistence of magnetic order and superconductivity has had a rich and interesting history. Early work revealed that magnetic ions substituted into a superconducting host generally had a very detrimental effect on superconductivity [8.1] due to the exchange-induced spin depairing of the Cooper electrons [8.2]. Typically less than a percent magnetic ion concentration was found to be sufficient to destroy superconductivity, and at such low concentrations the spin correlations between the magnetic impurities could to a good approximation be neglected. Thus, there was no chance for a cooperative magnetic state to develop.

The first exception to this rule was the binary $CeRu_2$ system in which large concentrations of heavy rare earths could be substituted for Ce before suppressing the superconductivity [8.3,4]. The reason for the extremely small pairbreaking interaction is the isolation of the magnetic ions from the Ru sublattice on which the superconducting electrons travel. The high concentration of magnetic ions afforded the first possibility for magnetic correlations to develop in a superconductor, which in turn initiated neutron scattering measurements on $(Ce_{1-x}Tb_x)Ru_2$ [8.5,6] and $(Ce_{1-x}Ho_x)Ru_2$ [8.7,8]. Ferromagnetic correlations were found to develop at low temperatures, but long-range order was not realized in the range of concentrations where superconductivity exists.

The exploration of the interplay between long-range magnetic order and superconductivity began with the discovery of a new class of superconductors with ternary crystal structures. The materials which have proven to be the most interesting so far are the rare-earth (R) molybdenum chalcogenides (RMo_6S_8, RMo_6Se_8) and the rare-earth rhodium borides (RRh_4B_4). The magnetic ions in these systems are also electronically isolated from the electrons which interact in the Cooper pair formation, which is of course essential if these systems are to be superconductors at all. Equally important, the magnetic ions reside on a (separate) periodic sublattice in the crystal structure, which avoids the complications of competing interactions that often occur in randomized magnetic alloys. Thus, at low temperatures spin-entropy must drive the magnetic ions to order. The observation of anomalies in the electric, magnetic and thermodynamic properties of these superconductors indicated they were undergoing magnetic transitions.

In this chapter we review the role that neutron scattering has taken in elucidating the nature of the magnetic order in these ternary superconducting systems. Of fundamental concern is the question of whether these systems do indeed develop *long-range* order, or whether they simply undergo a "spin freezing" phenomenon as in the $(Ce_{1-x}R_x)Ru_2$ alloy. Determination of the nature of the magnetic phase transitions can only be done unambiguously through neutron scattering observations. The results have shown that these systems do indeed exhibit long-range magnetic order. Most materials order in a compensated antiferromagnetic arrangement, and since there is no macroscopic (dipolar) magnetization associated with such a state, the superconductivity is relatively unaffected by the ordering. $DyMo_6S_8$ [8.9] provided the first unambiguous realization of the coexistence of superconductivity with long-range magnetic order.

Several materials (e.g., $ErRh_4B_4$ and $HoMo_6S_8$) undergo the more interesting case of ferromagnetic alignment, where the electromagnetic coupling to the superconductivity is dominant [8.10]. The competition between ferromagnetism and superconductivity produces a compromise long-wavelength oscillatory magnetization at intermediate temperatures. Further development of the magnetic state with decreasing temperature results in the destruction of superconductivity, and the compounds are pure ferromagnets at low T.

8.1 Neutron Scattering Theory

There are two basic types of scattering interactions that the neutron experiences in a solid; an interaction with the nuclei mediated by the strong nuclear force and an electromagnetic interaction due to the dipole moment of the neutron and any induction field $\underline{B}$ present in the sample. The nuclear interaction gives rise to Bragg diffraction, which yields information about the crystal structure. In addition to the equilibrium positions of the atoms, however, neutron scattering is also capable of measuring time-dependent phenomena such as lattice vibrations. The information which has been obtained on the phonon spectra in these systems is discussed by Bader et al. [Ref.8.11, Chap.7].

The magnetic interaction gives rise to *magnetic* Bragg diffraction which originates from the magnetic order in the crystal. Since the equivalent X ray magnetic scattering is very weak, neutron scattering is in fact the only technique routinely used to determine magnetic structures. This is the principle topic of the present chapter. The dynamics of the magnetic system can also be studied by inelastic neutron scattering, but since the characteristic collective excitations are expected to be at very low energies, there have been no investigations of the spin dynamics yet. On the other hand, the interaction of the crystalline electric field with the orbital part of the 4f wave function turns out to be large compared with exchange and dipolar energy in these systems (except Gd, which is in an S-state). Therefore only the crystal field ground state is occupied at low temperatures where the order takes place. The studies of the crystal field effects have been reviewed recently [8.12] and hence they will not be discussed in detail here. We will simply take as given that there is a magnetic moment on the rare-earth ions in these systems which can order at low temperatures, keeping in mind that the degeneracy of the magnetic ground state as well as the value of the magnetic moment can be considerably reduced from the free ion values.

The induction field $\underline{B}$ typically originates from the unpaired electron spins and orbital magnetic moments of the (rare-earth) ions in the crystal, but for superconductors the supercurrents can also give rise to a magnetic field which will scatter neutrons. The classic example of this type of scattering is a vortex lattice, which in practice is the only evidence of the superconducting state in the magnetic neutron scattering [8.13]. The neutron observations of the phase transitions in these magnetic superconductors have thus been restricted to the magnetic properties, and it is therefore essential to have other types of measurements such as resistivity, susceptibility and specific heat to characterize the superconductivity. It is particularly important to have these measurements on the identical samples studied by neutrons so that the measured magnetic and superconducting transition temperatures can be correlated. These types of measurements are reviewed in Chaps.4 and 5.

8.1.1 Coherent Bragg Diffraction

The cross section for scattering of unpolarized neutrons from an ordered array of magnetic ions is given by

$$\frac{d\sigma}{d\Omega} = \left(\frac{\gamma e^2}{2mc^2}\right)^2 |\underline{F}_M|^2 \delta(\underline{Q} - \underline{\tau}) \quad , \tag{8.1}$$

where

$$\left(\frac{\gamma e^2}{2mc^2}\right) = -\,0.2695 \cdot 10^{-12}\ \mathrm{cm} \tag{8.2}$$

is the neutron-electron dipole coupling constant [8.14]. The delta function ensures that the Bragg peaks occur at the magnetic reciprocal lattice points $\underline{\tau}$, with $\hbar\underline{Q}$ being the neutron momentum transfer. The magnetic structure factor is

$$\underline{F}_M(\underline{Q}) = \sum_j e^{i\underline{Q}\cdot\underline{r}_j}\, \hat{\underline{Q}} \times [\underline{M}_j(\underline{Q}) \times \hat{\underline{Q}}] \tag{8.3}$$

with $\hat{\underline{Q}}$ designating a unit vector in the direction of $\underline{Q}$, $\underline{M}(\underline{Q})$ is the vector form factor of the jth ion in the unit cell, and the sum is over all atoms in the magnetic unit cell. The Debye-Waller factor has not been included in (8.3) since all the magnetic transitions occur at very low temperatures. The vector cross product can be regarded as a magnetic scattering amplitude for the jth atom, and (8.3) then takes the form of the structure factor equation familiar to X ray scatterers. The more complicated form for (8.3) occurs because the dipole-dipole interaction is a vector coupling rather than a scalar coupling, and the atomic magnetization can change direction as well as magnitude. The neutrons scatter from the component of the magnetization perpendicular to $\underline{Q}$. For the present case of interest, $|F_M|^2$ can be written as

$$|\underline{F}_M(\underline{Q})|^2 = \langle\mu_z\rangle^2 f^2(\underline{Q})\langle 1 - (\hat{\underline{Q}}\cdot\hat{\underline{\eta}})^2\rangle \quad , \tag{8.4}$$

where $f(\underline{Q})$ is the magnetic form factor which is the Fourier transformation of the magnetization density, and $\hat{\underline{\eta}}$ is a unit vector in the direction of the magnetic moment μ_z. The angular brackets in the last term indicate that an average over all possible spin directions must be taken. In writing (8.4) we have taken advantage of the fact that the magnetic sublattices in these systems are simple Bravais lattices. In the rhombohedral Chevrel phase crystal structure ($R\overline{3}$), for example, there is only one rare-earth ion per unit cell. The lattice is therefore simple primitive, and in fact is nearly simple cubic since the rhombohedral angle is nearly 90°. Details of the crystal structure of these systems can be found in [Ref.8.11, Chaps.2 and 3].

Measurements of the intensities and positions of the magnetic Bragg peaks yield the magnetic structure, the spatial distribution of the atomic magnetization and the temperature and field dependence of the (sublattice) magnetization. Essentially all of the measurements on these magnetic superconductors, however, have been taken on polycrystalline materials. Thus, in (8.1) only powder Bragg peaks will be observed at values of $|\underline{\tau}|$, and information can be lost in this powder average. Thus, some values of the form factor will not be known and ambiguities in the determination of the magnetization density can result. In addition, observed intensities are much lower in powders because the scattering in reciprocal space is distributed over a sphere of radius $|\underline{\tau}|$, rather than being concentrated at a single position $\underline{\tau}$. Thus, the measurements are not sufficiently accurate to make critical comparisons with calculated form factors [8.15].

Information about the spin direction (8.4) can also be lost in a powder average. In a system which possesses a unique axis, however, one is assured that at least the angle with respect to this unique axis can be determined [8.16]. If the intensity of one of the magnetic reflections is found to vanish, then from (8.4) we see that $\underline{\eta}$ must be parallel to $\underline{\tau}$ and the direction can still be determined uniquely. In the antiferromagnetic systems we will discuss, there is a unique magnetic axis as well as a unique crystallographic axis, which has allowed a determination of the spin direction in some cases.

We see from (8.4) that the behavior of the magnetic moment $<\mu>$ can be determined from the intensity data. The average here denotes a time average, i.e., the expectation value of $\underline{\mu}$, as well as a spatial average over all equivalent magnetic sites in the lattice. In an alloy such as $(Er_{1-x}Ho_x)Rh_4B_4$, for example,

$$<\mu_z> = x<\mu_{Ho}>_T + (1 - x)<\mu_{Er}>_T \tag{8.5}$$

where we have explicitly assumed that the rare-earth site is fully occupied. We can thus study the spontaneous magnetization (or sublattice magnetization for an antiferromagnet) as a function of temperature, and of course determine the magnetic transition temperature T_M. We emphasize that the magnetic moment measured by neutrons is the z component, which for a free ion is gJ, and not the "effective moment" $g\sqrt{J(J + 1)}$.

Finally, to determine the saturation magnetization either at low temperatures or in a sufficiently large field, the magnetic intensities must be put on an absolute basis. The most convenient way to do this is to compare the magnetic intensities with the nuclear intensities given by

$$|F_N|^2 = |\sum_j b_j \exp(i\underline{Q} \cdot \underline{r}_j)|^2 \quad , \tag{8.6}$$

where the b_j are the (known) nuclear scattering amplitudes. For any given crystal structure, the positions $\underline{r}_j$ of the atoms in the unit cell are usually known very well and F_N can be readily evaluated. If the crystal structure has not been completely determined by X rays, then the F_N can be used to refine or solve the structure.

8.1.2 Phase Transitions and Critical Scattering

At sufficiently high temperatures the magnetic moments fluctuate as a function of time in a random, uncorrelated fashion. Each moment behaves independently so there is no coherence between the magnetic scatterers and the paramagnetic cross section is independent of $\underline{Q}$ [aside from the form factor $f(\underline{Q})$]. With decreasing temperature correlations begin to develop between the magnetic moments, leading to a wave vector dependence of the scattering. The cross section can be conveniently written as

$$\frac{d\sigma}{d\Omega} = \frac{2}{3}\left(\frac{\gamma e^2}{2mc^2}\right)^2 f^2(\underline{Q})\mu(\mu + 1)\chi(\underline{Q}) \quad , \tag{8.7}$$

where for simplicity we have neglected the inelasticity of the scattering. The wave vector-dependent susceptibility $\chi(\underline{Q})$ contains all the information about the spatial correlations in the system. Physically, $\chi(\underline{Q})$ relates the response of the system to an applied magnetic field which varies sinusoidally in space as $\sin(\underline{Q} \cdot \underline{r})$. Such a (small) sinusoidal field results from the scattering of a neutron of momentum change $\hbar\underline{Q}$.

For a conventional ferromagnet above T_C, $\chi(Q)$ at small wave vectors is given to a good approximation by the Ornstein-Zernike form

$$\chi(Q) = \frac{C}{Q^2 + \kappa^2} , \tag{8.8}$$

where C is a constant related to the strength of the correlations [8.17]. The Fourier inversion of (8.8) yields the spin-spin correlation function in real space, which has the familiar spatial dependence $\exp(-\kappa r)/r$. The range of the correlations in real space is denoted by ξ, with $\xi = 1/\kappa$. As the phase transition is approached from high temperatures the range of the correlations increases and $\xi \rightarrow \infty$ as $T \rightarrow T_C^+$, which is the characteristic of a phase transition to *long-range* order. Experimentally one can never see this divergence since the intrinsic width of the scattering given by (8.8) will at some temperature necessarily become small in comparison with the instrumental width. Thus experimentally, only a lower limit to ξ can be determined.

For the case of a ferromagnetic superconductor we should expect strong competition, as already noted. The basic reason for this is that the most favorable configuration for ferromagnetism is parallel alignment of the moments. The associated macroscopic dipolar field, however, will have an effect similar to an applied magnetic field, which of course is energetically unfavorable for the superconductor. It may then be desirable for the coupled system to compromise by producing an oscillation in the magnetization, the criterion being that the wavelength of the oscillation be longer than the magnetic stiffness, but short compared to the superconducting penetration depth. Above the magnetic transition we may then expect (8.8) to be modified such that the long wavelength (small Q) fluctuations are suppressed relative to those at smaller wavelengths, resulting in a $\chi(Q)$ which has a peak at finite Q. At lower temperatures these fluctuations may condense into an ordered oscillatory state characterized by a wave vector Q_C.

There has been considerable theoretical effort along these lines which indicates the possibility at intermediate temperatures of a peak in the neutron scattering cross section at small wave vectors. This peak could originate from either critical fluctuations, a long-range ordered oscillatory magnetic state, or the formation of a spontaneous vortex lattice. According to Ginzburg-Landau theory [8.18] there are the following possible states: superconductivity and no magnetic order; ferromagnetism with no superconductivity; the coexistence of an oscillatory magnetic state with superconductivity. This latter case includes two possibilities, one where the superconducting state is characterized by an order parameter ψ which is essentially uniform in space, and another where ψ is also oscillatory in nature (the vortex state). A review of the theory is given in Chapter 9.

8.2 Antiferromagnetic Superconductors

Most of the ternary rare-earth systems studied to date have turned out to order as compensated antiferromagnets. The order parameter in this case is the *staggered* magnetization, which does not couple strongly to the superconducting state since there is no associated macroscopic magnetic field. The superconducting state survives the magnetic transition and both phases are found to coexist on a microscopic scale down to low temperatures. Since only polycrystalline samples have been available for the materials discussed in this section, it is fortunate that most of the observed magnetic structures have been simple enough to be solved unambiguously from powder diffraction data alone.

8.2.1 Magnetic Structures

In addition to electronic, thermodynamic, and room temperature X ray diffraction measurements on each sample, neutron diffraction patterns were taken at low temperatures, but well above the magnetic transitions. An example of such a powder pattern for a sample of $TbMo_6S_8$ is shown in the lower part of Fig.8.1 [8.19]. Since the rhombohedral angle in these Chevrel phase systems is close to 90°, the splitting of the peaks due to the lowering from cubic symmetry is small. With the resolution used to obtain these data, only the {111} peak shows a resolved splitting. Therefore the peaks have been labeled by cubic indices. In addition to (nuclear) Bragg peaks from the $TbMo_6S_8$, there are a few additional peaks demonstrating that there are small amounts of impurity phases present in the sample. The presence of these impurity phases was of no significance in measuring the nuclear structure or in the determination of the magnetic component of the scattering at low temperatures. They can, however, contribute to measurements of the bulk properties.

The cross section for nuclear Bragg scattering (8.6) has been calculated for each peak by using the known coherent nuclear scattering lengths and the positional parameters of each atom in the unit cell [8.20]. These data confirm that the crystal structure is rhombohedral ($R\overline{3}$), with one formula unit per unit cell. They also provide an absolute calibration for obtaining the saturation magnetic moment μ_z.

The magnetic phase transition in $TbMo_6S_8$ occurs at $T_M = 1.05$ K and gives rise to new peaks (shown by the hatching) in the low temperature powder pattern of Fig.8.1. These peaks can be indexed with half-integral values based on the chemical unit cell of the Chevrel phase. There is no temperature dependence to any of the Bragg peaks at the nuclear positions, so there is no ferromagnetic component. The magnetic Bragg peaks have scattering vectors $\underline{Q} = \underline{\tau} + \underline{q}_m$, where $\underline{\tau} = (h,k,\ell)$ are the reciprocal lattice vectors for the Chevrel structure. The wave vector of the magnetization is $\underline{q}_m = (0,0,1/2)$ (and equivalent directions). This simple magnetic structure consists of alternate (001) planes with oppositely directed magnetic moments. Identical magnetic structures have been found for $GdMo_6S_8$ [8.21], $DyMo_6S_8$ [8.9] and

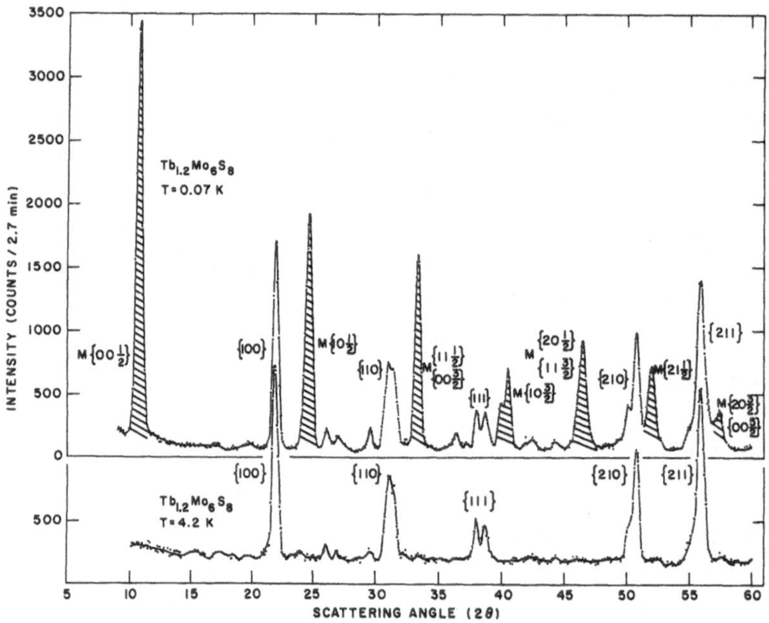

Fig.8.1. Powder neutron diffraction data for $TbMo_6S_8$ above (4.2 K) and below (0.07 K) the antiferromagnetic transition of T_M = 1.05 K [8.19]

$GdMo_6Se_8$ [8.22]. High resolution scans of selected peaks have provided a lower limit to the magnetic correlation length of ξ > 300 Å. It is thus concluded that these materials have conventional long-range antiferromagnetic order. The superconducting transition of $TbMo_6S_8$ (Table 8.1) is 2.05 K and the system remains superconducting below T_M. Thus the magnetic order coexists on a microscopic scale with superconductivity.

The simple magnetic structures found for the sulfides contrast with the situation for $ErMo_6Se_8$, which was the first coexistence compound to be studied [8.23]. Here long-range magnetic order was observed below T_M, but the ordering could not be uniquely linked to the Chevrel phase. Presumably the magnetic structure is not commensurate with the nuclear structure, which means the magnetic structure cannot be described by a simple vector such as $\underline{q}_m$ = (0,0,1/2). Complicated magnetic structures, of course, are more the rule than the exception for rare earths [8.24]. Induced moment studies [8.12,25] have shown the sample to be of high purity, and thus the magnetic order is very likely occurring in the Chevrel phase as anticipated [8.26], with the oscillatory state being characterized by a modulation vector along the [111] rhombohedral axis. There are a few weak magnetic reflections which this assumed magnetic structure does not explain, but these could be originating from very small concentrations of magnetic impurity phases as found in the sulfides. The more complicated behavior for the selenide systems is likely due to the more complicated crystal field effects in these materials [8.12].

Table 8.1. Magnetic ordering in superconducting ternary compounds. T_{C1} is the upper superconducting transition as determined by electrical and thermodynamic methods. T_{C2} is the re-entrant superconducting transition temperature. Actual values for the temperatures vary somewhat from sample to sample. T_{C2} also depends on whether it is measured on heating or cooling. T_M is the magnetic transition temperature and μ_z is the z-component of the magnetic moment as determined by the neutron scattering experiments

Compound	T_{C1} [K]	T_{C2} [K]	T_M [K]	Measured $\mu_z[\mu_B]$	Free ion moment [gJ]	Magnetic structure	Reference
$GdMo_6S_8$	1.4		0.84	6.5±0.3	7	AF	[8.21]
$TbMo_6S_8$	2.05		1.05	8.3±0.2	9	AF	[8.19,29]
$DyMo_6S_8$	2.05		0.4	8.8±0.2	10	AF	[8.9]
$HoMo_6S_8$	1.2	0.64	0.67	9.1±0.3	10	F	[8.32]
$ErMo_6S_8$	2.2		0.2	3.5±0.3	9	AF	[8.29]
$GdMo_6Se_8$	5.6		0.75		7	AF	[8.22]
$ErMo_6Se_8$	6.0		1.1		9	AF (complex)	[8.23]
$NdRh_4B_4$	5.3		1.55 1.2		36/11	AF (complex)	[8.28]
$TmRh_4B_4$	9.9		0.6		7	AF (complex)	[8.27]
$ErRh_4B_4$	8.5		1.0	5.6±0.2	9	F	[8.33,37]
$Er_{0.4}Ho_{0.6}Rh_4B_4$	7.2	3.6	3.6	5.0±0.5	9+	F	[8.44]
$Er_{0.11}Ho_{0.89}Rh_4B_4$	5.7	5.4	5.56	7.8±0.3	9+	F	[8.40]
$HoRh_4B_4$			6.8	8.7±0.3	10	F	[8.43]
$ErRh_{1.1}Sn_{3.6}$	1.22	0.34	0.61		9	F(?)	[8.45]

Two examples from the $(RE)Rh_4B_4$ class of materials have also been found to order magnetically at temperatures below T_c without destroying the superconducting state. $TmRh_4B_4$ [8.27] shows clear evidence for antiferromagnetic order. However, the details of the magnetic structure have not yet been solved. $NdRh_4B_4$ [8.28] exhibits two magnetic-ordering temperatures; first the development of antiferromagnetic order at the higher magnetic transition temperature, and then a distinct change in the magnetic symmetry at lower temperature. The details of these structures have also not been solved.

8.2.2 Magnetic Moment and Spin Direction

In addition to the basic configuration of magnetic moments, the direction of the spin as well as the magnitude of the moment as a function of temperature and magnetic field can also be determined from single crystal data. For powder data it may not be possible to uniquely determine the spin direction as discussed in Sect. 8.1.

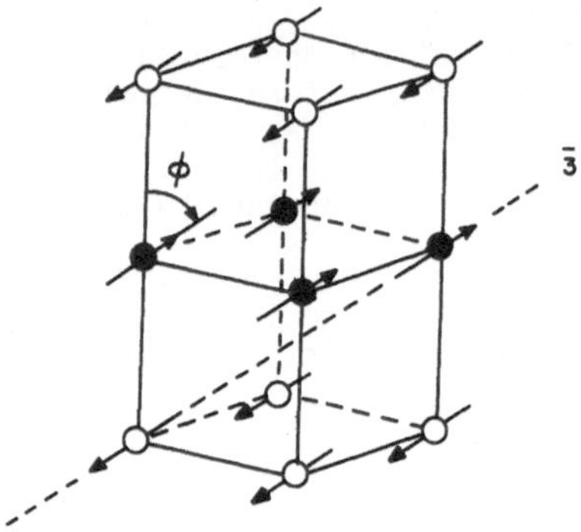

Fig.8.2

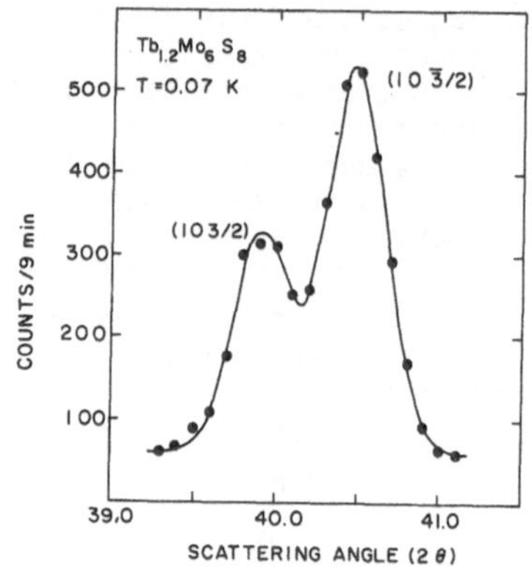

Fig.8.3 ▶

Fig.8.2. Antiferromagnetic structure for Dy, Er, Tb and Gd molybdenum sulfides and $GdMo_6Se_8$, consisting of alternating ferromagnetic planes of spins. The unit cell is doubled along this stacking axis. ϕ is the angle between the spin direction and the doubled axis. The spin direction shown ($\underline{\eta}$ ‖ [111]) is for $DyMo_6S_8$

Fig.8.3. High resolution scans through the {1 0 3/2} magnetic Bragg peaks of $TbMo_6S_8$. The splitting of these peaks originates from the rhombohedral crystal structure [8.29]

The magnetic structure of the Chevrel phase antiferromagnetic materials is approximately tetragonal as shown in Fig.8.2. With this assumption, the angle ϕ between the spin direction and the unique (001) direction can be determined and yields $\phi = 55^\circ$, 60° and 45° for $DyMo_6S_8$, $TbMo_6S_8$ and $ErMo_6S_8$ [8.29], respectively. The saturated value of the magnetic moment can then be calculated from (8.4,6). The resulting values for μ_z are given in Table 8.1. In all cases the values of μ_z obtained by neutron scattering are less than the free ion values gJ. Presumably this reflects the importance of crystal field effects in these materials.

The Chevrel phase materials have a small rhombohedral distortion which reduces the crystal and magnetic symmetry and results in a splitting of the magnetic (tetragonal) peaks. Measurements of the ratio of the intensities of these split peaks allow a unique determination of the moment direction. For example, a scan of the {1 0 3/2} peaks for $TbMo_6S_8$ is shown in Fig.8.3. The moment direction determined from these data is [hkℓ] = [0.85 0.15 0.5]. For $DyMo_6S_8$ the moment was found to lie along the [111] direction [8.9], which is the spin direction depicted in Fig.8.2. For the $ErMo_6S_8$, $GdMo_6S_8$ and $GdMo_6Se_8$, the counting statistics were insufficient to determine the spin direction reliably.

The magnetic-ordering temperatures listed in Table 8.1 have been determined by the temperature dependence of the magnetic intensity of selected magnetic peaks. For the antiferromagnetic superconductors the {0 0 1/2} peak intensity, which is proportional to the square of the staggered magnetization, was selected. The data can be plotted as $\mu_z(T)$ using the normalization to the low temperature saturated moments (Table 8.1). The temperature dependence of the magnetization for four

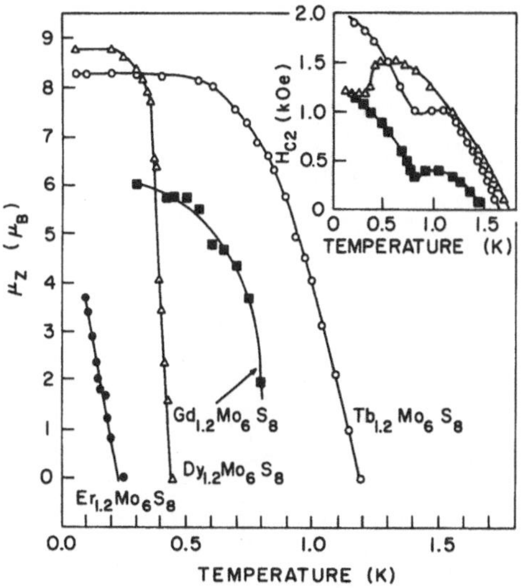

Fig.8.4. Temperature dependence of the sublattice magnetization for $REMo_6S_8$. The inset shows the measured upper critical field values [8.30]

rare-earth molybdenum sulfides is shown in Fig.8.4. No hysteresis was observed in the order parameters on warming and cooling, consistent with the transitions being of second order. The corresponding temperature dependence of the upper critical magnetic fields [8.30] is included in the inset for comparison. These data clearly demonstrate an important interaction between the magnetic order and the superconductivity.

$ErMo_6S_8$ was found to be an *incipient* antiferromagnet [8.29]. It was initially cooled to T = 0.11 K in a zero external applied magnetic field with no evidence of magnetic order developing. A magnetic field was then applied and increased to H = 20 kOe. As the field decreased to a nominal value of 250 Oe, antiferromagnetic order developed. Subsequently, the magnetization was reversible with both temperature and field as shown in Fig.8.4.

8.2.3 Magnetic Field Dependence

Experiments have been performed in applied magnetic fields on $TbMo_6S_8$, $DyMo_6S_8$ and $ErMo_6S_8$. The results for $TbMo_6S_8$ at T = 0.06 K are shown in Fig.8.5, where the intensity of the antiferromagnetic {1 0 1/2} peak is seen to monotonically decrease as the field increases. This behavior indicates that the field tends to align the moments at the expense of the antiferromagnetic state, but a field of 30 kOe was insufficient to completely align the moments. The magnetization measurements on $TbMo_6S_8$ also showed incomplete saturation fields up to 18.5 kOe [8.31]. The intensity of the {100} peak in Fig.8.5 consists of a field-independent nuclear Bragg peak plus the magnetic scattering due to the induced ferromagnetic alignment of the Tb moments in the applied field. Although there is a substantial reduction in the antiferromagnetic component, little ferromagnetic alignment develops for

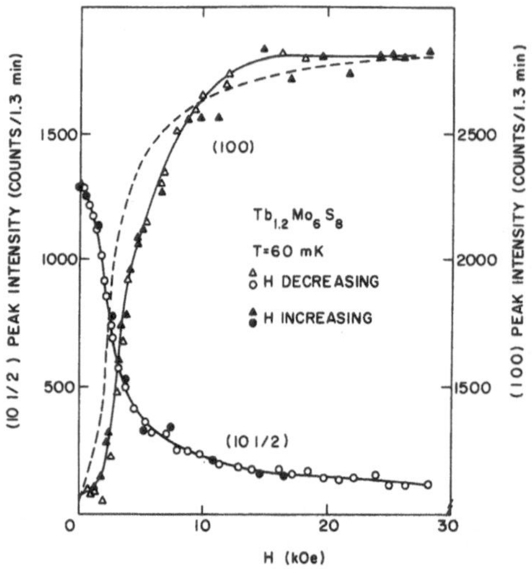

Fig.8.5. Field dependence of the antiferromagnetic {1 0 1/2} and ferromagnetic {100} peak intensities for $TbMo_6S_8$. The solid curves are simply guides to the eye. The dashed curve is the ferromagnetic intensity calculated from the measured antiferromagnetic intensity [8.29]

$H < 2.0$ kOe, which is near H_{c2}. For $DyMo_6S_8$, on the other hand, the ferromagnetic component increases substantially at low fields (below H_{c2} of 1.2 kOe). At an applied field near but less than H_{c2}, the ferromagnetic magnetization was already about 80% of the saturation value. In both cases the observed magnetization in the neutron experiment was in good agreement with the measured bulk magnetization data obtained in a field [8.31]. At sufficiently high fields, where the induced moment is nearly saturated, the value of the moment is the same (within experimental error) as the antiferromagnetic moment at low temperatures and zero field. Thus, the entire sample participates in the low-field antiferromagnetic ordering.

There may be a straightforward explanation for the qualitatively different behavior of $TbMo_6S_8$ and $DyMo_6S_8$. In anisotropic antiferromagnets there is a spin-flop transition at a finite applied field, H_{SF}. If $H_{SF} < H_{c2}$ for $DyMo_6S_8$ and $H_{SF} > H_{c2}$ for $TbMo_6S_8$, then the presence of superconductivity would lead to the different observed behavior in the following way. When the field is applied, flux vortices are created within which the field is H_{c2}. For $DyMo_6S_8$ this field is above H_{SF} and therefore a ferromagnetic magnetization can develop. As the field is increased, additional flux lines develop and hence the ferromagnetic intensity increases. For $TbMo_6S_8$, on the other hand, no ferromagnetism develops for $H < H_{SF}$ and increasing the field simply reduces the antiferromagnetic order parameter. One apparent weakness of this model is that H_{SF} would have to coincide with H_{c2} for the Tb compound, since the data indicate that the ferromagnetic component develops essentially at H_{c2}. Additional measurements will be needed to resolve this point.

8.3 Ferromagnetic Superconductors

For systems where the magnetic interactions favor parallel alignment there is strong competition with the superconducting state as discussed in Sect.8.1. This ferromagnetic-superconductor case has been studied in some detail for two systems, $ErRh_4B_4$ and $HoMo_6S_8$. These materials are re-entrant supercondcutors, that is, they are superconducting only over the temperature interval $T_{c2} < T < T_{c1}$. Moreover, the re-entry to the normal conducting state at T_{c2} is accompanied by a magnetic transition(s). At sufficiently low temperatures this order is purely ferromagnetic in nature, whereas near T_{c2} there is an interesting competition between the ferromagnetism and superconductivity which gives rise to an oscillatory magnetization as discussed below.

8.3.1 Ferromagnetic State

Figure 8.6 shows the magnetic diffraction pattern for $HoMo_6S_8$ [8.32] obtained by subtracting diffraction data taken well above the phase transition from the low temperature data. This subtraction procedure eliminates the nuclear Bragg scattering, which is temperature independent. The new peaks coincide in position with the Chevrel phase nuclear peaks demonstrating that the magnetic structure is ferromagnetic. The widths of the peaks are instrumentally limited, with the highest resolution results giving a lower limit of $\xi > 300$ Å. The absence of the diffuse paramagnetic scattering in the ordered state causes the decrease in the "background" between the Bragg peaks at low temperatures. Similar data have been obtained for $ErRh_4B_4$ [8.33].

If $HoMo_6S_8$ were cubic, then it would be impossible to determine the preferred direction (i.e., the easy magnetic axis) in the system from powder data. The small rhombohedral distortion, however, establishes a unique axis in the system, and it is then possible to determine the angle between the easy axis and the rhombohedral [111] direction as discussed in Sect.8.1. Data for the (111) and $(11\bar{1})$ peaks are shown in Fig.8.7. (If there were no rhombohedral distortion then the positions of these peaks would be identical). The subtraction of the nuclear scattering yields a single magnetic peak (lower part of the figure) at the $(11\bar{1})$ position, with no magnetic intensity at the (111) position. From (8.4) we see that the spin direction must coincide with the [111] rhombohedral axis. $ErRh_4B_4$ is tetragonal, and in this case it was possible to determine that the spin direction is perpendicular to the unique tetragonal axis, that is, the spins lie in the basal plane.

The temperature dependence of the magnetic intensity is shown in Fig.8.8 for both ferromagnetic materials, along with the re-entrant superconducting transition temperatures. It is clear that there is a close correspondence between the development of magnetic order and the disappearance of superconductivity, a subject to which we will return in the following section. Note that there is a clear differ-

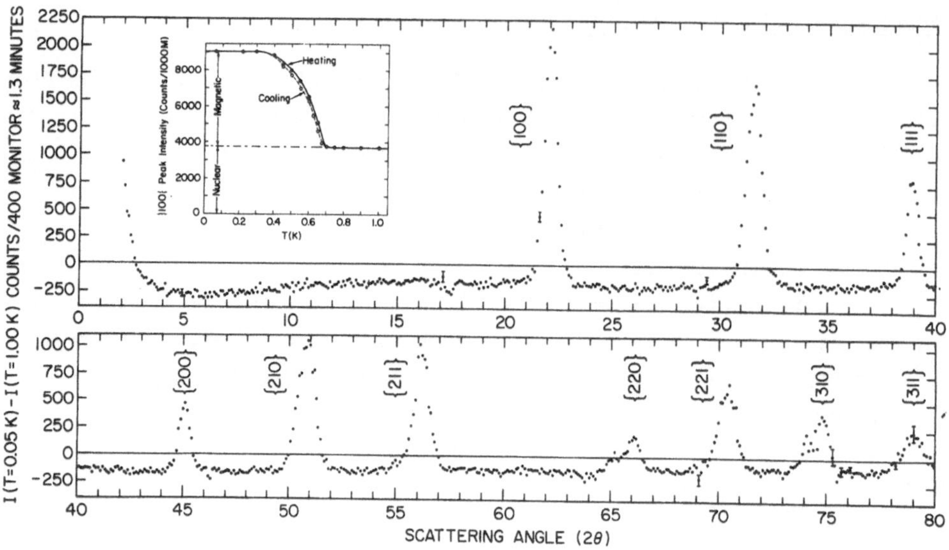

Fig.8.6. Magnetic diffraction pattern for $HoMo_6S_8$. Data taken above the transition have been subtracted from the low temperature data. The magnetic Bragg peaks which are observed coincide in position with the $HoMo_6S_8$ nuclear peaks, establishing that long-range ferromagnetic order exists. The inset shows the temperature dependence of the {100} peak intensity [8.32]

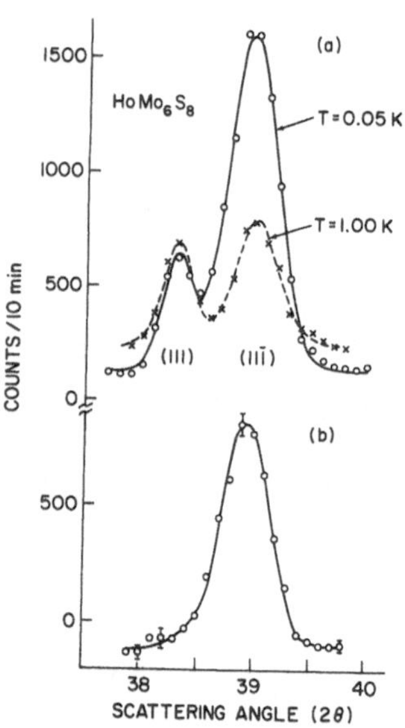

◄ Fig.8.7

► Fig.8.8

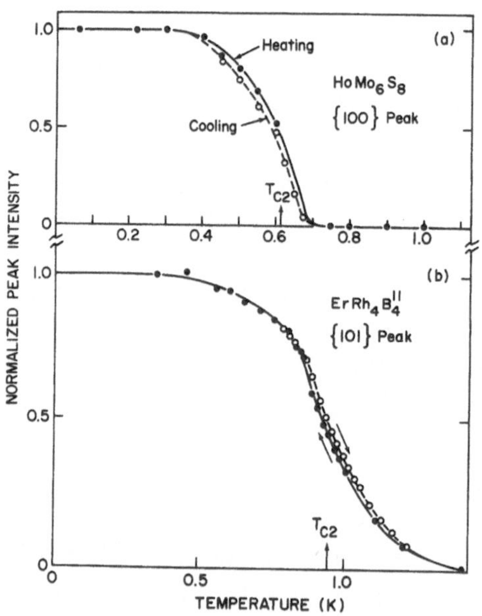

Fig.8.7. a) High resolution scans of the (111) and (11$\bar{1}$) reflections above and below the transition for $HoMo_6S_8$. The difference between these two scans (b) shows that the magnetic contribution occurs only for the (111) reflection, so that the easy magnetic axis is the rhombohedral [111] direction [8.32]

Fig.8.8. Temperature dependence of the magnetic peak intensity at the ferromagnetic Bragg position for $ErRh_4B_4$ and $HoMo_6S_8$. The development of magnetic order is seen to be in close correspondence to the disappearance of superconductivity at T_{C2} [8.32,33]

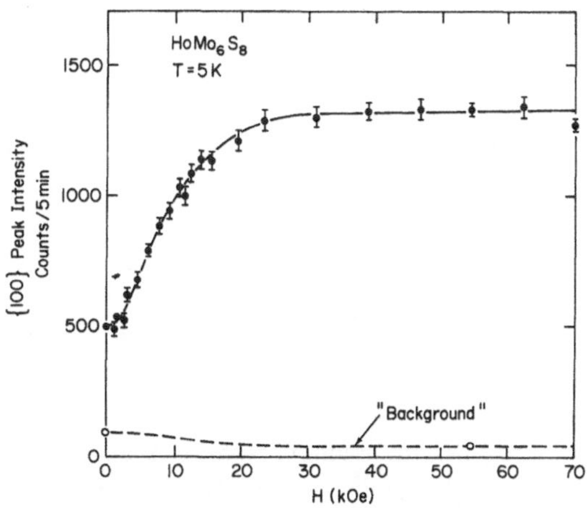

Fig.8.9. Peak intensity of the ferromagnetic component of $HoMo_6S_8$ induced by an applied field well above the magnetic and superconducting transition temperatures. The moment observed is the same as the spontaneous ferromagnetic moment measured at low temperatures [8.12,25]

ence in intensity on heating and cooling, which is presumably due to the competition between the ferromagnetism and superconductivity. The superconducting transition itself is clearly first order [8.34,35]. The magnetic transition is fairly well defined for $HoMo_6S_8$, whereas for the $ErRh_4B_4$ it appears to be smeared in temperature, but the width of the Bragg peak is instrument-limited at all temperatures. This smearing of the phase transition appears to be an intrinsic effect and not due to sample inhomogeneities.

The spontaneous magnetization is seen to saturate at low temperatures and from (8.4,6,7) we find that $\langle\mu_z\rangle_{T=0}$ is $(9.06 \pm 0.3)\mu_B$ and $(5.6 \pm 0.2)\mu_B$ for the Ho and Er, respectively. These values are considerably below the free-ion values (see Table 8.1), indicating that crystal field effects are important in these materials. For the Ho compound the same saturated value for the moment is found by applying a magnetic field well above T_M (and T_{C1}) as shown in Fig.8.9. Here the intensity in zero field is the nuclear Bragg peak, and the additional intensity at finite fields is due to the alignment of the moments by the field. The moment is seen to saturate above ~20 kOe, in contrast to the results for the selenide materials [8.12,25]. Since the moment obtained in these measurements is the same as the spontaneous moment in the ordered state, it is clear that the entire sample has ordered at low temperatures. For the $ErRh_4B_4$ compound, saturation of the moment is not found for fields up to 20 kOe [8.36], but the value observed at this field $(6.9\mu_B)$ is already substantially above the spontaneous moment observed at low temperatures. This indicates that the crystal field splittings are considerably smaller for $ErRh_4B_4$ than for $HoMo_6S_8$, and this has been confirmed directly by inelastic scattering measurements [8.25,37]. Of course, in sufficiently high fields the full free ion moment must be realized, but this may be well above the fields available in the laboratory. We should also point out that the Er moment measured by Mössbauer techniques [8.38] is not in agreement with the neutron data, and the source of this discrepancy is not yet known.

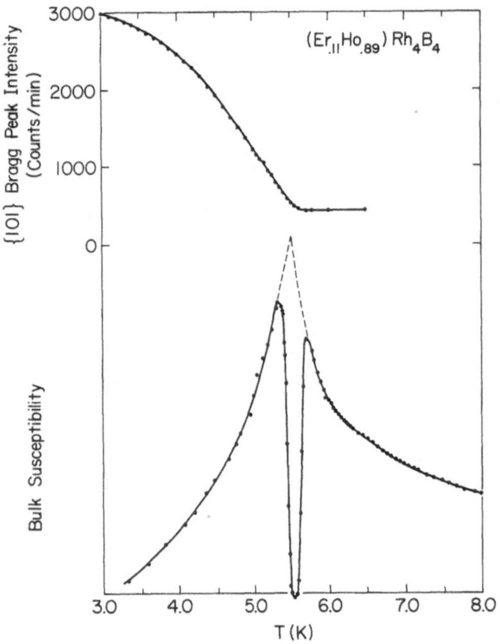

Fig.8.10. Magnetic phase transition in $(Er_{0.11}Ho_{0.89})Rh_4B_4$. The top part of the figure shows the Bragg peak intensity, which is proportional to the square of the magnetization. The ferromagnetic transition temperature is found to be 5.56 K, which is in good agreement with bulk susceptibility data extrapolated into the superconducting region [8.40]

For the $ErRh_4B_4$ system the superconducting and magnetic phase transitions can be adjusted by substituting Ho for Er [8.39]. Initially all three phase boundaries (T_{C1}, T_{C2}, and T_M) decrease slowly in temperature with increasing Ho concentration. Around 30% Ho, T_{C2} and T_M begin to increase and the three-phase boundaries meet at a multicritical point for a Ho concentration of 89% and a temperature of 6 K. Figure 8.10 shows neutron Bragg intensity data along with bulk susceptibility data on a sample which is slightly on the Er rich side of the multicritical point. The sample is superconducting over a small temperature interval [8.40,41]. The multicritical point may then be reached by the application of pressure [8.42]. The susceptibility shows the customary λ-type anomaly indicative of a magnetic phase transition, except that near T_M the sample becomes superconducting. The magnetic phase transition is of course masked in the susceptibility measurements by the Meissner effect. Neutrons, however, probe the atomic magnetic state directly and show the development of a ferromagnetic state at T_M = 5.56 K. This transition agrees well with the susceptibility data extrapolated into the superconducting region. These data demonstrate the complementary nature of the two experimental techniques applied to this problem.

The Bragg peak intensity in Fig.8.10 was found to be continuous and reversible, in contrast to the pure Er material [8.33] and $HoMo_6S_8$ [8.32], where a considerable difference on heating and cooling was observed. The transition for 89% Ho thus appears to be directly into the ferromagnetic state, with the spin direction along the tetragonal axis, as was found to be the case in the pure Ho material [8.43] (Table 8.1). This does not prove, however, the coexistence of ferromagnetism and superconductivity on a microscopic scale since we could have a mixed state with coexisting ferromagnetic and superconducting domains. Note that the Bragg intensity just below T_M is linear in reduced temperature, which would imply mean-field behavior if the transition is second order. Mean-field behavior has already been noted in the specific heat data [8.39]. At lower Ho concentrations, such as 60% Ho, the magnetic phase transition has been found to be strongly first order [8.44].

Finally we note that a new re-entrant system has recently been discovered, $ErRh_{1.1}Sn_{3.6}$ [8.45]. The re-entrant superconducting transition temperature in this material is accompanied by a "ferromagnetic transition", except that the widths of the Bragg peaks are considerably broader than the instrumental resolution. This indicates that the range of the ferromagnetic correlations is not truly of long range in this system, as was also found for the pseudobinary system $(Ce_{1-x}RE_x)Ru_2$ [8.5-7]. In the pseudobinary system, though, the superconducting state was not destroyed.

8.3.2 Oscillatory Behavior

The competition between ferromagnetism and superconductivity may give rise to an oscillatory magnetization in the superconducting state, and this has indeed turned out to be the case for both $ErRh_4B_4$ and $HoMo_6S_8$. Figure 8.11 shows the development of a peak in the scattering at a wave vector $Q_c = 0.03$ Å^{-1} for $HoMo_6S_8$ [8.46]. The width of this peak is limited by the instrumental resolution, indicating that a long-range oscillatory magnetic state has developed. The lower limit for ξ in this case is $\xi > 500$ Å, which is already considerably longer than the lower limit set for the ferromagnetic state at low temperatures. No higher-order peaks are observed, demonstrating that the magnetization in real space must be very nearly sinusoidal, with a characteristic wavelength $\lambda = 2\pi/Q_c$ of 200 Å. In addition, no precursor critical scattering which peaked around Q_c was observed.

The temperature dependence of the oscillatory magnetization is shown in Fig.8.12 [8.47], where a rather remarkable difference is found on heating and cooling. The wave vector dependence of this scattering shows that the oscillatory state forms only on cooling. Applying a magnetic field is seen to reduce the strength of the oscillatory scattering at this wave vector, and in general the wave vector dependence shows that the peak shifts to smaller Q with increasing field. Satellite peaks are also observed around the {100} powder line.

A peak in the small wave vector scattering was first observed in $ErRh_4B_4$ [8.33, 48] and this peak was clearly related to presence of the superconducting phase. Recent measurements [8.37] on a single crystal specimen of $ErRh_4B_4$ have shown that resolution limited satellites appear, demonstrating that there is long-range oscillatory order in both systems.

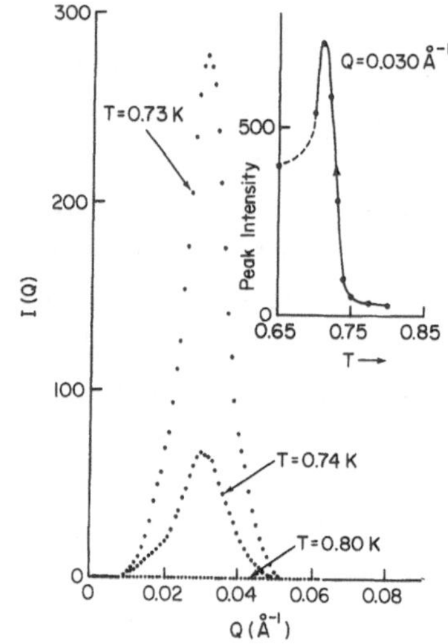

Fig.8.11 ▲

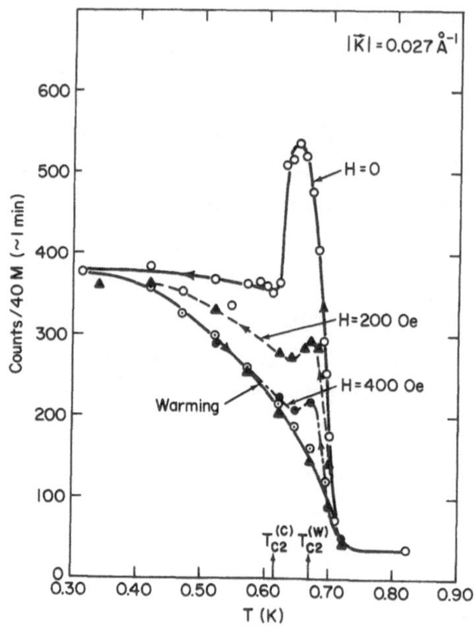

► Fig.8.12

Fig.8.11. Net intensity on cooling as a function of wave vector for $HoMo_6S_8$. The observed peak is limited in width by the instrumental resolution, indicating that the oscillatory magnetization which has developed is long range in nature [8.46]

Fig.8.12. Temperature dependence of the observed scattering intensity at a wave vector near the maximum for $HoMo_6S_8$. On cooling, the intensity rapidly rises as the oscillatory state develops, then decreases at lower temperature as ferromagnetism sets in. The application of a magnetic field is seen to suppress the scattering at this wave vector. On warming, no oscillatory state is observed and the field has little effect on the scattering [8.47]

The oscillatory component to the scattering in $ErRh_4B_4$ and $HoMo_6S_8$ could be due to the formation of a spiral component of the magnetization, a linearly polarized oscillatory state, or the spontaneous formation of a vortex state as discussed in Sect.8.1. The vortex possibility is apparently ruled out for these two systems since the Q_c is too large and has the opposite dependence with field to that expected. However, most of the measurements have only recently been taken and therefore should be regarded as preliminary. The nature of the oscillatory state is currently being actively investigated.

Acknowledgments. We would like to thank our many collaborators who have worked with us on various aspects of these studies. Work at Brookhaven supported by the Division of Basic Energy Studies, DOE, under Contract No. DE-AC02-76-CH00016. Work at Maryland supported by the NSF, DMR 79-00908 and DMR 82-07958.

References

8.1 See, for example, G.T. Rado, H. Suhl (eds.): *Magnetism*, Vol.5 (Academic, New York 1973) Chaps.10,11,12;
M.B. Maple: Appl. Phys. *9*, 179 (1976)
8.2 A.A. Abrikosov, L.P. Gorkov: Zh. Eksp. Teor. Fiz. *39*, 1781 (1960) [Sov. Phys. JETP *12*, 1243 (1961)]
8.3 B.T. Matthias, H. Suhl, E. Corenzwit: Phys. Rev. Lett. *1*, 449 (1958)
8.4 M. Wilhelm, B. Hillenbrand: Z. Naturforsch. Teil A*26*, 141 (1971); J. Phys. Chem. Solids *31*, 559 (1970)
8.5 See, for example, S. Roth: Appl. Phys. *15*, 1 (1978)
8.6 J.A. Fernandez-Baca, J.W. Lynn: J. Appl. Phys. *52*, 2183 (1981)
8.7 J.W. Lynn, D.E. Moncton, L. Passell, W. Thomlinson: Phys. Rev. B*21*, 70 (1980)
8.8 J.W. Lynn, C.J. Glinka: J. Mag. Mag. Mat. *14*, 179 (1979)
8.9 D.E. Moncton, G. Shirane, W. Thomlinson, M. Ishikawa, Ø. Fischer: Phys. Rev. Lett. *41*, 1133 (1978)
8.10 E.I. Blount, C.M. Varma: Phys. Rev. Lett. *42*, 1079 (1979)
8.11 Ø. Fischer, M.B. Maple (eds.): *Superconductivity in Ternary Compounds I*, Topics in Current Physics, Vol. 32 (Springer, Berlin, Heidelberg, New York 1982)
8.12 J.W. Lynn: In *Crystalline Electric Field and Structural Effects in f-Electron Systems*, ed. by J.E. Crow, R.P. Guertin, T. Mihalisin (Plenum, New York 1980) p.547
8.13 See, for example, D.K. Christen, H.R. Kerchner, S.T. Sekula, P. Thorel: Phys. Rev. B*21*, 102 (1980)
8.14 M. Blume: Phys. Rev. *124*, 96 (1961)
8.15 M. Blume, A.J. Freeman, R.E. Watson: J. Chem. Phys. *37*, 1345 (1962)
8.16 G. Shirane: Acta Cryst. *12*, 282 (1959)
8.17 W. Marshall, S.W. Lovesey: *Theory of Thermal Neutron Scattering* (University Press, Oxford 1971)
8.18 C.M. Varma, E.I. Blount, H.S. Greenside, T.V. Ramakrishnan: In *Ternary Superconductors*, ed. by G. Shenoy, B. Dunlap, F. Fradin (North-Holland, Amsterdam 1981) p.261
8.19 W. Thomlinson, G. Shirane, D.E. Moncton, M. Ishikawa, Ø. Fischer: J. Appl. Phys. *50*, 1981 (1979)
8.20 K. Yvon: In *Current Topics in Materials Science*, Vol.3, ed. by E. Kaldis (North-Holland, Amsterdam 1979) Chap.2
8.21 C.F. Majkrzak, G. Shirane, W. Thomlinson, M. Ishikawa, Ø. Fischer, D.E. Moncton: Solid State Commun. *31*, 773 (1979);
S. Quêzel, F. Tchéou, J. Rossat-Mignod, R. Chevrel, M. Sergent: Solid State Commun. *38*, 1003 (1981)
8.22 M.B. Maple, L.D. Woolf, C.F. Majkrzak, G. Shirane, W. Thomlinson, D.E. Moncton: Phys. Lett. *77*A, 487 (1980)
8.23 J.W. Lynn, D.E. Moncton, G. Shirane, W. Thomlinson, J. Eckert, R.N. Shelton: J. Appl. Phys. *49*, 1389 (1978)
8.24 W.C. Koehler: J. Appl. Phys. *36*, 1078 (1965);
J.J. Rhyne, T.McGuire: IEEE Trans. Mag. *8*, 105 (1972);
D.E. Cox: IEEE Trans. Mag. *8*, 161 (1972)
8.25 J.W. Lynn, R.N. Shelton: J. Mag. Mag. Mat. *15*, 1577 (1980); J. Appl. Phys. *50*, 1984 (1979)
8.26 R.W. McCallum, D.C. Johnston, R.N. Shelton, W.A. Fertig, M.B. Maple: Solid State Commun. *24*, 501 (1977)
8.27 C.F. Majkrzak, S.K. Satija, G. Shirane, H.C. Hamaker, H.B. MacKay, L.D. Woolf, M.B. Maple: Private communication
8.28 C.F. Majkrzak, H.A. Mook, G. Shirane, H.C. Hamaker, L.D. Woolf, H.B. MacKay, Z. Fisk, M.B. Maple: Phys. Rev. B, to be published
8.29 W. Thomlinson, G. Shirane, D.E. Moncton, M. Ishikawa, Ø. Fischer: Phys. Rev. B*23*, 4455 (1981)
8.30 M. Ishikawa, Ø. Fischer: Solid State Commun. *24*, 747 (1977)
8.31 M. Ishikawa, J. Muller: Solid State Commun. *27*, 761 (1978)

8.32 J.W. Lynn, D.E. Moncton, W. Thomlinson, G. Shirane, R.N. Shelton: Solid State Commun. *26*, 493 (1978); Phys. Rev. B*24*, 3817 (1981)
8.33 D.E. Moncton, D.B. McWhan, J. Eckert, G. Shirane, W. Thomlinson: Phys. Rev. Lett. *39*, 1164 (1977);
D.E. Moncton, D.B. McWhan, P.H. Schmidt, G. Shirane, W. Thomlinson, M.B. Maple, H.B. MacKay, L.D. Woolf, Z. Fisk, D.C. Johnston: Phys. Rev. Lett. *45*, 2060 (1980)
8.34 M. Ishikawa, Ø. Fischer: Solid State Commun. *23*, 37 (1977);
L.D. Woolf, M. Tovar, H.C. Hamaker, M.B. Maple: Phys. Lett. *71*A, 137 (1979)
8.35 W.A. Fertig, D.C. Johnston, L.E. DeLong, R.W. McCallum, M.B. Maple, B.T. Matthias: Phys. Rev. Lett. *38*, 987 (1977)
8.36 H.A. Mook, M.B. Maple, Z. Fisk, D.C. Johnston: Solid State Commun. *36*, 287 (1980)
8.37 S.K. Sinha, G.W. Crabtree, D.G. Hinks, H. Mook: Phys. Rev. Lett. *48*, 950 (1982)
8.38 G.K. Shenoy, B.D. Dunlap, F.Y. Fradin, C.W. Kimball, W. Potzel, F. Probst, G.M. Kalvius J. Appl. Phys. *50*, 1872 (1979);
G.K. Shenoy, B.D. Dunlap, F.Y. Fradin, S.K. Sinha, C.W. Kimball, W. Potzel, F. Probst, G.M. Kalvius: Phys. Rev. B*21*, 3886 (1980)
8.39 H.B. MacKay, L.D. Woolf, M.B. Maple, D.C. Johnston: Phys. Rev. Lett. *42*, 918 (1979)
8.40 J.W. Lynn, R.N. Shelton: J. Mag. Mag. Mat., to be published
8.41 See, for example, J.W. Lynn: In *Ternary Superconductors*, ed. by G. Shenoy, B. Dunlap, F. Fradin (North-Holland, Amsterdam 1981) p.51
8.42 R.N. Shelton, C.U. Segre, D.C. Johnston: Solid State Commun. *33*, 843 (1980)
8.43 G.H. Lander, S.K. Sinha, F.Y. Fradin: J. Appl. Phys. *50*, 1990 (1979)
8.44 H.A. Mook, W.C. Koehler, M.B. Maple, Z. Fisk, D.C. Johnston: In *Superconductivity in d and f Bands*, ed. by H. Suhl, M.B. Maple (Academic, New York 1980) p.427
8.45 J.P. Remeika, G.P. Espinosa, A.S. Cooper, H. Barz, J.M. Rowell, D.B. McWhan, J.M. Vandenberg, D.E. Moncton, Z. Fisk, L.D. Woolf, H.C. Hamaker, M.B. Maple, G. Shirane, W. Thomlinson: Solid State Commun. *34*, 923 (1980)
8.46 J.W. Lynn, J.L. Ragazzoni, R. Pynn, J. Joffrin: J. Physique Lettre *42*, L45 (1981)
8.47 J.W. Lynn, G. Shirane, W. Thomlinson, R.N. Shelton: Phys. Rev. Lett. *46*, 368 (1981)
8.48 See, for example, D.E. Moncton: J. Appl. Phys. *50*, 1880 (1979);
D.E. Moncton, G. Shirane, W. Thomlinson: J. Mag. Mag. Mat. *14*, 172 (1979)

9. Theory of Magnetic Superconductors

P. Fulde and J. Keller

With 12 Figures

The problem of magnetic superconductors is a long standing one [9.1,2]. After the BCS theory was formulated in which electrons are paired with opposite spins and momenta, it was realized that this kind of electron correlation is in conflict with ferromagnetic order since the latter requires an electron-spin polarization. YOSHIDA [9.3] calculated explicitely the vanishing of the spin susceptibility in a BCS superconductor as the temperature T approaches zero. Stimulated by the experimental findings of a nonvanishing Knight shift [9.4], it was first realized by FERRELL [9.5] that a spin-orbit interaction of the conduction electrons will lead to a finite spin susceptibility in a superconductor even at T = 0. ANDERSON [9.6] cast the theory for spin-orbit scattering into the concept of a generalized pairing prescription in terms of time-reversed states. This concept was further extended by BALTENSPERGER and STRÄSSLER [9.7] who showed that an antiferromagnet can be a superconductor. The field gained new impact by the development of Green's function techniques by ABRIKOSOV and GOR'KOV [9.8] who applied them to the treatment of the influence of magnetic impurities on superconductivity. Treating the exchange interaction between conduction electrons and impurities in the Born approximation, they predicted the phenomenon of gapless superconductivity which was later found by REIF and WOOLF [9.9]. Earlier pionering work on that problem by SUHL, ANDERSON and others is summarized in [9.10].

The Green's function method allowed readily for the inclusion of a homogeneous exchange field, spin-orbit scattering and an applied magnetic field acting on electron orbit and spin. The Clogston-Chandrasekhar limit played a prominent role. These developments are reviewed, e.g., in [9.11-16] which also discuss other extensions such as spin-compensation effects [9.17], finite momentum pairing [9.18, 19] or the influence of the Kondo effect [9.15,16].

Since the discovery of the strong interplay of superconductivity and magnetism in ternary compounds by FERTIG et al. [9.20], and ISHIKAWA and FISCHER [9.21], an unprecedented burst of theoretical work has resulted. Of particular stimulation were also the neutron-scattering experiments done at Brookhaven, Oak Ridge and Grenoble [9.22,23] which provided detailed information on the nature of the magnetic order in the presence of superconductivity. Not only was the coexistence of superconductivity and antiferromagnetic order confirmed, but it was also demonstrated that a cryptoferromagnetic state can exist in a superconductor. The origin of such a long wavelength order can be due to different reasons. The long-range part of the RKKY interaction is modified in the superconducting state and this may lead to cryptoferromagnetic order as proposed by ANDERSON and SUHL [9.24]. Alternately, the screening currents may be responsible for it, which are set up in the superconductor in order to screen the field of the magnetization. This situation was investigated by BLOUNT and VARMA [9.25], FERRELL et al. [9.26], and TACHIKI et al. [9.29]. The latter mechanism is believed to be of importance in the ternary compounds since the exchange interaction between the magnetic ions and the conduction electrons is particularly weak in these systems. There is also the possibility that the field of the magnetization induces a vortex structure in the superconductor as studied by KREY [9.27], TACHIKI and coworkers [9.71,75], and KUPER et al. [9.28]. We shall discuss here our present understanding of the mutual influence of superconducting and magnetic order. The different topics which we shall be dealing with can be read off the generalized phase diagrams shown in Figs.9.1a,b. There is first the influence of the paramagnetic ions, including their short-range correlations, on superconductivity [line (a)]. Then there is the problem of the change in the magnetic ordering temperature due to superconductivity [line (b)]. Furthermore, one wants to understand under what circumstances superconductivity may persist after magnetic order has set in. Thereby one has to distinguish between antiferromagnetic order which is compatible with superconductivity [line (d)] and ferromagnetic order which competes with and may be modified by superconducting order. Thus there is a regime between lines (b) and (c) where superconductivity coexists with cryptoferromagnetism, spinal or self-induced vortex structures. The point MC requires special attention. In its vicinity, as well as close to line (b), one may assume that the superconducting as well as the magnetic-order parameter are both small.

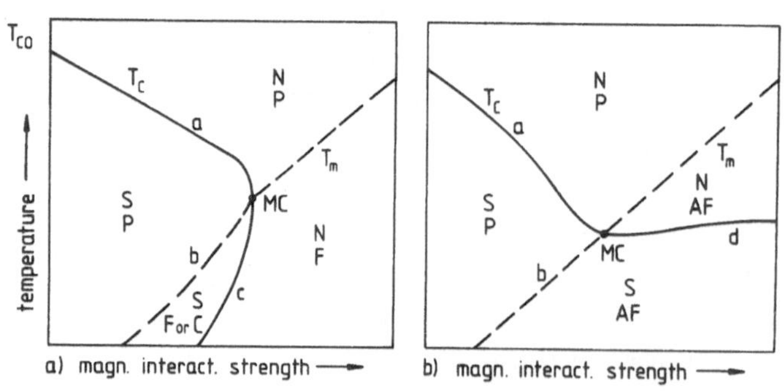

Fig.9.1a,b. Schematic phase diagram for magnetic superconductors with (a) ferromagnetic and (b) antiferromagnetic order. N, S, P, F, AF, C denote normal, superconducting, paramagnetic, ferromagnetic, antiferromagnetic and cryptoferromagnetic states, respectively

9.1 Influence of Spin Scattering on Superconductivity

The exchange interaction between conduction electrons and magnetic ions is conventionally written as

$$H_{int} = -2 \sum_n \underline{S}_n \int d^3\underline{r} I(\underline{r} - \underline{R}_n)\underline{s}(\underline{r}) \quad . \tag{9.1}$$

Here $I(\underline{r} - \underline{R}_n)$ is the exchange energy of conduction electrons with spin density $\underline{s}(\underline{r})$ and ions at positions $\underline{R}_n$ and spins $\underline{S}_n$. Its $\underline{r}$ dependence usually serves only as a short wave cut-off. For rare-earth ions, one may replace $\underline{S}_n$ by $(g - 1)\underline{J}_n$ where g is the Landé factor and $\underline{J}_n$ is the total angular momentum of the incomplete 4f shell.

The exchange interaction has a strong effect on superconductivity. Thereby one can distinguish between several different situations. The simplest one is when the magnetic ions are uncorrelated with each other. This is the situation dealt with by ABRIKOSOV and GOR'KOV [9.8] and will not be discussed here. When the magnetic ions are *correlated* with each other, their effect will be different in the paramagnetic phase and in the magnetically-ordered phase. In the study of this effect, the spin susceptibility χ_s of the superconducting electrons plays a prominent role. It describes the response of the electron system to a wave number dependent exchange field which is set up by the ordered ions. For that reason we shall first discuss the behavior of χ_s as function of temperature and wave vector $\underline{q}$ before we turn to the different cases of correlated ionic spins. Later, in Sect.9.3, we shall study how the correlations between the magnetic ions and the form of their ordering depend on the superconducting order of the conduction electrons.

9.1.1 Electron-Spin Susceptibility

The following definition will be used for the electron-spin susceptibility:

$$\chi(\underline{q},T) = \frac{\partial\langle\sigma^z_{\underline{q}}\rangle}{\partial h_{\underline{q}}} \quad . \tag{9.2}$$

Here $\sigma_{\underline{q}}$ is the Fourier transform of the electron-spin density $\underline{\sigma}(\underline{r}) = 2\underline{s}(r)$ and $\langle\ldots\ldots\rangle$ denotes a thermal average. $h_{\underline{q}}$ is the transform of a field which is supposed to couple to the spin density by

$$H_{int} = - \int d^3\underline{r}\sigma^z(\underline{r})h(\underline{r}) \quad . \tag{9.3}$$

Due to the formation of Cooper pairs the spin susceptibility χ_s in a superconductor is drastically different from that in a normal metal χ_n. Changes in the Knight shift are a measure of this difference [9.30].

The earliest treatment of $\chi_s(T)$ is due to YOSHIDA [9.3]. He considered the original BCS case where electrons are paired in states $(\underline{k},\sigma)$ and $(-\underline{k},-\sigma)$. In that case, $\chi_s(T = 0) = 0$ since the electron spins can align in a magnetic field only by the breaking of Cooper pairs. This requires a finite energy. The situation is changed when the electrons acquire a finite spin-orbit mean free path due to impurity or crystallite-surface scattering. Spin-orbit scattering is conventionally introduced by assuming n_i scattering centers per volume with a scattering potential of the form

$$v(\underline{k},\underline{k}') = v_1(\underline{k} - \underline{k}') + i \frac{v_{so}}{k_F^2} [\underline{k} \times \underline{k}']\underline{\sigma}$$

(k_F is the Fermi momentum).

The spin-orbit scattering rate is then

$$\frac{1}{\tau_{so}} = \frac{1}{2} n_i N(0) \int_{|\underline{k}'|=k_F} d\Omega_{\underline{k}'} |v_{so}|^2 \sin^2\left(\frac{(\underline{k} \cdot \underline{k}')}{k_F^2}\right) \tag{9.4}$$

and is usually considered as a parameter. $N(0)$ is the electron-density of states per spin direction and per volume. In the presence of spin-orbit scattering, $0 \leq \chi_s(T = 0)/\chi_n \leq 1$ depending on the size of τ_{so}^{-1}. Frequently the question is raised as to whether the spin-orbit parameter which appears in an electronic band-structure calculation can play a similar role to τ_{so}^{-1} resulting from impurities and imperfections. This is not the case, as has been pointed out by GOR'KOV [9.31]. A simple way of seeing this is to express $\chi_s(T = 0)$ in terms of the single-electron band states $|n\rangle$ as

$$\chi_s(T = 0) = \sum_{n \neq n'} \frac{|\langle n'|\sigma^z|n\rangle|^2}{E_n + E_{n'}} (u_n v_{n'} - u_{n'} v_n)^2 \quad .$$

The E_n are the excitation energies of the superconductor and $(u_n v_{n'} - u_{n'} v_n)$ is the BCS coherence factor. $\chi_s(0) \neq 0$ requires nonvanishing matrix elements $\langle n'|\sigma^z|n\rangle$ between states belonging to different energies E_n, $E_{n'}$ (otherwise the coherence factor is zero). Within a band, $\langle n'|\sigma^z|n\rangle = 0$, since Bloch functions belonging to different momenta are orthogonal to each other. Thus, $|n\rangle$ and $|n'\rangle$ must belong to different bands but to the same momentum vector. Since the band splitting D is much larger than the energy gap, this contribution to the susceptibility is unchanged when going from the normal to the superconducting state. However, it constitutes only a small fraction of χ and hence $\chi_s(0)/\chi_n \ll 1$.

Similarly, as spin-orbit scattering leads to a finite $\chi_s(0)$, so does spin-flip scattering from magnetic impurities. The spin-flip scattering rate is defined by

$$\frac{1}{\tau_s} = \frac{1}{2} n_i \; N(0)S(S+1) \int_{|\underline{k}'|=k_F} d\Omega_{\underline{k}'} |I(\underline{k} - \underline{k}')|^2$$

$$= 2\pi n_i N(0)S(S+1)I^2 \quad . \tag{9.5}$$

Here $I(\underline{k} - \underline{k}')$ is the Fourier transform of $I(r - \underline{R}_n)$, i.e., $I(\underline{k}) = \int d^3\underline{r} I(\underline{r})\exp(i\underline{k}\underline{r})$. Furthermore we have assumed a contact interaction $I(\underline{r}) = I\delta(\underline{r})$. We discuss in the following the different forms of $\chi_s(\underline{q},T)$ depending on τ_{so}^{-1} and τ_s^{-1}.

a) *No Spin-Orbit and Exchange Scattering* ($\tau_{so}^{-1} = \tau_s^{-1} = 0$)

Usually in a superconductor the mean-free path ℓ is much smaller than the coherence length $\xi_o = v_F/\pi\Delta(0)$ (dirty limit). Here v_F is the Fermi velocity and $\Delta(0)$ is the superconducting order parameter at T = 0. Then it is found [9.32] that

$$\chi_s(\underline{q},T)/\chi_n(\underline{q}) = 1 - (q\ell)^{-1} \arctan(q\ell) \frac{\pi T}{\Delta} \sum_n (1 + \omega_n^2/\Delta^2)^{-1}$$

$$\times [(1 + \omega_n^2/\Delta^2)^{1/2} + (2\tau\Delta)^{-1}(1 - (q\ell)^{-1} \arctan(q\ell)]^{-1} \quad , \tag{9.6}$$

where $\omega_n = \pi T(2n+1)$ and n is an integer. Furthermore, $\tau_o = \ell/v_F$ and $\Delta = \Delta(T)$. The normal-state susceptibility for $q/k_F \ll 1$ is

$$\chi_n(\underline{q}) = 2\,N(0)(1 - q^2/12k_F^2) \quad . \tag{9.7}$$

The functions $\chi_s(\underline{q},T=0)$ and $\chi_n(\underline{q})$ are sketched in Fig.9.2. More generally, $\chi_n(\underline{q})$ is given by twice the Lindhard function. Equation (9.6) has the following limiting behavior:

(i) $q\ell \gg 1$

$$\chi_s(\underline{q},T)/\chi_n = 1 - \pi(2\xi(T)q)^{-1} \tanh\left(\frac{\Delta}{2T}\right) - \frac{q^2}{12k_F^2} \quad , \tag{9.8}$$

where $\chi_n = \chi_n(q=0)$ and $\xi(T) = v_F/\pi\Delta(T)$.

At T = 0 the function has a maximum when q equals

$$Q_o = (3\pi k_F^2 \xi_o^{-1})^{1/3} \quad . \tag{9.9}$$

(ii) $q\ell \ll 1$

In the limit T = 0 one obtains

$$\chi_s(\underline{q},0)/\chi_n = \frac{\pi^2}{24}(q\ell)^2(\xi_o/\ell) - \frac{q^2}{12k_F^2} \quad . \tag{9.10}$$

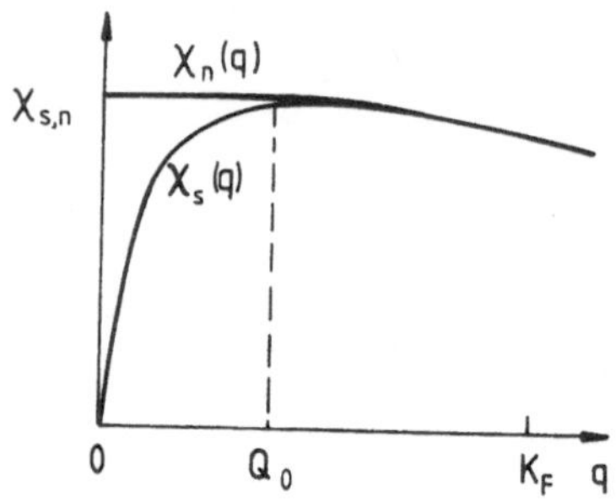

Fig.9.2. Electronic susceptibility χ_s and χ_n in the superconducting and normal state at T = 0 (schematic)

b) *Finite Spin-Orbit and Exchange Scattering* ($\tau_{so}^{-1} \neq 0$, $\tau_s^{-1} \neq 0$)

In the case where $\tau_s^{-1} \neq 0$, one has to introduce quantities u_n which are related to the ω_n through

$$\frac{\omega_n}{\Delta} = u_n[1 - (\tau_s\Delta)^{-1}(1 + u_n^2)^{-1/2}] \quad . \tag{9.11}$$

We shall consider $\chi_s(\underline{q},T)$ in the limit q = 0. In that case [9.32,33]

$$\chi_s(T)/\chi_n = 1 - \frac{\pi T}{\Delta}\sum_n (1 + u_n^2)^{-1}[(1 + u_n^2)^{1/2} - (3\tau_s\Delta)^{-1} \times (1 + 2u_n^2)(1 + u_n^2)^{-1} + 2/3\tau_{so}\Delta]^{-1} \quad . \tag{9.12}$$

The following limits are of interest

(i) $\tau_{so}^{-1} \neq 0$, $\tau_s^{-1} = 0$ (9.13)

$$\chi_s(T)/\chi_n = 1 - \frac{\pi T}{\Delta}\sum_n (1 + \omega_n^2/\Delta^2)^{-1}[(1 + \omega_n^2/\Delta^2)^{1/2} + 2/3\tau_{so}\Delta]^{-1} \quad .$$

For T = 0 the function is plotted in Fig.9.3a for different values of $[\tau_{so}\Delta(0)]^{-1}$. In the limit of a short spin-orbit mean free path ($\tau_{so}\Delta \ll 1$), it is found [9.34] that

$$\chi_s/\chi_n = 1 - \frac{3\pi}{4}(\tau_{so}\Delta)\tanh(\Delta/2T) \quad . \tag{9.14}$$

(ii) $\tau_{so}^{-1} = 0$, $\tau_s^{-1} \neq 0$

This is the case considered in [9.35]. We have evaluated $\chi_s(T = 0)/\chi_n$ as a function of $[\tau_s\Delta(T = 0, \tau_s^{-1} = 0)]^{-1}$ as shown in Fig.9.3b.

The generalization of the results to finite values of q is straightforward. For $q\ell \ll 1$ one finds an expression of the form of (9.10), but with $\pi^2/24$ replaced by

$$\frac{\pi^2}{6}\frac{T}{\Delta}\sum_n (1 + u_n^2)^{-1}[(1 + u_n^2)^{1/2} - (3\tau_s\Delta)^{-1}(1 + 2u_n^2)(1 + u_n^2)^{-1} + 2/3\tau_{so}\Delta]^{-2} \quad .$$

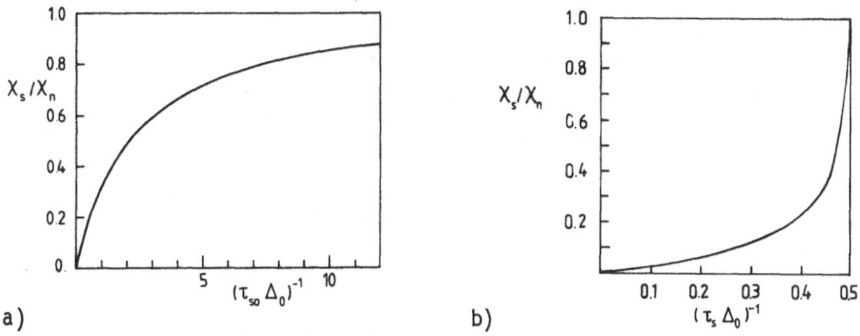

a) b)

Fig.9.3. (a) Normalized electronic susceptibility χ_s in the superconducting state as a function of the spin-orbit scattering rate $(\tau_{so})^{-1}$ for T = 0 [$\Delta_0 = \Delta$ (T = 0)]. (b) Normalized electronic susceptibility χ_s in the superconducting state as a function of the magnetic impurity scattering rate $(\tau_s)^{-1}$ for T = 0 [$\Delta_0 = \Delta$ (T = 0, $\tau_s^{-1} = 0$)]

Magnetic impurity scattering ($\tau_s^{-1} \neq 0$) and spin-orbit scattering ($\tau_{so}^{-1} \neq 0$) influence superconductivity differently as the former is Cooper pairbreaking while the latter is not. For that reason, Figs.9.3a,b look rather different. For example, the sharp increase in Fig.9.3b of χ_s/χ_n near $(\tau_s\Delta_0)^{-1} = 0.5$ is due to the rapid change of Δ/Δ_0 in that region because of the pairbreaking.

Finally we discuss how the spin susceptibility changes when the exchange interaction between the conduction electrons as well as the renormalization due to electron-phonon interactions are taken into account [9.36]. For that purpose we define the function $\bar{u} = [2N(0)]^{-1}\chi_s$. Then in the presence of the above interactions, the susceptibility $\tilde{\chi}_s$ is

$$\tilde{\chi}_s = 2\,\frac{N^*(0)\bar{u}}{1 - [N^*(0)V_{ex} - \lambda]\bar{u}} \quad .$$

Here V_{ex} denotes the exchange integral and $N^*(0) = N(0)(1 + \lambda)$ is the density of states as modified by the electron-phonon coupling λ. It is seen that the latter modifies the spin susceptibility in the superconducting state but not in the normal state where $\bar{u} = 1$.

9.1.2 Paramagnetic Phase

The influence of spin correlations on the superconducting transition temperature has been the subject of numerous investigations [9.37-44]. Thereby one has to distinguish between spatial correlations which are reflected in the static spin susceptibility $\chi_m(\underline{q})$ (per ion) of the magnetic ions and dynamical correlations described by $\chi_m(\underline{q},\omega)$. Considerable insight into the effect of correlations is gained by applying a sum-rule argument as suggested by RAINER [9.38]. We shall discuss it below in detail.

In fact, frequently in the literature not enough attention was paid to that sum rule and has resulted in the prediction of spurious effects.

When calculating the effects of magnetic ions on the superconducting transition temperature T_c, one has to solve the self-consistency equation for the order parameter Δ in the limit $\Delta \to 0$, i.e., (see, e.g., [9.11])

$$\ln(T_c/T_{co}) = 2\pi T_c \sum_{\omega_n>0} \left(\frac{1}{\Delta}\frac{\tilde{\Delta}_n}{\tilde{\omega}_n} - \frac{1}{\omega_n}\right) \quad . \tag{9.15}$$

T_{co} is the transition temperature in the absence of the exchange interaction and $\tilde{\omega}_n$ and $\tilde{\Delta}_n$ are renormalized frequencies and order parameter, respectively.

In order to relate them to the ion susceptibility $\chi_m(q,i\omega_n)$, we define $\tau = \tau_s S(S+1)$. Furthermore, we denote by τ_0^{-1} the scattering rate due to Coulomb scattering from ordinary, i.e., nonmagnetic impurities, grain boundaries, etc. Then,

$$\tilde{\omega}_n = \omega_n + \frac{1}{2\tau_0}\,\mathrm{sgn}\{\omega_n\} + \frac{3}{2\tau} T_c \sum_m \int d^3\underline{q}\; g(q)\chi_m[\underline{q},i(\omega_n - \omega_m)]\,\mathrm{sgn}\{\omega_m\}$$

$$\tilde{\Delta}_n = \Delta + \frac{1}{2\tau_0}\frac{\tilde{\Delta}_n}{\tilde{\omega}_n}\,\mathrm{sgn}\{\omega_n\} - \frac{3}{2\tau} T_c \sum_m \int d^3\underline{q}\; g(q)\chi_m[\underline{q},i(\omega_n - \omega_m)]\,\frac{\tilde{\Delta}_m}{\tilde{\omega}_m}\,\mathrm{sgn}\{\omega_m\} \quad . \tag{9.16}$$

One notices that in (9.16) the nonmagnetic scattering rate τ_0^{-1} appears in both equations with the same sign while the magnetic scattering rate τ^{-1} enters with opposite sign. This sign change is due to the lack of time-reversal invariance of the magnetic interaction and leads to pairbreaking and gapless superconductivity.

Both equations can be combined to form one equation for $\tilde{\Delta}_n/(\Delta\tilde{\omega}_n)$ which enters (9.15). In this equation, τ_0^{-1} is no longer contained. The weighting function $g(q)$ was calculated by RAINER [9.38]. It is normalized such that

$$\int d^3\underline{q}\, g(q) = 1 \quad .$$

In general it is also dependent on frequency but in the dirty limit $\tau_0^{-1} \gg T$, it can be approximated by

$$g(q) = \frac{1}{4\pi^2 k_F^2}\frac{1}{q}\begin{cases} \arctan(2q\ell) & ; \quad q < 2k_F \\ \arctan(2q\ell) - \arctan[2\ell(q^2 - 4k_F^2)^{1/2}] & ; \quad q > 2k_F \end{cases} \tag{9.17}$$

where as before, ℓ is the electron-mean free path. As $g(q)$ has a maximum for small values of q, ferromagnetic correlations are weighted more strongly than antiferromagnetic correlations or free spins since they imply that $\chi_m(\underline{q},i\omega_n)$ is largest for $q = 0$. Therefore they result in a larger pairbreaking effect than the latter. Furthermore, in (9.16) terms with low frequencies make the largest contributions.

Therefore, spin systems with low excitation energies will be particularly effective in breaking Cooper pairs.

The equations in (9.16) have to be solved together with (9.15) in order to obtain T_c. For a meaningful comparison of the effect on T_c of different forms of $\chi_m(\underline{q},i\omega_m)$, one has to observe the following sum rule [9.38] relating the dynamic susceptibility to the local correlation function:

$$T \sum_n \int_{BZ} d^3q \chi_m(\underline{q},i\omega_n) = \frac{1}{3} S(S+1) \int_{BZ} d^3\underline{q} \quad . \tag{9.18}$$

The integral extends over the first Brillouin zone. In the following we want to discuss some special cases and forms of $\chi_m(\underline{q},i\omega_m)$.

a) *Static Approximation*

When the characteristic frequency of the ion-spin motion is small compared with $(k_BT_c)^{-1}$, one may replace

$$\chi_m(\underline{q},i\omega_n) = \chi(\underline{q})\delta_{\omega_n 0} \quad . \tag{9.19}$$

Then the theory reduces to the one of ABRIKOSOV-GOR'KOV [9.8] but with a temperature-dependent scattering rate:

$$1/\tau_s = 3/\tau \int d^3\underline{q} g(q)\chi_m(\underline{q},T) \quad . \tag{9.20}$$

Thus, (9.15) becomes

$$\ln(T_c/T_{co}) + \Psi\left[\frac{1}{2} + (2\pi T_c\tau_s)^{-1}\right] - \Psi\left(\frac{1}{2}\right) = 0 \quad , \tag{9.21}$$

where $\Psi(x)$ is the digamma function. From this equation one can obtain the influence of ferromagnetic or antiferromagnetic correlations on T_c [9.38,42] by using an Ornstein-Zernicke form for $\chi(\underline{q},T)$, i.e.,

$$\chi(\underline{q},T) = \sum_{i=1}^{r} \frac{1}{r} \frac{\chi_0(T)}{\gamma(T) + a^2(T)(\underline{q} - \underline{q}_i)^2} \quad . \tag{9.22}$$

The $\underline{q}_i$ denote the wave vectors at which the susceptibility has its maximum and eventually diverges at the ordering temperature T_m. The quantity $\chi_0(T) = S(S+1)/3T$ is the susceptibility of free spins. Reference [9.38] uses a special ansatz for $\gamma(T)$ which determines T_m as function, e.g., of the magnetic-ion concentration or the coupling constant I. The term a(T) denotes an effective range of the magnetic ion-ion interactions and is chosen such that the sum rule (9.15) is fulfilled. Results for T_c are shown in Fig.9.4.

Ferromagnetic and antiferromagnetic correlations influence the superconducting transition temperature T_c rather differently. In the case of *ferromagnetic* corre-

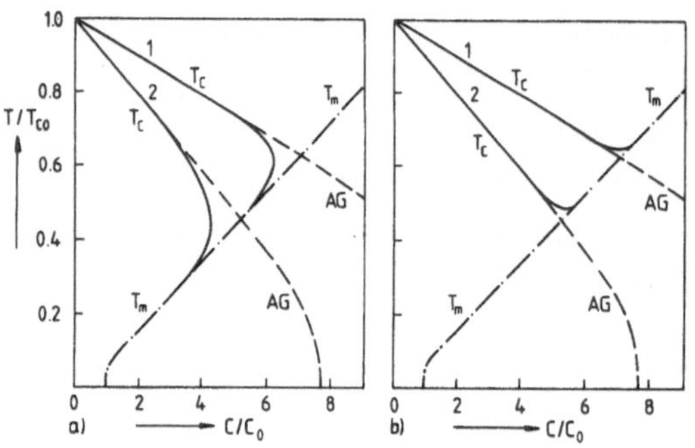

Fig.9.4. Superconducting transition temperature T_c vs magnetic ion concentration c for systems with (a) ferromagnetic and (b) antiferromagnetic spin correlations and given magnetic transition temperature $T_m(c)$. Curves 1 and 2 correspond to different values of the exchange interaction. Curves marked AG correspond to uncorrelated spins (from [9.38])

lations, superconductivity may be suppressed by spin fluctuations in the vicinity of the ferromagnetic transition (Fig.9.4a). This happens when the interaction is sufficiently strong and the electronic mean-free path is not too short. The nature of the second (lower) normal-superconducting phase transition has been studied in more detail in [9.45]. It was found that at that transition, the jump in the specific heat becomes extremely large. This signals the closeness to a first-order phase transition. *Antiferromagnetic* correlations have a reverse effect (Fig.9.4b). This is a consequence of the form of the weighting factor $g(\underline{q})$ as explained above. Similar results were obtained in [9.42], without the restriction to the dirty limit ($\tau_0^{-1} \gg T_c$), and also in [9.41].

b) *Dynamic Correlations*

When the characteristic energy of the fluctuating spins is of the order or less than T, then the frequency-dependent form of $\chi_m(\underline{q}, i\omega_n)$ has to be used. There are two approaches which have been taken.

The first consists of again deriving a temperature-dependent effective scattering rate $(\tau_s^{eff})^{-1}$ which is set into (9.21). In [9.40] an ansatz of the following form was made:

$$\frac{1}{\tau_s^{eff}} = \frac{3}{\tau} \int d^3\underline{q}\, g(q)\, \frac{1}{\pi} \int_{-\infty}^{+\infty} d\omega \beta\omega e^{\beta\omega} (e^{\beta\omega} - 1)^{-2}\, \mathrm{Im}\{\chi_m(\underline{q},\omega)\} \tag{9.23}$$

where $\beta = T^{-1}$, with a diffusion type of ansatz for $\chi_m(\underline{q},\omega)$, i.e.,

$$\chi_m(\underline{q},\omega) = \chi(\underline{q})\, \frac{iDq^2}{\omega + iDq^2} \quad , \quad (\mathrm{Im}\{\omega\} > 0) \quad . \tag{9.24}$$

For the static susceptibility see (9.22).

It was found that for large values of the diffusion constant D, the dynamical effects reduce the pairbreaking. But as pointed out in [9.45], for realistic values of D the effects are too small in order to lead to any appreciable changes as compared with the case of static correlations only. The second approach consists of

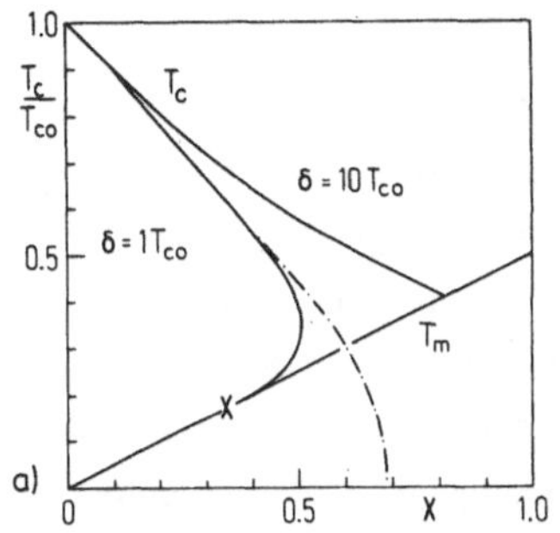

Fig.9.5. Superconducting transition temperature T_c vs magnetic ion concentration x for a ferromagnetic-coupled ion system with crystal field split energy levels (2 singlets separated by energy δ). The dotted line corresponds to uncorrelated spins with $\delta = 0$. All curves are normalized such that the initial slope is the same

numerically solving (9.15,16) for a given suceptibility $\chi_m(\underline{q}, i\omega_n)$. This is the one used when one is dealing with magnetic excitations of high energies and at low temperatures where the approximation (9.23) becomes inadequate. In [9.45], a spin system with crystal field split energy levels and a nonmagnetic ground state was examined. If was found that for excitation energies $\delta \gg T$, magnetic fluctuations above the magnetic-ordering temperature were not strong enough in order to suppress superconductivity (Fig.9.5).

Finally we discuss the influence of high energy excitations. When the excitation energies of the spin system are sufficiently large, the ω-dependence of $\chi_m(q, i\omega_n)$ in (9.16) can be completely neglected. This is in analogy to the phonon case where high-frequency phonons contribute to Δ but cause only a negligible frequency dependence of the order parameter. Then those equations reduce to

$$\tilde{\omega}_n = \omega_n(1 + Z_s) + \frac{1}{2\tau_o}\,\mathrm{sg}\{\omega_n\} \tag{9.25}$$

$$\tilde{\Delta}_n = \Delta[1 - Z_s/V_{ph}N(0)] + \frac{1}{2\tau_o}\,\frac{\tilde{\Delta}_n}{\tilde{\omega}_n}\,\mathrm{sg}\{\omega_n\}$$

with

$$Z_s = \frac{3}{2\pi\tau}\int d^3\underline{q}\,g(q)\,\chi_m(\underline{q}, 0) \quad . \tag{9.26}$$

The transition temperature is then found to be

$$T_c = 1.13\omega_D \exp[-(1 + Z_s)(N(0)V_{ph} - Z_s)^{-1}] \quad , \tag{9.27}$$

where ω_D is the Debye frequency and V_{ph} the electron-electron interaction energy due to phonons, respectively. Such a functional dependence of T_c on Z_s (or χ_m) is typical for pair-*weakening* instead of pair-*breaking* (see, e.g., [9.67]). It appears also in systems with electron exchange enhancement [9.101]. The above approximation may be applicable even near a magnetic phase transition when the low frequency soft or diffusive mode is confined to a small phase-space volume in the Brillouin zone.

9.1.3 Magnetically-Ordered Systems

In the following we shall consider systems with long-range ferromagnetic or antiferromagnetic order. It will be seen that in both cases the influence of magnetic order on superconductivity is drastically different. The ferromagnetic order is simulated by a uniform exchange field. This implies leaving out screening currents which are set up in a superconductor in order to keep $\underline{B} = 0$ when a magnetization is present [9.46]. A discussion of those currents is the subject of Sect.9.3. The exchange scattering is also not included when a uniform exchange field is considered, but it is easy to add it afterwards [9.47,48]. Also discarded here is the problem of what the reaction of the magnetic state will be to the onset of superconducting order. The possible development of a cryptoferromagnetic spin structure will be discussed in Sect.9.2.

a) *Uniform Exchange Field*

Let us assume that in addition to the BCS Hamiltonian H_{BCS} there is a uniform exchange field h along the z-axis acting on the electron spins so that

$$H = H_{BCS} - \sum_{\underline{k},\sigma=\pm 1} c^+_{\underline{k}\sigma} c_{\underline{k}\sigma} \sigma h \quad . \tag{9.28}$$

For example, we imagine that the uniform field results from ferromagnetically aligned magnetic ions. The $c^+_{\underline{k}\sigma}$, $c_{\underline{k}\sigma}$ are electron creation and annihilation operators. In order to study the influence of h on superconductivity, one must determine the coefficients $f_0(h,T)$ and $f_1(h,T)$ which appear in the Ginzburg-Landau equation

$$\ln(T/T_{co}) + f_0(h,T) + \frac{1}{2} f_1(h,T)\left(\frac{\Delta}{2\pi T}\right)^2 = 0 \quad . \tag{9.29}$$

This equation is obtained, e.g., by minimizing the free-energy difference between the superconducting and normal state. For details we refer, e.g., to [9.11].

The transition temperature $T_c(h)$ is determined through

$$\ln(T_c/T_{co}) + f_0(h,T_c) = 0 \tag{9.30}$$

while the sign of $f_1(h,T_c)$ determines the order of the phase transition, i.e., first order for $f_1(h,T_c) < 0$ and second order for $f_1(h,T_c) > 0$.

It has been shown by SARMA [9.49] that

$$f_0(h,T) = \mathrm{Re}\left\{\Psi\left(\frac{1}{2} + i\,\frac{h}{2\pi T}\right) - \Psi\left(\frac{1}{2}\right)\right\} \quad . \tag{9.31}$$

Furthermore,

$$f_1(h,T) = -\frac{1}{2}\,\mathrm{Re}\left\{\Psi^{(2)}\left(\frac{1}{2} + \frac{ih}{2\pi T}\right)\right\} \quad . \tag{9.32}$$

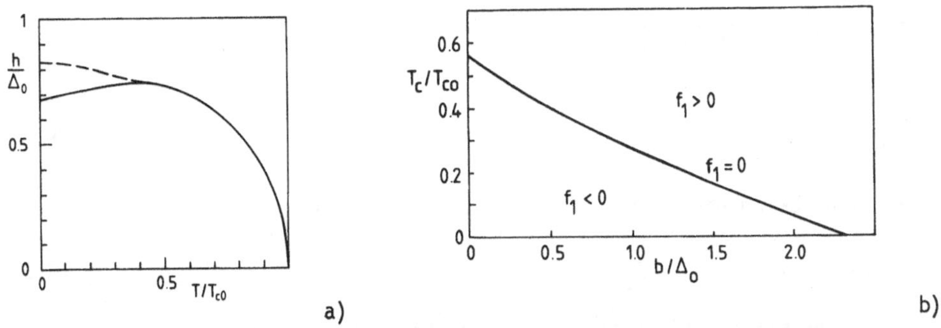

Fig.9.6. (a) Critical exchange field h(T) for systems with finite spin-orbit interaction $b = 0.2\ \Delta_0$. The broken curve denotes a first-order phase transition. (b) Change from second ($f_1 > 0$) to first-order phase transition denoted by $f_1[h(T_c),\ T_c, b] = 0$ vs spin-orbit scattering rate $b = (3\tau_{so})^{-1}$ (from [9.55])

$f_1(h,T_c)$ changes sign at $[T_1/T_{co};\ h^{(1)}/\Delta(0)] = (0.56;\ 0.62)$ implying that for $T_c/T_{co} < 0.56$, the phase transition is of first order. A plot of the solutions of (9.30) can be found in [9.49] or [9.11]. This result is qualitatively changed if one allows for spin-orbit scattering. With the notation

$$b = (3\tau_{so})^{-1} \quad , \tag{9.33}$$

one obtains in the dirty limit $\ell \ll \xi$ [9.47,50],

$$\begin{aligned} f_0(h,T_c,b) &= \frac{1}{2}\,[1 + b(b^2 - h^2)^{-1/2}]\Psi\left(\frac{1}{2} + \rho_-\right) \\ &\quad + \frac{1}{2}\,[1 - b(b^2 - h^2)^{-1/2}]\Psi\left(\frac{1}{2} + \rho_+\right) \end{aligned} \tag{9.34}$$

$$\begin{aligned} f_1(h,T_c,b) &= (2\pi T)^3\ \mathrm{Re}\left\{\sum_{n=0}^{\infty}\left[\left(\frac{\omega_n - ih + 2b}{(\omega_n + b)^2 - b^2 + h^2}\right)^3\right.\right. \\ &\quad \left.\left. - 2h^2 b\,\frac{(\omega_n + 2b)^2 + h^2}{[(\omega_n + b)^2 - b^2 + h^2]^4}\right]\right\} \quad . \end{aligned} \tag{9.35}$$

Hereby,

$$\rho_\pm = (2\pi T)^{-1}[b \pm (b^2 - h^2)^{1/2}] \quad . \tag{9.36}$$

$T_c(h,b)$ is obtained from (9.30) as before (Fig.9.6a). The curve $f_1[h(T_c),\ T_c,b] = 0$ is plotted in Fig.9.6b and defines the change in the order of the phase transition from second to first [9.51].

It is seen that for $b > 2.32\ \Delta_0$, the phase transition is always of second order [here $\Delta_0 = \Delta(T = 0,\ h = 0)$]. The expressions for $f_0(h,T,b)$ simplify considerably in the limit of a large spin-orbit scattering rate b (i.e., $\tau_{so}\Delta_0 \ll 1$). In that case,

$$f_0(h,T,b) = \Psi\left[\frac{1}{2} + \alpha(2\pi T)^{-1}\right] - \Psi(\tfrac{1}{2})$$

and

$$\ln(T_c/T_{co}) + \Psi\left[\frac{1}{2} + \alpha(2\pi T_c)^{-1}\right] - \Psi(\tfrac{1}{2}) = 0 \quad , \tag{9.37}$$

where the pairbreaking parameter α is given by $\alpha = \alpha_{so}$ with

$$\alpha_{so} = \tau_{so}h^2/2 \quad . \tag{9.38}$$

In this limit the effect of a uniform exchange field is equivalent to that of magnetic impurities with freely rotating spins [9.8]. In particular, the critical exchange field beyond which superconductivity is destroyed is given by $\alpha_{so}^c = \Delta_o/2$.

It is now easy to include the pairbreaking effect of additional external perturbations on the conduction electrons. For that purpose we use the equivalence and additivity of different pairbreaking mechanisms [9.32,47].

(i) When in addition an external magnetic field $\underline{B}$ is applied which acts on the electron orbits, then (9.37) again applies but with α given by [9.47]

$$\alpha = \alpha_{so} + \alpha_{orb} \quad ,$$

where

$$\alpha_{orb} = \frac{eDB}{3} \tag{9.39}$$

and D is the diffusion constant $D = v_F^2\, \ell/3$.

(ii) When the external field is sufficiently large so that its effect on the electron spins is much more important than its effect on the electron orbits [9.50,52, 53], then in (9.37) α is given by

$$\alpha = \frac{\tau_{so}}{2}(h + \mu_B B)^2 \quad . \tag{9.40}$$

This demonstrates the possibility of a "spin-compensation effect" first pointed out by JACCARINO and PETER [9.17] (see also [9.46]). The exchange field h set up by spin polarized magnetic ions and the external magnetic field $\underline{B}$ can have opposite directions and hence reduce the total effect on the electron spins. The effect was found experimentally by FISCHER and coworkers [9.54].

(iii) When there is, in addition, scattering from magnetic ions, then (9.37) again holds provided that the characteristic frequency of the spin motion is much less than T_c. In that case α is given by [9.47]

$$\alpha = \alpha_{so} + \alpha_{orb} + \alpha_{sp} \quad , \tag{9.41}$$

where α_{sp} is the magnetic scattering rate

$$\alpha_{sp} = 1/\tau_s \tag{9.42}$$

as given by (9.20). When the above condition is not fulfilled, one may in certain cases replace τ_s^{-1} by $(\tau_s^{eff})^{-1}$ as discussed in connection with (9.23).

Within the simple picture of a homogeneous exchange field, the following explanation would apply for the vanishing of superconductivity with the onset of ferromagnetic order as observed in the ternary systems $ErRh_4B_4$ and $HoMo_6S_8$. Above the Curie temperature T_m, the exchange scattering is not strong enough in order to destroy superconductivity. As T_m is approached, critical fluctuations develop and below T_m a uniform exchange field h is set up which adds to the pairbreaking. Thus, in (9.37),

$$\alpha = \tau_s^{-1} + \frac{\tau_{so}}{2} h^2 \tag{9.43}$$

where

$$\underline{h} = n_i \langle \underline{S} \rangle I \quad .$$

Below T_m, $\langle \underline{S} \rangle$ increases rapidly with $(T_m - T)$. Thus the critical value $\alpha^c = \Delta_0/2$ is reached either slightly above T_m when the spin fluctuations are sufficiently strong or slightly below T_m when $\langle \underline{S} \rangle$ becomes sufficiently large. Whether α^c is reached at $T \gtrless T_m$ also depends to some extent on the electron-mean free path [9.38]. This is due to the increase of the pairbreaking effect with increasing mean-free path of the long-range spin fluctuations.

b) *Antiferromagnetic Order*

From an experimental point of view there are many more substances which show an interplay of superconductivity and antiferromagnetism than, e.g., of superconductivity and ferromagnetic order. Therefore a theoretical understanding of the former case is of particular importance. One especially wants to understand the experimentally observed coexistence of both phenomena (Chaps.4,5, and 8).

An additional reason for treating the case of antiferromagnetic order in rather great detail here is the fact that in the past it has not obtained the attention which it deserves.

The first theoretical treatment of superconductivity and antiferromagnetic order was given by BALTENSPERGER and STRÄSSLER [9.7]. In fact, those authors predicted the possible coexistence of both ordering phenomena by slightly modifying the usual pairing condition by which electrons are paired in time-reversed states. They also pointed out that as a result, the effective electron-electron interaction via phonons is weakened when antiferromagnetic order sets in. We shall discuss in the following this theory. Thereby it is advantageous to reformulate parts of it in a way which makes it easy to apply to various experimental situations such as the inclusion of impurity scattering or the computation of the critical magnetic fields.

The starting point is the Hamiltonian (9.1) in a slightly rewritten form

$$H_{int} = -I \sum_{\substack{n \\ \mu\nu}} \sum_{\underline{k}\underline{k}'} \underline{S}_n \underline{\sigma}_{\mu\nu} c^+_{\underline{k}\mu} c^+_{\underline{k}'\nu} \exp[i(\underline{k} - \underline{k}')R_n] \quad . \tag{9.44}$$

For simplicity, only s-wave scattering is assumed. The $\underline{\sigma}_{\mu\nu}$ are the Pauli matrices. Consider two sublattices with magnetization $<\underline{S}_i> = m\hat{z}$ and $<\underline{S}_i> = -m\hat{z}$, respectively, when referred to a z-axis. Then H_{int} can be decomposed into a molecular-field part $H^{(1)}_{int}$ and a fluctuating part $H^{(2)}_{int}$. In the following discussion, we will first consider $H^{(1)}_{int}$ only, and later comment on the influence of $H^{(2)}_{int}$.

Molecular-field part

The molecular field part of the interaction Hamiltonian can be generally written as

$$H^{(1)}_{int} = - \sum_{\underline{k}'} \sum_{\underline{Q}} h_{\underline{Q}} (c^+_{\underline{k}\uparrow} c_{\underline{k}+\underline{Q}\uparrow} - c^+_{\underline{k}\downarrow} c_{\underline{k}+\underline{Q}\downarrow}) \tag{9.45}$$

$$h_{\underline{Q}} = \frac{mI}{2} \sum_{\underline{K}} \delta_{\underline{Q}\underline{K}}$$

where the $\underline{K}$'s are the reciprocal lattice vectors of the magnetic lattice restricted to the first Brillouin zone.

The electron-pairing prescription in the presence of $H^{(1)}_{int}$ is found as follows. In the absence of interactions which break time-reversal symmetry (e.g., magnetic interactions), the Hamiltonian commutes with the time-reversal operator T, i.e., [H,T] = 0. Then time-reversed single electron states have the same energy and can combine to form Cooper pairs. In the absence of impurity scattering, this implies pairing of states $(\underline{k},\sigma)$ and $(-\underline{k},-\sigma)$. In the present case $[H^{(1)}_{int},T] \neq 0$. However, the operator $Y = T \cdot R$ has the property that [H,Y] = 0, where R denotes a translation by a vector connecting the two sublattices. This implies that when $\psi_{\underline{k}\sigma}(r)$ is an eigenfunction of $H = H_0 + H^{(1)}_{int}$, then $Y \cdot \psi_{\underline{k}\sigma}(r)$ is also an eigenfunction with the same energy (we denote by $H_0 = \sum_k \varepsilon_{\underline{k}} c^+_{\underline{k}\sigma} c_{\underline{k}\sigma}$ the unperturbed Hamiltonian). The pairing prescription for an antiferromagnet is then to pair electrons in states $\psi_{\underline{k}\sigma}(\underline{r})$ and $\exp(i\tilde{\Phi})Y\psi_{\underline{k}\sigma}(\underline{r})$ where $\tilde{\Phi}$ is an arbitrary phase which is chosen for convenience.

To make this prescription more transparent, we assume a cubic system with antiferromagnetically ordered planes along the [100] direction. This is a good model for the Chevrel phases. We write $\psi_{\underline{k}\sigma}(\underline{r})$ in Bloch form

$$\psi_{\underline{k}\sigma}(\underline{r}) = u_{\underline{k}\sigma}(\underline{r}) \exp(i\underline{k}\underline{r})\eta_\sigma \quad , \tag{9.46}$$

where η_σ is the spin function. The function $u_{\underline{k}\sigma}(\underline{r})$ has the periodicity of the magnetic lattice, i.e., $u_{\underline{k}\sigma}(\underline{r}) = u_{\underline{k}\sigma}(\underline{r} + 2\underline{a})$. Here $\underline{a}$ is a lattice vector along [100]. The operator T implies taking the complex conjugate and changing $\sigma \to -\sigma$. The operator R when applied on any function $f(\underline{r})$ changes $Rf(\underline{r}) = f(\underline{r} + \underline{a})$. Thus,

$$Y\psi_{\underline{k}\sigma}(\underline{r}) = \exp(-i\underline{k}(\underline{r}+\underline{a}))u^*_{\underline{k}\sigma}(\underline{r} + \underline{a})\eta_{-\sigma} \quad . \tag{9.47}$$

It is $u^*_{\underline{k}\sigma}(\underline{r}+\underline{a}) = u_{-\underline{k}-\sigma}(\underline{r})$ (see below) and hence

$$Y\psi_{\underline{k}\sigma}(\underline{r}) = \exp(-i\underline{k}\underline{a})\exp(-i\underline{k}\underline{r})u_{-\underline{k}-\sigma}(\underline{r})\eta_{-\sigma} \quad . \tag{9.48}$$

This suggests choosing the phase factor $\exp\tilde{\phi} = \exp(i\underline{k}\underline{a})$ and defining

$$\psi_{-\underline{k}-\sigma}(\underline{r}) = \exp(i\underline{k}\underline{a})Y\psi_{\underline{k}\sigma}(\underline{r}) \quad .$$

The operators which create electrons in states $\psi_{\underline{k}\sigma}(\underline{r})$ and $\exp(i\tilde{\phi})\, Y\psi_{\underline{k}\sigma}$ will in the following be denoted by $a^+_{\underline{k}\sigma}, a^+_{-\underline{k}-\sigma}$. The amplitudes of $\psi_{\underline{k}\sigma}(\underline{r})$ and $\psi_{-\underline{k}-\sigma}(\underline{r})$ are sketched in Fig.9.7 and from that figure it is apparent why both states have the same single-particle energy $E_{\underline{k}}$. The presence of $h_{\underline{Q}}$ as given by (9.45) causes gaps to appear in the one-electron spectrum at points $\underline{p} = \pm\underline{K}/2$. It is of the form

$$E_{\underline{k}} = (\varepsilon_{\underline{k}} + \varepsilon_{\underline{k}-\underline{K}})/2 \pm [(\varepsilon_{\underline{k}} - \varepsilon_{\underline{k}-K})^2/2 + |h_{\underline{K}}|^2]^{1/2} \quad . \tag{9.49}$$

The two signs correspond to the two different branches of the single-particle spectrum due to the folding of the Brillouin zone at $\underline{k} = \pm\underline{K}/2$. The upper branch in the [100] direction belongs to

$$\psi^{ex}_{\underline{k}\sigma}(\underline{r}) = u^{ex}_{\underline{k}\sigma}(\underline{r}) \exp(i\underline{k}\underline{r})\eta_\sigma \tag{9.50}$$

where $u^{ex}_{\underline{k}\sigma}(\underline{r})$ is orthogonal to $u_{\underline{k}\sigma}(\underline{r})$. It corresponds to a distribution of electrons such that in regions of large electron density, their magnetic moments are antiparallel to the molecular field. For rare-earth systems the gaps $|h_{\underline{K}}|$ are expected to be much larger than T_c $[|h_{\underline{K}}|/\varepsilon_F \simeq (g_J - 1)I \cdot N(0) \simeq 10^{-2}]$. On the other hand, the coupling between different magnetic ions $J_{eff} \simeq (g_J - 1)^2 J(J + 1)I^2 N(0)$ and hence the Néel temperature may be rather small. In order to study the influence of antiferromagnetic order on superconductivity, we introduce a Green's function formalism which incorporates the pairing condition described above [9.56]. We define the matrix-Green's function

$$\underline{\underline{G}}(\underline{r},\underline{r}';\tau) = \begin{pmatrix} \langle T_t\Phi_\uparrow(\underline{r},\tau)\Phi^+_\uparrow(\underline{r}',0)\rangle & \langle T_t\Phi_\uparrow(\underline{r},\tau)\Phi_\downarrow(\underline{r}',0)\rangle \\ \langle T_t\Phi^+_\downarrow(\underline{r},\tau)\Phi^+_\uparrow(\underline{r}',0)\rangle & \langle T_t\Phi^+_\downarrow(\underline{r},\tau)\Phi_\downarrow(\underline{r}',0)\rangle \end{pmatrix} \quad , \tag{9.51}$$

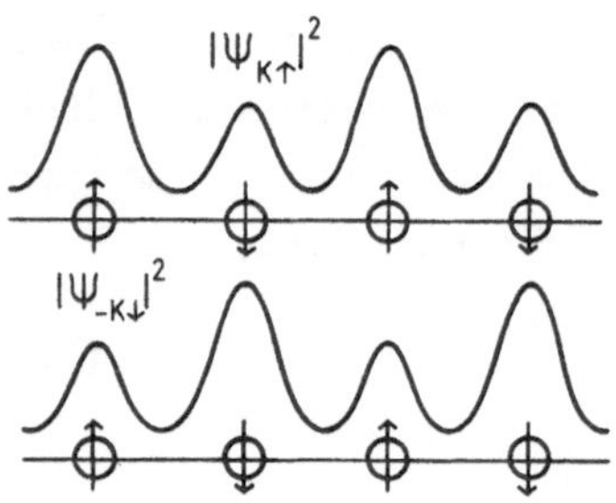

Fig.9.7. Electronic wave functions in an antiferromagnetic lattice belonging to the same energy $E_{\underline{k}}$

where the operators $\Phi_\sigma(\underline{r})$ are expressed in terms of the Bloch functions through

$$\Phi_\sigma(\underline{r}) = \sum_n \sum_{\underline{k}} a^n_{\underline{k}\sigma}\psi^n_{\underline{k}\sigma}(\underline{r})$$

$$\psi^n_{-\underline{k}-\sigma}(\underline{r}) = e^{i\underline{k}\underline{a}}[\psi^n_{\underline{k}\sigma}(\underline{r}+\underline{a})]^* \quad . \tag{9.52}$$

Here n is the band index and the sum over $\underline{k}$ extends over the first Brillouin zone of the antiferromagnetic lattice. T_t is the time-ordering operator and τ is the fictitious "imaginary time" τ which varies from 0 to 1/T [9.57].

Due to Bloch's theorem the diagonal parts of the Green's function matrix must be diagonal in the wave vector $\underline{k}$. By making use of (9.52) we can rewrite $\underline{\underline{G}}(\underline{r},\underline{r}';\tau)$ in the form

$$\underline{\underline{G}}(\underline{r},\underline{r}';\tau) = \sum_{nn'} \sum_{\underline{k}} \underline{\underline{\Psi}}^n_{\underline{k}}(\underline{r})\underline{\underline{G}}^{nn'}(\underline{k},\tau)[\underline{\underline{\Psi}}^{n'}_{\underline{k}}(\underline{r}')]^+ \quad , \tag{9.53}$$

where

$$\underline{\underline{G}}^{nn'}(\underline{k},\tau) = \begin{pmatrix} \langle T_t a^n_{\underline{k}\uparrow}(\tau)a^{n'+}_{\underline{k}\uparrow}(0)\rangle & \langle T_t a^n_{\underline{k}\uparrow}(\tau)a^{n'}_{-\underline{k}\downarrow}(0)\rangle \\ \langle T_t a^{n+}_{-\underline{k}\downarrow}(\tau)a^{n'+}_{\underline{k}\uparrow}(0)\rangle & \langle T_t a^{n+}_{-\underline{k}\downarrow}(\tau)a^{n'}_{-\underline{k}\downarrow}(0)\rangle \end{pmatrix} \tag{9.54}$$

and

$$\underline{\underline{\Psi}}^n_{\underline{k}}(\underline{r}) = \begin{pmatrix} \psi^n_{\underline{k}\uparrow}(t) & 0 \\ 0 & [\psi^n_{-\underline{k}\downarrow}(\underline{r})]^* \end{pmatrix} \tag{9.55}$$

is the matrix of the wave functions. The latter is proportional to the unit matrix when the molecular field $h_{\underline{Q}} \equiv 0$ since then $\psi^n_{\underline{k}\uparrow}(\underline{r}) = [\psi^n_{-\underline{k}\downarrow}(\underline{r})]^*$.

Now we want to show how one can calculate the changes in the quasi-particle interactions when $h_Q \neq 0$. For that purpose we have to calculate the phonon contribution to the matrix $\underline{\underline{G}}$. This requires the knowledge of the phonon propagator

$$D_{ij}(\underline{R}_a,\underline{R}_b;\omega_m) \quad .$$

It describes the dynamical displacements of the ions at positions $\underline{R}_a$, $\underline{R}_b$ in the directions i and j, respectively. $D_{ij}(\underline{R}_a, \underline{R}_b; \omega_m)$ is a solution of the dynamical matrix equation

$$M_a\omega_m^2 D_{ij}(\underline{R}_a,\underline{R}_b;\omega_r - \omega_s) + \sum_{l,\underline{R}_c} \Phi_{il}(\underline{R}_a,\underline{R}_c)D_{lj}(\underline{R}_c,\underline{R}_b;\omega_r - \omega_s)$$

$$= \delta_{ij}\delta_{\underline{R}_a,\underline{R}_b} \quad . \tag{9.56}$$

The M_a are the ionic masses and ω's are again the fermion Matsubara frequencies. The matrix of the force constants $\Phi_{il}(\underline{R}_a,\underline{R}_b)$ is assumed to be known and independent of antiferromagnetic order. The same applies then for the matrix D_{ij}. With D_{ij} given, one can calculate the Green's function $\underline{\underline{G}}$ from the well-known Dyson equation (in abbreviated notation)

$$\underline{\underline{G}} = \underline{\underline{G}}_o + \underline{\underline{G}}_o\underline{\underline{\Sigma}}\underline{\underline{G}} \tag{9.57}$$

where $\underline{\underline{G}}_o$ is the Green's-function matrix in the absence of electron-phonon interactions. The self-energy $\underline{\underline{\Sigma}}$ is given by

$$\Sigma^{nn'}(\underline{k},\omega_r) = \pi T \sum_{\underline{k}',\omega_s} \sum_{mm'} \sum_{ab} \sum_{ij} [g_i(\underline{R}_a)]^{nm}_{\underline{k}\underline{k}'}$$

$$\times\, \underline{\underline{G}}^{mm'}(\underline{k}',\omega_s)[g_j(\underline{R}_b)]^{m'n'}_{\underline{k}'\underline{k}} D_{ij}(\underline{R}_a,\underline{R}_b;\omega_r - \omega_s) \quad . \tag{9.58}$$

Here $\underline{\underline{G}}^{mm'}(\underline{k},\omega_n)$ is the Fourier transform of $\underline{\underline{G}}^{mm'}(\underline{k}',\tau)$. The electron-phonon coupling matrix $[g_i(\underline{R}_a)]^{nn'}_{\underline{k}\underline{k}}$ is defined by

$$[g_i(\underline{R}_a)]^{nn'}_{\underline{k}\underline{k}'} = \begin{pmatrix} \langle n\underline{k}\uparrow|g_i(\underline{R}_a)|n'\underline{k}'\uparrow\rangle & 0 \\ 0 & \langle n(-\underline{k})\downarrow|g_i(\underline{R}_a)|n'(-\underline{k}')\downarrow\rangle \end{pmatrix} \quad . \tag{9.59}$$

The $|n\underline{k}\sigma\rangle$ denote the Bloch functions $\psi^n_{\underline{k}\sigma}(\underline{r})$. The matrix elements, e.g., $\langle n\underline{k}\uparrow|g_i(\underline{R}_a)|n'\underline{k}'\uparrow\rangle$ change as the Bloch functions change below the antiferromagnetic ordering temperature. With them the effective interaction changes between electrons and between electrons and holes. Neglecting interband pairing results in the Baltensperger-Strässler pairing prescription. In that case these interactions are characterized by two functions $\lambda(\omega_n)$ and $\tilde{\lambda}(\omega_n)$,

$$\left.\begin{array}{l}\tilde{\lambda}^{nn'}_{\underline{kk}'}(\omega_r-\omega_s)\\ \\ \lambda^{nn'}_{\underline{kk}'}(\omega_r-\omega_s)\end{array}\right\} = \sum_{ab}\sum_{ij} \langle n\underline{k}\uparrow|g_i(\underline{R}_a)|n'\underline{k}'\uparrow\rangle D_{ij}(\underline{R}_a,\underline{R}_b;\omega_r-\omega_s)$$
$$\times\begin{cases}\langle n'\underline{k}'\uparrow|g_j(\underline{R}_b)|n\underline{k}\uparrow\rangle\\ \langle n(-\underline{k})\downarrow|g_j(\underline{R}_b)|n'(-\underline{k}')\downarrow\rangle\end{cases} \quad . \tag{9.60}$$

The self-consistency equation for the superconducting order parameter and the equation for the renormalized frequencies $\tilde{\omega}_n$ which enter it can be expressed in terms of those functions. One finds

$$\Delta(n\underline{k},\omega_m) = \pi T \sum_{\hat{\underline{k}}'m'n'} \left[\lambda^{nn'}_{\underline{kk}'}(\omega_m-\omega_{m'})-\mu^* N(0)^{-1}\right] \frac{\Delta(n'\underline{k}',\omega_{m'})}{[\tilde{\omega}^2_{m'}(n'\underline{k}')+\Delta^2(n'\underline{k}',\omega_{m'})]^{1/2}}$$
$$\tilde{\omega}_m(n\underline{k}) = \omega_m+\pi T \sum_{\hat{\underline{k}}'m'n'} \tilde{\lambda}^{nn'}_{kk'}(\omega_m-\omega_{m'}) \frac{\tilde{\omega}_{m'}(n'\underline{k}')}{[\tilde{\omega}^2_{m'}(n'\underline{k}')+\Delta^2(n'\underline{k}',\omega_{m'})]^{1/2}} \tag{9.61}$$

where μ^* is the Coulomb pseudopotential. The summation over $\hat{\underline{k}}'$ is restricted to the Fermi surface.

These two equations may serve for a calculation of T_c as a function, e.g., of the field h_Q but also of $\Delta(T)$. They can also be generalized to include a finite electron-mean free path or an external magnetic field and can then be used to calculate the temperature dependence of the critical field. Figure 9.8 shows the result of such a calculation for $TbMo_6S_8$. It does not contain any adjustable parameters but only experimentally determined quantities. For details we refer to [9.56].

From Fig.9.7 one expects that the local contribution $\underline{R}_a = \underline{R}_b$ of the phonon matrix $D_{ij}(\underline{R}_a, \underline{R}_b; \omega_m)$ to $\tilde{\lambda}(\omega_n)$ and $\lambda(\omega_n)$ and hence to $\Delta(\underline{k},\omega_n)$ is reduced, as compared with the case of vanishing molecular field h_Q. On the other hand, one expects that the contributions $\underline{R}_a \neq \underline{R}_b$ are somewhat enlarged by the onset of antiferromagnetic order when $\underline{R}_a$ and $\underline{R}_b$ belong to different sublattices. This change in the attractive electron-electron interaction is the main effect of the antiferromagnetic order [9.56]. For local interactions it results in a weakening of superconductivity and antiferromagnetism. In accordance with Fig.9.1b, T_c remains finite even for large IN(0).

It is instructive to study how the $\hat{\underline{k}}$ integration over the Fermi surface is affected when $E_{\underline{k}}$ replaces $\epsilon_{\underline{k}}$ for $h_Q \neq 0$, see (9.61). Since $E_{\underline{k}}$ has gaps in the spectrum near $\underline{k} = \pm\underline{K}/2$, there is a reduction in phase space for virtual pair scattering provided that $|\epsilon_F - E_{K/2}| \lesssim \omega_D$ which is the Debye energy. This reduction can be simply estimated for a half-filled band ($\epsilon_F = E_{\underline{K}/2}$) where it turns out to be of order $(h_{\underline{K}/2})/\epsilon_F \simeq 10^{-2}$.

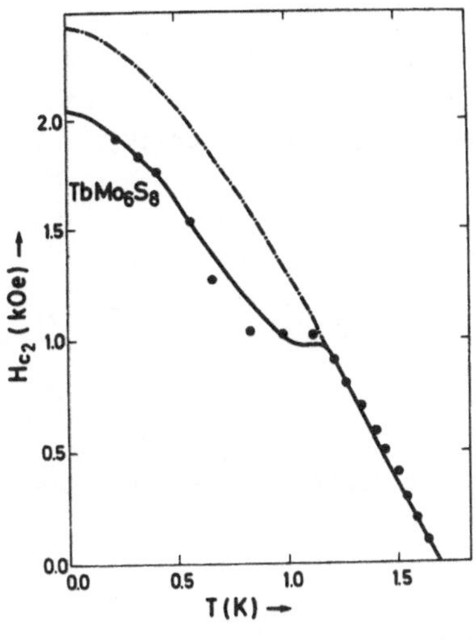

Fig.9.8. Theoretical and experimental results for the upper critical field $H_{c2}(T)$ of $TbMo_6S_8$. The calculated influence of antiferromagnetic order is without adjustable parameters (from [9.56])

In that particular case one can show that both the weakening of the electron-electron attraction and the reduction of phase space lead to comparable effects on T_c although the effect of the former is still larger. But as $|E_{K/2} - \varepsilon_F|$ increases, the reduction in phase space becomes less important compared with the reduction in the electron attraction [9.58]. A more careful estimate of the phase-space reduction is found in [9.59], but since those authors have paired the electrons in states which contain time-reversed components, their calculations cannot otherwise be compared with the present outline of the theory. Other attempts to describe the influence of antiferromagnetic order on superconductivity are found in [9.60-62] and are discussed in [9.63]. They vary from the present approach either in the pairing prescription or the treatment of the changes of the electron interaction via phonons. Finally it should be noted that in antiferromagnetic superconductors, the presence of nonmagnetic scattering centers may influence T_c since at the impurity site, electrons of both spin directions will have the same energy and there will be no reduction in the electron interaction via phonons.

Fluctuations

What remains are some comments on the influence of antiferromagnetic magnons on the superconducting state. As can be inferred from the discussion given in Sect. 9.1.2, their effect on superconductivity is weaker than that of paramagnetic, freely rotating spins. When the characteristic frequency of magnons is sufficiently large (i.e., $|h_{\underline{Q}}| \gg T_c$), then their effect can be treated as a repulsive electron-electron interaction. Two situations may arise:

$$(\alpha) \quad |\varepsilon_F - E_{\underline{K}/2}| \gg |h_{\underline{K}/2}| \quad .$$

In this case, (9.25-27) apply. For an estimate of the repulsive interaction [9.102] which we shall denote by Z_s, it suffices to calculate the scattering of two electrons due to the exchange of magnons. This can be done in close analogy to the case of an exchange of phonons. It is found that approximately [9.7]

$$Z_s \simeq \frac{N}{\mathrm{Vol}} \frac{N(0)I^2C^2S}{2\omega_M} \quad .$$

Here ω_M is the Debye energy of the magnons, C is a constant of order unity and Vol denotes the volume. By using some reasonable values

$$N(0)I = 5\cdot 10^{-3} \quad , \quad C^2S(N/\mathrm{Vol})I/\omega_M = 4, \; N(0)V_{ph} = 0.1 \quad ,$$

one obtains

$$N(0)V_{ph}/Z_s = 10 \quad ,$$

i.e., the magnetic fluctuations are not strong enough to suppress superconductivity.

(β) $|\varepsilon_F - E_{\underline{k}/2}| \lesssim \omega_D$

As discussed before, the gap in the one-electron excitation spectrum $E_{\underline{k}}$ reduces the available phase space for virtual pair scattering. This applies to the scattering due to phonons as well as to magnons. Consequently, both interactions have to be multiplied by a reduction factor $\alpha \lesssim 1$. A more detailed study of the influence of magnons on T_c has to involve a numerical analysis.

9.2 Magnetic Structures in Superconductors Resulting From Exchange Interactions

Until now the influence of exchange scattering on the superconducting state was considered. In the following a discussion will be given of the changes in the magnetic order which result from superconductivity. The first theoretical treatment of that problem goes back to ANDERSON and SUHL [9.24] who considered dilute alloys, which in the absence of superconductivity were assumed to order ferromagnetically. In the superconducting state, the effective interaction (RKKY) between the magnetic ions is changed due to changes in the electron-spin susceptibility. As a consequence, long-wave spin structures induced by superconductivity were predicted. Stimulated by the experimental findings in ternary systems, that treatment was recently generalized to the case where the magnetic ions form a lattice. The resulting cryptoferromagnetism (or spiral magnetic state) is discussed in the following. First we shall outline the basic ideas for T = 0. Next a treatment is presented for a Ginzburg-Landau regime in which both the superconducting and the magnetic-order parameter are small [9.64]. Finally, the complete phase diagram is discussed by making use of a treatment in which only the magnetic-order parameter is assumed to be small [9.65]. In this section we shall not discuss modifications of the magnetic state which result from superconducting screening currents in the presence of a magnetization. Those effects are the subject of Sect.9.3.

9.2.1 Anderson-Suhl Cryptoferromagnetism

In Sect.9.1.1 it was shown that in a superconductor the zero-temperature electron spin susceptibility $\chi_s(\underline{q})$ differs from $\chi_n(\underline{q})$ for $q \lesssim \xi_0^{-1}$ and that it has a maximum at $Q_0 = (3\pi^2 k_F^2 \xi_0^{-1})^{1/3}$. The space-dependent susceptibility $\chi(\underline{r})$ is related to the interaction energy U of the magnetic ions through

$$U = -\frac{I^2}{2} \sum_{i \neq j} \chi(|\underline{R}_i - \underline{R}_j|)\underline{S}_i\underline{S}_j \quad . \tag{9.62}$$

Thus, a change in $\chi(\underline{q})$ for small values of $\underline{q}$ implies a change in the long-range tail of the ion-ion interaction. In a normal metal $\chi(\underline{r})$ is on the average positive, i.e., $\chi_n(\underline{q} = 0) > 0$. Since $\chi_s(\underline{q} = 0) = 0$ at T = 0, the change $\Delta\chi(\underline{r})$ due to superconductivity is negative and of range ξ_0^{-1}. The magnetic ions respond to this by forming domains of size $Q_0^{-1} \simeq 50$ Å. This results in so-called cryptoferromagnetism.

A different way of seeing this is by writing

$$U = -\frac{1}{2}\frac{N^2}{\mathrm{Vol}} \sum_{\underline{q}} J_{\mathrm{eff}}(\underline{q})\underline{S}_{\underline{q}}\underline{S}_{-\underline{q}} \tag{9.63}$$

where we assume that N magnetic ions form a lattice of volume Vol so that

$$\underline{S}_{\underline{q}} = N^{-1} \sum_{i=1}^{N} \underline{S}_i \exp(-i\underline{q}\underline{R}_i) \quad . \tag{9.64}$$

Furthermore,

$$J_{\mathrm{eff}}(\underline{q}) = I^2 \left[\sum_{\underline{G}} \chi_s(\underline{q} + \underline{G}) - \frac{1}{N}\sum_{\underline{k}} \chi_s(\underline{k})\right] \tag{9.65}$$

where $\underline{G}$ denotes the reciprocal lattice vectors. The second term in the bracket excludes the self-interaction of a spin with itself. For simplicity we shall consider only the simplified form $J_{\mathrm{eff}}(\underline{q}) = I^2\chi_s(\underline{q})$ as the remaining parts are only weakly $\underline{q}$ dependent. Since $\chi_s(\underline{q})$ is largest at $|\underline{q}| = Q_0$, the interaction energy U will take its minimum value when all the weight of $\underline{S}_{\underline{q}}$ is contained in the Fourier component $\underline{S}_{\underline{Q}_0}$. This again implies a cryptoferromagnetic spin structure. For an analysis of the precise form of this structure, the free energy of the system has to be studied. This is done below. A detailed energy analysis also requires the inclusion of additional effects such as spin-orbit scattering, pairbreaking due to exchange scattering, etc. As shown in Sect.9.1.1, the latter lead to changes in $\chi_s(\underline{q})$ and hence influence the effective ion-ion interaction.

9.2.2 Ginzburg-Landau Regime

When the superconducting order parameter $\Delta(\underline{r})$ and the magnetization density are both small, one can study the interplay of superconductivity and magnetism by a Ginzburg-Landau type of theory. For this purpose one constructs a free-energy functional in powers of the two-order parameters. In the following a simplified derivation of such a functional will be given which resembles the one of [9.66]. A more sophisticated treatment can be found in [9.64] and also in Sect.9.2.3. For the description of the spatial variation of the local moments, let us define a continuous spin field $\underline{m}(\underline{r})$ with the property that $\underline{m}(\underline{R}_i) = \underline{S}_i$. Then the free energy which depends on $\underline{m}(\underline{r})$ and the superconducting order parameter $\Delta(\underline{r})$ is decomposed into three terms:

$$F[\Delta,\underline{m}] = F_S^0[\Delta] + F_M^0[\underline{m}] + F_{SM}[\Delta,\underline{m}] \quad . \tag{9.66}$$

The first two terms are the free-energy functionals of the superconducting and magnetic system in the absence of mutual interactions, i.e.,

$$F_S^0[\Delta] = \int d^3\underline{r}\left[\frac{\alpha_s}{2}|\Delta(\underline{r})|^2 + \frac{\beta_s}{4}|\Delta(\underline{r})|^4 + \frac{\gamma_s}{2}|\underline{\nabla}\cdot\Delta(\underline{r})|^2\right] \tag{9.67a}$$

$$F_M^0[m] = \int d^3\underline{r}\left\{\frac{\alpha_m}{2}\underline{m}^2(\underline{r}) + \frac{\beta_m}{4}\underline{m}^4(\underline{r}) + \frac{\gamma_m}{2}[\underline{\nabla}\cdot\underline{m}(\underline{r})]^2\right\} \quad . \tag{9.67b}$$

The form of the coupling term $F_{SM}[\Delta,\underline{m}]$ can be derived by realizing that in the presence of a fixed $\Delta(\underline{r})$, the free-energy functional of the spin system is

$$F_M[\underline{m},\Delta] = F_M^0[\underline{m}] + F_{SM}[\Delta,\underline{m}] \quad . \tag{9.68}$$

For a uniform $\Delta(\underline{r}) = \Delta$, it can be expressed quite generally as

$$F_M[\underline{m},\Delta] = N\left[\frac{1}{2}\sum_{\underline{q}}\chi_m^{-1}(\underline{q})|\underline{m}_{\underline{q}}|^2 + O(\underline{m}^4)\right] \tag{9.69}$$

where

$$\underline{m}_{\underline{q}} = \frac{1}{\mathrm{Vol}}\int d^3\underline{r}\,\underline{m}(\underline{r})\exp(-i\underline{q}\underline{r})$$
$$= \frac{1}{N}\sum_i \underline{S}_i \exp(-i\underline{q}\underline{R}_i)$$

and $\chi_m(\underline{q})$ is the mean-field susceptibility of the spin field $\underline{m}(\underline{r})$. The latter can be written as

$$\chi_m(\underline{q}) = \frac{\chi_0(T)}{1 - J_{\mathrm{eff}}(\underline{q})n\chi_0(T)} \tag{9.70}$$

where $n = N/\mathrm{Vol}$ and $J_{\mathrm{eff}}(\underline{q})$ describes the effective interaction between the different spins as given by (9.65). As before, $\chi_0(T) = S(S + 1)/3T$. Note that $\chi_m(\underline{q})$ corresponds to the susceptibility per ion used earlier in Sect.9.1.2. Using (9.70) it is then easy to extract the coupling term $F_{SM}[\Delta,\underline{m}]$ from (9.68) as

$$F_{SM}[\Delta,\underline{m}] = - \mathrm{Vol}\left\{n^2 \frac{I^2}{2} \sum_{\underline{q}} [\chi_s^{-1}(\underline{q},\Delta) - \chi_n^{-1}(\underline{q})]|\underline{m}_{\underline{q}}|^2\right\} \quad . \tag{9.71}$$

It depends on the difference of the conduction electron susceptibility in the superconducting and normal state. The latter are given by (9.6,7).

Minimization of $F_M[\underline{m},\Delta]$ with respect to $\underline{m}_{\underline{q}}$ yields a spin structure with wave vector $\underline{Q}$ which is determined by the maximum of $\chi_s(\underline{q},\Delta)$. One finds as a function of temperature

$$Q = Q_o\left[\frac{\Delta(T)}{\Delta(0)} \tanh \frac{\Delta(T)}{2T}\right]^{1/3} \quad , \tag{9.72}$$

with Q_o given by (9.9). For a complete determination of the magnetic structure, the polarization of the magnetization has to be known. It is found by minimizing the $\underline{m}^4(\underline{r})$ term in (9.67b). In the absence of anisotropies it is found that a helical structure of the form

$$\underline{m}(\underline{r}) = m(\cos Qz, \sin Qz, 0) \tag{9.73}$$

is favored.

Until now Δ was kept arbitrarily large. By expanding in powers of Δ one finds for the coupling term

$$F_{SM}[\Delta,\underline{m}] = \mathrm{Vol}\ \frac{\Delta^2}{2} \eta(2\pi)^3 \sum_{\underline{q}} \tilde{g}(q)|\underline{m}_{\underline{q}}|^2 \quad , \tag{9.74}$$

where

$$\eta = I^2N^2(0)\pi^2n^2/T \tag{9.76}$$

and

$$\tilde{g}(q) = (8\pi^2 k_F^2 q)^{-1} \quad . \tag{9.76}$$

The q-dependence of F_{SM} is a consequence of the nonlocal character of the interaction. This point has been stressed by SUHL [9.64]. In real space F_{SM} can be written as

$$F_{SM}[\Delta,\underline{m}] = \frac{\eta}{2} \int d^3\underline{r}\Delta^2(\underline{r}) \int \frac{d^3\underline{\rho}}{(2\pi)^3}\, \underline{m}(\underline{r} + \underline{\rho}/2)\underline{m}(\underline{r} - \underline{\rho}/2) \times \frac{\sin^2(\rho k_F)}{(\rho k_F)^2} \quad . \tag{9.77}$$

Since this expression allows for a spatial variation of $\Delta(\underline{r})$, it is slightly more general than (9.74).

We want to complete the derivation of the Ginzburg-Landau functional by listing the various quantities which appear in the functional within the present calculation scheme:

$$\alpha_s = 2N(0)(T - T_{co})/T_{co}$$

$$\beta_s = \frac{7}{4}\frac{\zeta(3)}{\pi^2 T_{co}^2} N(0) \quad ; \quad \zeta(3) = 1.202$$

$$\gamma_s = \frac{7}{24}\frac{\zeta(3)}{\pi^4 T_{co}^2}\frac{k_F^3}{m} = 1.08 N(0)\xi_o^2$$

$$\alpha_m = 2N(0)I^2 n^2 (T - T_{mo})/T_{mo}$$

$$\beta_m \simeq nT_{mo}$$

$$\gamma_m = 2N(0)I^2 n^2 a^2 \quad ; \quad a = (12k_F^2)^{-1} \tag{9.78}$$

$$T_{mo} = 2N(0)I^2 nS(S+1)/3 \quad .$$

It is instructive to notice that the same Ginzburg-Landau functional can also be used to study the influence of spin fluctuations on the superconducting state. From (9.67a,71) one can derive an effective functional for the superconducting order parameter Δ [for simplicity we consider only homogeneous $\Delta(\underline{r})$] simply by replacing $|m_{\underline{q}}|^2$ by its thermal average $\langle|m_{\underline{q}}|^2\rangle$:

$$F[\Delta] = \mathrm{Vol}\left[\frac{\alpha_s}{2}\Delta^2 + \frac{\beta_s}{4}\Delta^4 + \frac{\eta}{2}\Delta^2 \int d^3\underline{q}\tilde{g}(q) \quad \langle|m_{\underline{q}}|^2\rangle \quad \mathrm{Vol}\right] \quad . \tag{9.79}$$

The changes in T_c due to $\langle|m_{\underline{q}}|^2\rangle \neq 0$ are obtained from

$$\alpha_s(T_c) + \eta\mathrm{Vol} \int d^3\underline{q}\tilde{g}(q) \quad \langle|m_{\underline{q}}|^2\rangle = 0 \quad . \tag{9.80}$$

The correlation function $\langle|m_{\underline{q}}|^2\rangle$ is related to the local spin susceptibility [see (9.22,70)] through

$$\langle|m_{\underline{q}}|^2\rangle = 3T\chi_m(\underline{q},T)/N \quad . \tag{9.81}$$

By using (9.78) one then obtains for the superconducting transition temperature

$$T_c = T_{co} - nI^2 N(0)\frac{3\pi^2}{2}\int d^3\underline{q}\tilde{g}(q)\chi_m(\underline{q},T_c) \quad . \tag{9.82}$$

This expression coincides with the one in (9.21) in the limit of weak pairbreaking. This connection between the two different approaches shows that the coupling term $F_{SM}[\Delta,\underline{m}]$ is correct to lowest order in Δ and $\underline{m}$. It also suggests that mean-free path effects of the conduction electrons can be taken into account by replacing the function $\tilde{g}(q)$ in (9.79) by the more general weighting function $g(q)$ given by (9.17). For sufficiently strong exchange coupling constant I, (9.82) may have two solutions (see Fig.9.4). In that case, superconductivity is quenched above the magnetic transition temperature due to fluctuations in the spin system and a development of the spiral structure is also inhibited. However, for sufficiently weak exchange coupling, the cryptoferromagnetic state should be realized in a certain temperature range below the magnetic transition temperature T_{mo} before, at lower temperatures, a ferromagnetic normal phase eventually becomes more stable.

In [9.45] it was shown that a quenching of superconductivity by magnetic fluctuations produces a large jump in the specific heat at the superconducting transition temperature T_c. This implies that also fluctuations of the superconducting order parameter are important in that regime. The latter have not been considered so far. For a correct treatment of the fluctuations one has to consider all quartic terms in the Ginzburg-Landau functional. In Sect.9.3 we shall discuss similar problems in connection with electromagnetic effects.

9.2.3 Phase Diagram for Cryptoferromagnetic Spin Structures

In the following we want to discuss in more detail the possible occurrence of a cryptoferromagnetic spin structure in a superconducting system. Of particular interest is thereby the relative stability of the cryptoferromagnetic phase as compared with a simple ferromagnet and a superconductor with paramagnetic ions. Hereby we shall follow the work of BULAEVSKII et al. [9.65] who also calculated a complete phase diagram. We begin by outlining the procedure which leads to a phase diagram before we become more specific.

Our starting point is a free energy function $F(\Delta,m,\underline{Q})$ for the coupled system of conduction electrons and local moments which are assumed to order cryptoferromagnetically with wave vector $\underline{Q}$. m is the magnetization amplitude defined in (9.73) and is a variational parameter as is Δ and $\underline{Q}$. It is important *not* to expand $F(\Delta,m,\underline{Q})$ in powers of Δ since at the onset of magnetic order Δ is usually not small. Minimization of $F(\Delta,m,\underline{Q})$ with respect to Δ and $\underline{Q}$ leads to an effective free-energy functional $F_{SC}(m,T)$ for the superconducting cryptoferromagnetic state. Generally it can be written as

$$F_{SC}(m,T) = \frac{A}{2} m^2 + \frac{B}{4} m^4 + F_o(T) \tag{9.83}$$

and $A(T) = 0$ determines the transition temperature T_1 from the paramagnetic to the cryptoferromagnetic state. The sign of B determines the order of the phase transition, i.e., first or second order. The possible transition to a purely ferromagnetic state at a temperature $T_2 < T_1$ is investigated by equating the free energies at equilibrium value for the two phases.

In order to be more specific, the following assumptions are made. The cryptoferromagnetic structure is supposed to be of the form given by (9.73) and the superconducting order parameter is assumed to be homogeneous, i.e., Δ = const. Furthermore, the exchange interaction between the conduction electrons and the spin system is treated like an external field $\underline{h}(\underline{r})$ acting on the electron-spin density $\underline{m}(\underline{r})$. Thus,

$$\underline{h}(\underline{r}) = h(\cos \underline{Q}\underline{r}, \sin \underline{Q}\underline{r}, 0) \qquad h = nmI \quad , \tag{9.84}$$

where, as before, n is the number of spins per volume.

This approximation neglects the effects of magnetic scattering. On the other hand it enables one to write down a complete expression for the free-energy functional $F[\Delta,m,Q]$ by using the quasi-particle energies of the superconducting electrons in the field $\underline{h}(\underline{r})$ and by making a mean-field approximation for the spin system. Yet the functional is not of simple form. This changes close to the magnetic transition when $h \ll qv_F$ and when, furthermore, $T \ll T_{co}$. In this case one obtains [9.65]

$$F(\Delta,m,Q) = \mathrm{Vol}\ N(0)n^2I^2\left\{\left(1 - \frac{T}{T_{mo}} + \frac{Q^2}{k_F^2} + \frac{\pi^2\Delta}{2Qv_F}\right)m^2 \left[f(S)\frac{I}{T_{mo}} - \frac{\pi^2\Delta}{2Qv_F}\frac{1}{16}\left(\frac{nI}{\Delta}\right)^2\right]m^4\right\} - \mathrm{Vol}\ N(0)\Delta^2(1 + \ln \Delta_o^2/\Delta^2) \tag{9.85}$$

with $f(S) = \frac{3}{10}\frac{S^2 + S + 1/2}{(S+1)^2S^2}$ and $\Delta_o = 1.76\ T_{co}$.

The last term is the free energy of a BCS superconductor. The first two terms can be compared with (9.69). With the help of (9.70) [see also (9.71)] and with the electron-spin susceptibility given by (9.7), one obtains immediately the term proportional to m^2. The Δ dependence of the m^4 term is more difficult to explain and we shall not attempt this here. By minimizing F with respect to Q and Δ, one finds as a function of the magnetization,

$$Q(m) = Q_o\left(1 - \frac{n^2}{48}\frac{I^2}{\Delta_o^2}m^2\right) \qquad \Delta(m) = \Delta_o - \frac{(nIm)^2Q_o}{2\cdot 3^{1/2}\Delta_o^2k_F} \tag{9.86}$$

with $|Q_o|$ given by (9.9). One notices that for m = 0 the value of Q(m) coincides with Q_o which is the wave vector found in Sect.9.2.1. When Q(m) and Δ(m) are inserted into (9.85), the effective free-energy functional for the magnetization of the cryptoferromagnetic state takes the following form:

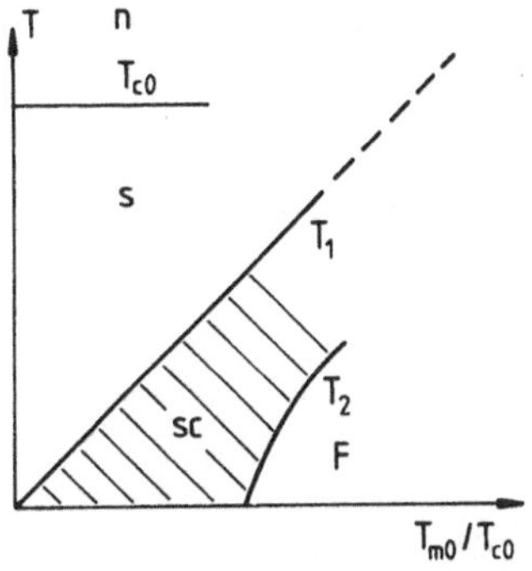

Fig.9.9. Phase diagram for an exchange coupled magnetic superconductor (without magnetic field effects). T_{mo} is the ferromagnetic transition temperature in the normal state. T_1 denotes the transition into a cryptoferromagnetic coexistence state and T_2 the (first order) transition into the ferromagnetic normal state (from [9.65])

$$F_{SC}(m,T) = \mathrm{Vol}\; N(0)n^2I^2[m^2(T - T_1)/T_1 + m^4 f(S)(1 - T_1/T_1^*)] - \frac{1}{2} N(0)\Delta_o^2 \;\mathrm{Vol} \quad . \tag{9.87}$$

Here, T_1 is

$$T_1 = T_{mo}\left(1 - \frac{1}{4}\frac{Q_o^2}{k_F^2}\right) \tag{9.88}$$

and T^* is a characteristic temperature of the order

$$T^* \sim \Delta_o(\Delta_o/\epsilon_F)^{1/3} \quad . \tag{9.89}$$

For strong magnetic coupling the phase transition from the paramagnetic (superconducting) state to the cryptoferromagnetic state becomes of first order ($T_1 > T^*$).

For the investigation of the possible (first order) transition from the cryptoferromagnetic to the ferromagnetic normal state, the free energies for the two phases have to be compared. Thereby one cannot expand F_{SC} in powers of m. From such a calculation for which we refer to the literature [9.65], the phase diagram shown in Fig.9.9 is obtained with the transition curve T_2. It turns out that the amplitude of the magnetization is almost unchanged in such a transition. It is primarily the direction of polarization which changes.

9.3 The Magnetic State in the Presence of Electromagnetic Coupling

9.3.1 Ginzburg-Landau Functional

For many problems with a spatially-varying order parameter, a Ginzburg-Landau expansion of the free energy is a good starting point for a theory. For a superconductor in an applied magnetic field and coupled to a magnetic system, such a free-energy functional was formulated by KREY [9.27] and subsequently used by different authors in a slightly modified version. It is expected to hold when one is in the vicinity of both the superconducting and the magnetic phase transition temperature. Here we shall make the following ansatz for the free energy as a functional of the superconducting order parameter $\Delta(\underline{r})$ and the magnetization $\underline{M}(\underline{r})$ (in this section we will write down explicitly factors of $\hbar$ in combination with the elementary charge e for clarity)

$$F = F_S + F_M + F_H \tag{9.90}$$

with

$$F_S = \int_K d^3\underline{r}\left[\frac{1}{2}\alpha_s|\Delta(\underline{r})|^2 + \frac{1}{4}\beta_s|\Delta(\underline{r})|^4 + \frac{1}{2}\gamma_s|(-i\underline{\nabla} - \frac{2e}{\hbar}\underline{A})\Delta(\underline{r})|^2\right]$$

$$F_M = \int_K d^3\underline{r}\left\{\frac{1}{2}\alpha_M\underline{M}^2(\underline{r}) + \frac{1}{4}\beta_M\underline{M}^4(\underline{r}) + \frac{1}{2}\gamma_M[\underline{\nabla}\underline{M}(r)]^2\right\}$$

$$F_H = \int_V d^3\underline{r}\,\frac{\mu_0}{2}H^2(\underline{r}) \quad . \tag{9.91}$$

Here K and V indicate integrations over the sample and all space, respectively. Furthermore, $\underline{H} = \underline{B}/\mu_0 - \underline{M}$ and $\underline{B} = \underline{\nabla}\times\underline{A}$. A local ansatz has been made for the free-energy contributions of both the superconducting and magnetic system including their mutual interaction. The microscopic origin of the different terms and the limitations of this ansatz will be discussed below. The ansatz (9.90) for the free energy is limited to an electromagnetic interaction between the magnetic ions and superconducting electrons. For a realistic system with appreciable exchange interactions between the magnetic ions and the conduction electrons, F would have to be supplemented by a coupling term as given by (9.77). In the following such a coupling will not be considered since we are here exclusively interested in effects resulting from the electromagnetic coupling. The electromagnetic interactions between the magnetic and superconducting systems are produced by the vector potential in the expression for F_S. The interaction between the magnetic moments is partially contained in F_M and in the magnetic field energy F_H. Note that $\underline{M}$ is the magnetization of the magnetic ions only. The magnetization produced by the supercurrents j_s is contained in $\underline{H}$, i.e., $j_s = \underline{\nabla}\times\underline{H}$ inside the sample. In this respect our expression for F differs from the one used, e.g., in [9.67]. There is also a difference to [9.25] in the notation used for the magnetic field energy. This is compensated for by a difference in the definition of the coefficient α_M. In the presence of an external magnetic field $\underline{H}_e$ one has to consider the free enthalpy

$$G = F - \int d^3\underline{r}\,\underline{H}_e(\underline{r})\underline{B}(\underline{r}) \quad . \tag{9.92}$$

The equilibrium state for a given $\underline{H}_e(\underline{r})$ is then obtained by minimizing G with respect to $\Delta(\underline{r})$, $\underline{A}(\underline{r})$ and $\underline{M}(\underline{r})$. In order to keep $\underline{H}_e(\underline{r})$ fixed it is conceptionally more appropriate to vary the vector potential $\underline{A}_s(\underline{r})$ of the supercurrents instead of the total vector potential $\underline{A}(\underline{r})$. The results, however, are the same in both cases inside the sample and lead to the following Ginzburg-Landau equations:

$$\alpha_s\Delta(\underline{r}) + \beta_s|\Delta(\underline{r})|^2\Delta(\underline{r}) + \gamma_s\left[-i\underline{\nabla} - \frac{2e}{\hbar}\underline{Ar}\right]^2\Delta(\underline{r}) = 0$$

$$\underline{\nabla}\times\underline{H}(\underline{r}) = \underline{j}_s(\underline{r}) = \frac{\gamma_s}{2}\cdot\frac{2e}{\hbar}\left[\Delta^*(\underline{r})\left[-i\underline{\nabla} - \frac{2e}{\hbar}\underline{A}(\underline{r})\right]\Delta(\underline{r}) + \text{c.c.}\right] \tag{9.93}$$

$$\mu_o\underline{H}(\underline{r}) = \alpha_M\underline{M}(\underline{r}) + \beta_M[\underline{M}(\underline{r})]^2\underline{M}(\underline{r}) - \gamma_M\nabla^2\underline{M}(\underline{r}) \quad .$$

The last equation has to be considered as an equation determining $\underline{M}$ for a given field $\underline{H}$. The field $\underline{H}$, on the other hand, is determined by the supercurrent density ($\underline{\nabla}\times\underline{H} = \underline{j}_s$) and the spin magnetization ($\nabla\underline{H} = -\underline{\nabla}\underline{M}$). The GL equations have to be solved with the following boundary conditions for the normal components of $\underline{j}_s$ and $\underline{\nabla}\underline{M}_i$ at the sample boundary: $j_{s\perp} = 0$ and $(\underline{\nabla}M_i)_\perp = 0$. We want to stress that there are various ways to write down Ginzburg-Landau functionals which may differ with respect to the energy densities which they contain [9.68]. This is possible since integrals of the form $\int d^3\underline{r}\underline{F}(\underline{r})\underline{G}(\underline{r})$ vanish when div $\underline{F} = 0$ and rot $\underline{G} = 0$ provided that the regions of the sources of F and G (i.e., rot $\underline{F} \neq 0$ and div $\underline{G} \neq 0$) are finite.

When comparing the parameters of the Ginzburg-Landau functional in (9.91) with those in (9.67b), one has to observe that $\underline{M} = g\mu_B n\underline{m}$, where g is the Landé factor and n is the number of spins per volume with $\langle S^z\rangle = m$. This implies the following relations:

$$\alpha_m = \alpha_M(g\mu_B n)^2 \quad , \quad \beta_m = \beta_M(g\mu_B n)^4 \quad , \quad \gamma_m = \gamma_M(g\mu_B n)^2 \quad .$$

For the following it is useful to define transition temperatures T_{co} and T_{mo} by the vanishing of the parameters $\alpha_S(T)$ and $\alpha_M(T)$, i.e., one sets $\alpha_S(T) = \alpha_S'(T - T_{co})/T_{co}$ and $\alpha_M = \alpha_M'(T - T_{mo})/T_{mo}$. For systems with dominating magnetic dipole interaction, $\alpha_M' \simeq \mu_o$ and $T_{mo} \simeq \mu_o\mu_B^2 n$.

In the following section we want to give a survey of possible solutions of the Ginzburg-Landau equations in the absence of an external magnetic field [$\underline{H}_e(\underline{r}) = 0$]. This is followed by a more detailed discussion of two special inhomogeneous magnetic and superconducting structures which are of particular relevance for bulk samples of ternary superconductors.

9.3.2 Survey of Solutions of the Ginzburg-Landau Equations

First we consider phases with *constant-order parameters* $\Delta(\underline{r})$ *and* $\underline{M}(\underline{r})$. When $\Delta(\underline{r})$ = const., the Meissner effect requires $\underline{B} \equiv 0$ in the bulk of the sample. This is fulfilled either when $\underline{M} = 0$ or $\underline{M} = -\underline{H}$. The latter case corresponds to a coexistence phase between ferromagnetism and superconductivity. $\underline{M} = -\underline{H}$ with constant $\underline{M}$

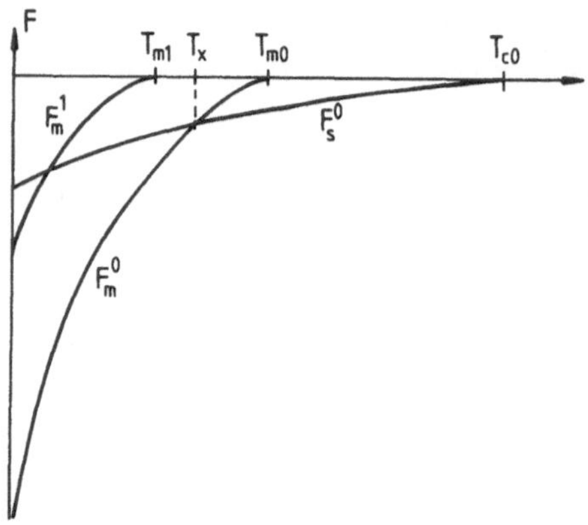

Fig.9.10. Schematic free energy of the superconducting state F_s^o and the ferromagnetic normal state F_m^o. The function F_m^1 is the total magnetic energy in a phase with $\underline{H} = -\underline{M}$

can be realized in two different ways. The first is by a thin film which is magnetized perpendicularly to the surface. Then $\underline{H}$ is the demagnetizing field. The second is by a thin rod which is magnetized parallel to the rod axis. In that case $\underline{B} = 0$ is achieved by superconducting screening currents which are set up in the surface of the sample. In both cases the magnetic field energy is large in the sample and negligible outside the sample. The situation is illustrated in Fig.9.10 which shows the free energy of a superconductor in the Meissner state ($\underline{B} = 0$), i.e., $F_s^o(T) = -\mathrm{Vol}\{\alpha_s^2/4\beta_s\}$ and the free energy of the spin system for $\underline{H} = 0$, i.e., $F_M^o(T) = -\mathrm{Vol}\{\alpha_M^2/4\beta_M\}$. Both quantities are obtained by setting the solutions of (9.93) for the order parameters Δ and $\underline{M}$ into (9.91). In Fig.9.10 it has been assumed that $T_{co} > T_{mo}$ as is usually the case for ternary superconductors. At $T = 0$ the free energy of the spin system is much lower than the superconducting condensation energy. T_x denotes the temperature below which the Meissner state ($\underline{B} = 0$) becomes unstable with respect to the ferromagnetic normal state in the absence of demagnetization effects ($\underline{H} = 0$). Also shown in Fig.9.10 is the function $F_M^1(T) = -\mathrm{Vol}\{(\alpha_M + \mu_o)^2/4\beta_M\}$ which is the free energy of the magnetic system $(F_M + F_H)$ for a phase with $\underline{H} = -\underline{M}$. The corresponding transition temperature is given by $T_{m1} = T_{mo}(1 - \mu_o/\alpha_M')$. The total free energy of a Meissner state in coexistence with ferromagnetism is then given by the sum $F_s^o(T) + F_M^1$. In principle, the temperature T_{m1} can be larger or smaller than T_x (see Fig.9.10). The latter case is realized for systems with predominantly dipolar coupling.

Now let us turn to *phases with constant* Δ *and spatially varying* $\underline{M}(\underline{r})$. When one allows for spatial variations of $\underline{M}(\underline{r})$, the system can avoid part of the magnetic field energy. The resulting phases may have transition temperatures much closer to T_{mo} than found before. Note that the condition $\underline{B} \equiv 0$ is now relaxed, but there must still be $\langle\underline{B}(\underline{r})\rangle = 0$ when $\Delta = \mathrm{const}$. Inhomogeneous magnetic structures may occur in the form of cryptoferromagnetic (spiral) states [9.25,26,29], or in the form of magnetic domains [9.27,69], in particular, in the presence of anisotropies and surfaces. While the cryptoferromagnetic state is discussed in detail in Sect.9.3.3, we want to comment here on the work which has been done on magnetic domain walls. In [9.69] the magnetic domain structure of a thin sheet with uniaxial anisotropy was investigated and the domain width as a function of sample thickness was calculated. This was done by a minimization of the free energy (9.91) with a special ansatz for the field energy of the sample surface. The order parameter Δ was kept constant. In [9.70] the influence of superconductivity on a Bloch wall in a bulk ferromagnet was studied. It was found that a Bloch wall is more stable in a superconductor than in a normal metal due to partial screening of the magnetic interactions. Therefore, it was speculated that superconductivity may persist in Bloch walls even when the rest of the sample is in the ferromagnetic normal phase.

Finally, we have to consider *phases with spatial variations of* $M(\underline{r})$ *and* $\Delta(\underline{r})$. Phase of this kind are expected when one is dealing with type II superconductors. Then a superconducting vortex state may develop in the magnetic field set up by the spin magnetization [9.28,71,75]. An upper critical value $M_{c2} = H_{c2}$ of the magnetization exists at which the system undergoes a second-order phase transition into a ferromagnetic normal state [9.71,72]. The same situation has been studied in [9.73] under the assumption of a spatially homogeneous spin magnetization. There is, however, one important difference to an ordinary type II superconductor in an external field. Because of the large field energy in a Meissner state the system will undergo a direct first-order phase transition from a paramagnetic superconducting state to a magnetically-ordered vortex state, i.e., the diamagnetic regime with $H < H_{c1}$ will usually be suppressed. This transition should occur between T_x and T_{mo}. Close to the onset of magnetic order one must also consider the alternative development of a spiral state (Sect.9.3.4).

9.3.3 Inhomogeneous Phases

As pointed out before, coexisting phases of ferromagnetic ions and superconducting electrons are difficult to achieve with constant order parameters Δ and $\underline{M}$. However, when one allows for spatially varying solutions of (9.93), the situation is more favorable. In that case the system can decrease the field energy in the sample considerably. We shall discuss and compare in the following two inhomogeneous solutions. One corresponds to a cryptoferromagnetic spin structure of the ions while the second corresponds to superconducting screening currents and magnetization forming a vortex lattice. Both solutions may be realized experimentally for different regimes of the relevant parameters. Particular attention is devoted to fluctuations of the order parameters around their mean values since they may influence considerably the stability of the coexistence phases.

a) *Blount-Varma Theory*

When investigating oscillatory solutions of the Ginzburg-Landau equations, it is advantageous to use the Fourier transform of the Ginzburg-Landau functional (9.91). For this purpose one sets

$$\underline{M}(\underline{r}) = \sum_{\underline{q}} \underline{M}_{\underline{q}}\, e^{i\underline{q}\underline{r}} \quad ; \quad (\underline{M}_{-\underline{q}} = \underline{M}_{\underline{q}}^{+}) \tag{9.94}$$

and similarly for $\underline{A}(\underline{r})$ and $\underline{B}(\underline{r})$. In the Coulomb gauge ($\Delta\underline{A} = 0$) one has for $\underline{q} \neq 0$,

$$\underline{A}_{\underline{q}} = i(\underline{q} \times \underline{B}_{\underline{q}})/q^2 \quad . \tag{9.95}$$

Furthermore, it is assumed that $\Delta(\underline{r}) = \Delta_o$ = const. Dropping the term $\underline{M}^4(r)$ which is unessential in the present context, one rewrites (9.91) for the free-energy densities as

$$\begin{aligned}
f &= f_s + f_M + f_H \\
f_s &= \frac{1}{2}\alpha_s\Delta_o^2 + \frac{1}{4}\beta_s\Delta_o^4 + \frac{1}{2}\gamma_s\Delta_o^2 \sum_{\underline{q}\,0} \left(\frac{2e}{\hbar q}\right)^2 |\underline{B}_{\underline{q}}|^2 \\
f_M &= \frac{1}{2}\alpha_M \underline{M}_o^2 + \sum_{\underline{q}\neq 0} \frac{1}{2}(\alpha_M + q^2\gamma_M)|\underline{M}_{\underline{q}}|^2 \\
f_H &= \frac{1}{2}\mu_o \sum_{\underline{q}} |\underline{B}_{\underline{q}}/\mu_o - \underline{M}_{\underline{q}}|^2 \quad .
\end{aligned} \tag{9.96}$$

The Meissner effect requires that $\underline{B}_o = 0$. By minimizing f with respect to $\underline{B}_{\underline{q}}$ one obtains

$$\frac{\partial f}{\partial \underline{B}_{\underline{q}}^*} = -\underline{M}_{\underline{q}}^{\perp} + \frac{1}{\mu_o}[1 + (\lambda^2 q^2)^{-1}]\underline{B}_{\underline{q}} = 0 \quad , \tag{9.97}$$

where

$$\lambda^{-2} = \left(\frac{2e}{\hbar}\right)^2 \mu_o\gamma_s\Delta_o^2 \quad .$$

It determines the equilibrium value $\underline{B}_{\underline{q}}$ of the magnetic field which is set up by the magnetization $\underline{M}_{\underline{q}}$ and the supercurrents. $\underline{M}_{\underline{q}}^{\perp}$ denotes the transverse part of $\underline{M}_{\underline{q}}$, i.e., the component perpendicular to $\underline{q}$. With its help one can eliminate $\underline{B}_{\underline{q}}$ from the Ginzburg-Landau function (9.96) and obtain

$$f[\Delta_o, \underline{M}_{\underline{q}}] = \frac{1}{2}\alpha_s\Delta_o^2 + \frac{1}{4}\beta_s\Delta_o^4 + \frac{\mu_o}{2}\sum_{\underline{q}} [\chi_{\perp}^{-1}(q)\,|\underline{M}_{\underline{q}}^{\perp}|^2 + \chi_{\parallel}^{-1}(q)\,|\underline{M}_{\underline{q}}^{\parallel}|^2] \quad , \tag{9.98}$$

where

$$\mu_o \chi_\perp^{-1}(\underline{q}) = \alpha_M + q^2\gamma_M + \mu_o \frac{1}{1 + \lambda^2 q^2}$$

$$\mu_o \chi_{||}^{-1}(\underline{q}) = \alpha_M + q^2\gamma_M + \mu_o \quad . \tag{9.99}$$

The quantities $\chi_\perp$, $\chi_{||}$ can be interpreted as transverse and longitudinal spin susceptibilities in the presence of a superconducting electron gas. The superconducting properties enter into $\chi_\perp(\underline{q})$ through the penetration depth λ and lead to partial screening of the magnetic interactions. It is seen that for $\lambda > \bar{\ell} = (\gamma_M/\mu_o)^{1/2}$, $\chi_\perp(\underline{q})$ has a maximum at $q = q_o$ where

$$q_o^2 = (\bar{\ell}\lambda)^{-1} - \lambda^{-2} \quad . \tag{9.100}$$

$\chi_\perp(\underline{q})$ diverges at a temperature T_s which is given by

$$T_s = T_{mo}\,[1 - (2\bar{\ell}/\lambda - \bar{\ell}^2/\lambda^2)\mu_o/\alpha_M'] \quad . \tag{9.101}$$

Thereby the system goes over into a cryptoferromagnetic state with $\langle \underline{M}(\underline{q}_o)\rangle \neq 0$ [9.25]. For a realization of this phase, T_s must satisfy the condition $T_{mo} > T_s > T_x$ (Fig.9.11a). Otherwise the normal ferromagnetic state is more stable. In addition, the transition temperature T_{m1} for the onset of a coexistence phase with $\Delta(\underline{r})$ = const. and $\underline{M}(r)$ and $\underline{M}_o$ = const. has to be lower than T_s. Both conditions are expected to be fulfilled for systems with predominantly dipolar coupling. Then $T_{m1} \ll T_{mo}$ and the factor μ_o/α_M' is of order one. When the cryptoferromagnetic phase with $\langle \underline{M}(\underline{q}_o)\rangle \neq 0$ is formed, the most probable structure is a spiral structure with $\underline{M}_{\underline{q}} \perp \underline{q}$ since a structure with $|\underline{M}|$ = const. minimizes the free-energy expression (9.91) with the M^4 term. This is not true, though, for samples with strong anisotropies. In that case phases with spatial variation of the amplitude of $\underline{M}(\underline{r})$ may be favorable [9.74]. The wave vector of the spiral structure is given by $(\lambda\bar{\ell})^{-1/2}$ where $\bar{\ell}$ is the magnetic interaction length introduced above. This holds as long as $(\bar{\ell}\lambda)^{1/2}$ is large as compared with the superconducting coherence length $\xi(T)$. Otherwise nonlocal effects have to be taken into account in the relationship between supercurrent and vector potential. This has been done in [9.26,29]. The resulting value of q_o is larger than the value given by (9.105) and helps to stabilize the spiral state [9.26].

b) *Vortex States*

The spontaneous formation of superconducting vortices in magnetic superconductors was first considered by KREY [9.27] and more recently by TACHIKI et al. [9.71,75], and KUPER et al. [9.28]. Thereby one has to distinguish between two different situations. The first one is an instability of the coexistence phase with $\Delta(\underline{r})$ = const. and $\underline{M}(\underline{r})$ = const. with respect to vortex formation in the magnetic field induced by the spin polarization. This situation resembles that of type II superconductors,

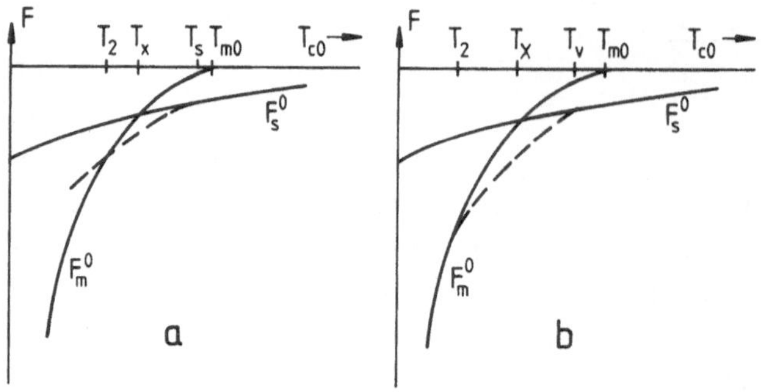

Fig.9.11. Free energy comparison of the spiral state [dashed curve in (a)] and the vortex state [dashed curve in (b)], and the pure superconducting state (F_s^0) and the ferromagnetic normal state (F_m^0)

the difference being that the role of the external field is taken up by the spin polarization $\underline{M}(\underline{r})$ [9.27]. The second situation is one in which the superconducting state with $\underline{M} = 0$ becomes unstable with respect to the formation of vortices with a finite spin polarization in the core. Generally one expects that for systems with predominantly dipolar coupling the latter case is more important since then the homogeneous coexistence phase is energetically unfavorable as initial phase. The ternary magnetic superconductors do not appear to fall into this category.

The mathematical treatment is the same in both cases. The form of an isolated vortex is obtained from the Ginzburg-Landau equations (9.93) by assuming cylindrical symmetry for $\underline{M}(\underline{r})$ and $\underline{H}(\underline{r})$ and making the ansatz $\Delta(\underline{r}) = |\Delta(r)| \exp[\pm i\varphi(r)]$. The two situations described above differ merely in the boundary conditions which are important when a single vortex is considered, but are irrelevant for a vortex lattice. It is $\lim_{r\to\infty} \underline{M}(\underline{r}) \neq 0$ in the first case and $\lim_{r\to\infty} \underline{M}(\underline{r}) = 0$ in the second case. For the latter it is found that a vortex solution may exist in a temperature range $T_x < T < T_{mo}$ for reasonable values of the parameters. It then has a lower free energy than the pure superconducting state. However, the interaction energy between different vortices is negative. Therefore, with the formation of one vortex, a whole vortex lattice is created [9.28,75]. Consequently the phase transition at T_v from the pure superconducting state to the vortex state is of first order (Fig. 9.11b). When the temperature is lowered the spin polarization and hence the density of vortices increases. At T_2, where $M = H_{c2}$, the system undergoes a second-order phase transition into the ferromagnetic normal state [9.74,75].

In the vicinity of T_{mo} the field distribution inside a vortex is strongly influenced by the high polarizability of the spin system [9.75]. The penetration depth of the magnetic field is reduced from λ to a value of order $(\lambda\bar{\ell})^{1/2}$ where $\bar{\ell}$ is the magnetic interaction length introduced in the previous section. This leads to a strong confinement of the quantized flux. As a further consequence, type II superconductors with large $\kappa = \lambda/\xi$ behave more like systems for which κ exceeds only slightly the critical value $1/\sqrt{2}$. This also explains the strong first-order nature of the transition into the vortex state.

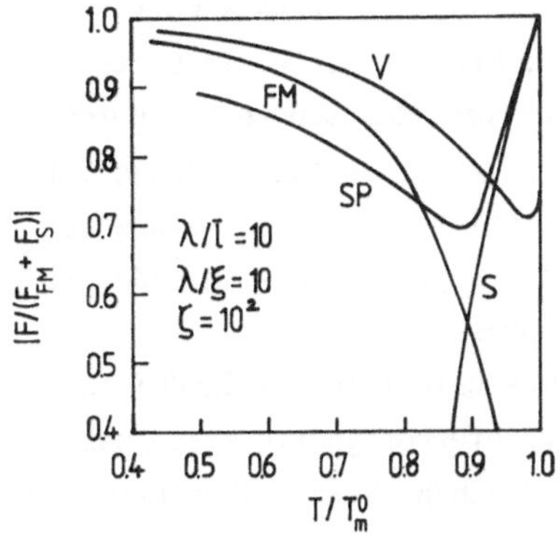

Fig.9.12. Calculated free energy of the spiral state (SP), vortex state (V), ferromagnetic state (FM) and pure superconducting state (S) below the magnetic transition temperature T_m^0. Parameters are chosen such that at low T the vortex state is the most stable phase (from [9.74])

Finally one must compare the stability of the vortex state with respect to the spiral state [9.74]. It is found that for large correlation energies of the spin system $[F_{mo}(0)/F_{so}(0) \equiv \zeta = 10^2]$ and large values of $\lambda/\bar{\ell} = 10^2$, the spiral state is more favorable. For smaller values of the parameters ($\zeta = 10^2$, $\lambda/\bar{\ell} = 10$), the vortex state wins out at low temperatures (see Fig.9.12). Both states also differ in the temperature dependence of the wavelength of their structures. At low temperatures, the periods of the vortex lattice and the spiral state are of the same order of magnitude. But close to T_V, the distance between vortices increases whereas the wavelength of the spiral state remains almost unchanged up to T_s. The experiments on $HoMo_6S_8$ [9.22,23] indicate that the spiral state is realized in that system. The rather long period of 250 Å suggests that it is induced by electromagnetic effects. The corresponding wavelength of the Anderson-Suhl cryptoferromagnetic state [see (9.9)] would presumably be much shorter.

c) *Influence of Fluctuations*

Until now we have discussed equilibrium states which are obtained by minimizing the Ginzburg-Landau functional. This procedure corresponds to a mean-field approximation. However, it is known that near a phase transition fluctuations may strongly influence the thermodynamic properties of the system. In the theory of magnetic superconductors, one has to distinguish between fluctuations in the amplitude and phase of the superconducting order parameter, the magnetization and the magnetic field.

Fluctuations in $\Delta(\underline{r})$ are known to have an extremely small effect on the thermodynamic properties in three dimensions. This is connected with the large coherence length in a superconductor [9.76].

Somewhat more important are fluctuations in the vector potential $\underline{A}(\underline{r})$. When integrated out they lead, for a pure superconductor, to the appearance of a term $|\Delta(\underline{r})|^3$ in the free energy with negative sign and as a consequence to a phase transition which is weakly first order [9.77]. But this effect is very small and has not yet been observed.

On the other hand, spatial fluctuations in the magnetization have a more pronounced influence on the stability of the cryptoferromagnetic state. In Sects. 9.1.2,9.2.2, it was demonstrated how critical fluctuations may suppress superconductivity when they are coupled to the conduction electrons via the exchange interaction. Here we shall discuss the effect of the fluctuating magnetic field which is due to fluctuations in the magnetic moments and the supercurrents in order to see whether it can suppress the cryptoferromagnetic phase.

The fluctuation contribution to the free energy was calculated by BLOUNT and VARMA [9.25]. Its effect on the stability of the cryptoferromagnetic state was investigated by FERRELL et al. [9.26], and by GREWE and SCHUH [9.78]. The influence of magnetic fluctuations on the superconducting state with constant $\Delta(\underline{r}) = \Delta_0$ can be treated as follows (see also KELLER [9.63]).

First, one eliminates the magnetic field in the Ginzburg-Landau functional (9.96) by replacing the field components $\underline{B}_q$ by their equilibrium values for a fixed value of $\underline{M}_{\underline{q}}$. This was done in the section on the Blount-Varma theory. The resulting functional $f[\Delta_0, \underline{M}_{\underline{q}}]$ is given by (9.98). Note that the coupling between the two order parameters Δ_0 and $\underline{M}_{\underline{q}}$ is contained in the penetration depth λ which appears in $\chi_\perp(\underline{q})$. From $f[\Delta_0, \underline{M}_{\underline{q}}]$ one can derive an effective function $\tilde{f}[\Delta_0]$ of Δ_0 alone by integrating out the fluctuating magnetization, i.e.,

$$\exp(-\beta\tilde{f}[\Delta_0]) = \int \prod_{\underline{q}} d^3\underline{M}_{\underline{q}} \exp(-\beta f[\Delta_0, \underline{M}_{\underline{q}}]) \quad . \tag{9.102}$$

The resulting free-energy difference $f_s - f_n$ between normal and superconducting phase then takes the form [9.26,78]

$$\begin{aligned} f_s - f_n &= \tilde{f}[\Delta_0] \\ &= \frac{1}{2}\alpha_s\Delta_0^2 + \frac{1}{4}\beta_s\Delta_0^4 + \Delta f_{\text{fluct}} \\ \Delta f_{\text{fluct}} &= -\frac{T}{\text{Vol}}\sum_{\underline{q}} \ln\left[\chi_\perp(\underline{q},\lambda)/\chi_\perp(\underline{q},\lambda=\infty)\right] \\ &= \frac{T}{\text{Vol}}\sum_{\underline{q}} \ln\left(1 + \mu_0 \frac{1}{1+\lambda^2 q^2} \frac{1}{\alpha_M + q^2\gamma_M}\right) \quad . \end{aligned} \tag{9.103}$$

The integration over $\underline{q}$ can be performed analytically. The results become simple at the magnetic phase transition ($\alpha_M = 0$) when, in addition, the penetration depth λ is large compared with the magnetic interaction length $\bar{\ell}$, i.e., $\lambda \gg \bar{\ell} = (\gamma_M/\mu_0)^{1/2}$. In this case,

$$\Delta f_{\text{fluct}} \simeq \frac{1}{\pi} T \lambda^{-3}(\lambda/\bar{\ell})^{3/2} \quad . \tag{9.104}$$

Assuming $\lambda = 100$ a, $\bar{\ell} = a$ where a is the lattice constant, one obtains for the fluctuation energy in a unit cell of volume a^3 at $T = T_{mo}$,

$$\Delta F_{fluct} \simeq T_{mo} \cdot 10^{-3} \quad . \tag{9.105}$$

This value has to be compared with the mean-field value of the superconducting condensation energy

$$(F_s - F_n)_{MF} = \left(\frac{1}{2}\, \alpha_s \Delta_o^2 + \frac{1}{4}\, \beta_s \Delta_o^4\right) a^3 \simeq -T_{co}^2/\varepsilon_F \tag{9.106}$$

and one notices that both energies are of the same order of magnitude. This implies that magnetic fluctuations may affect the stability of the superconducting phase.

For a more quantitative statement the problem has to be studied numerically. This was done by GREWE and SCHUH [9.78] who found that the cryptoferromagnetic phase should be suppressed by the fluctuations. Differentiating from the description given above, the authors also included the nonlinear coupling of the fluctuations, i.e., the $\underline{M}^4$ term and, furthermore, assumed that the magnetic field is an independent fluctuating variable. Both effects will be discussed in the following.

The treatment of the magnetic field $\underline{B}_{\underline{q}}$ as an independent fluctuating variable is open to objections. In doing so, one derives an additional term to the fluctuation energy density [see (9.103)] of the form

$$\Delta f_{fluct}^{(1)} = \frac{T}{\mathrm{Vol}} \sum_{\underline{q}} \ln \frac{\lambda^{-2} + q^2}{q^2} \quad . \tag{9.107}$$

It is obtained by considering that part of the free energy which contains the magnetic field, i.e.,

$$f[\underline{B}_{\underline{q}}] = \frac{1}{2\mu_o} \sum_{\substack{\underline{q} \neq 0 \\ \sigma}} \left(1 + \frac{1}{\lambda^2 q^2}\right) B_{\underline{q}\sigma}^2 \tag{9.108}$$

and integrating classically over the variables $B_{\underline{q}\sigma}$ [compare with (9.102)] (σ denotes the two polarization directions).

However, (9.107) cannot be correct since it would imply that fluctuations in $\underline{B}_{\underline{q}}$ with q values as short as $q_1 = \lambda^{-1}$ contribute to the fluctuation energy. But the corresponding excitation energies $\omega(q_1) = c/\lambda$ are much too large in order to be thermally excited except very close to the superconducting phase transition. The result is changed when the fluctuations are treated quantum mechanically. In this case, one replaces $\underline{B}_{\underline{q}}^2/2\mu_o$ by $\omega_q n_q/\mathrm{Vol}$ where n_q is the photon occupation number. Thereby it has been tacitly assumed that the electric field fluctuations behave similarly and can be added. Otherwise this replacement would not hold. Equation (9.108) is then rewritten as

$$f[\underline{B}_q] = \frac{1}{\mathrm{Vol}} \sum_{\substack{q \neq 0 \\ \sigma}} \tilde{\omega}_q n_q \quad , \tag{9.109}$$

where $\tilde{\omega}_q = cq[1 + (\lambda^2 q^2)^{-1}]$. This shows that a minimum energy of order $2\omega(\underline{q}_1)$ is required for the excitation of an independent $\underline{B}_q$ fluctuation. Therefore, in the superconducting state the corresponding fluctuation energy can be neglected. In the normal state the fluctuation energy is the usual "black body" radiation energy, which is very small compared with the superconducting condensation energy. From this, one may conclude that the independent $\underline{B}_q$ fluctuations may be safely neglected in a superconductor except when one is very close to a second-order superconducting phase transition [9.77] where $\lambda \to \infty$ and a classical treatment is appropriate.

In [9.78] the $\underline{M}^4$ term is factorized in a Hartree-like way and results in a renormalization of the coefficient α_M in (9.96), and hence of the susceptibilities as given by (9.99). One finds

$$\alpha_M \to \alpha_M + \beta_M(\underline{M}_o^2 + \Delta\underline{M}^2) \tag{9.110}$$

where ΔM^2 is a measure of the fluctuations in $\underline{M}(\underline{r})$. It is

$$\begin{aligned} \Delta\underline{M}^2 &= \frac{1}{\mathrm{Vol}} \int d^3\underline{r} \langle \underline{M}^2(\underline{r}) \rangle - M_o^2 \\ &= \sum_{\underline{q} \neq 0} [\langle |\underline{M}_q^{\perp}|^2 \rangle + \langle |\underline{M}_q^{\parallel}|^2 \rangle] \quad . \end{aligned} \tag{9.111}$$

The thermal average is performed with $\exp[-\beta f(\underline{M}_q)]$ as statistical operator. Thereby fixed values for $\underline{M}_o$, Δ_o and $\Delta\underline{M}^2$ are assumed. One obtains

$$\langle |\underline{M}_q^{\perp}|^2 \rangle = 2T\chi_{\perp}(\underline{q},\lambda)\mu_o^{-1}$$

with

$$\mu_o \chi_{\perp}^{-1}(\underline{q},\lambda) = \alpha_M + \beta_M(\underline{M}_o^2 + \Delta\underline{M}^2) + \gamma_m q^2 + \mu_o \frac{1}{1 + \lambda^2 q^2} \tag{9.112}$$

which can be compared with (9.99). Similarly, one can recalculate $\chi_{\parallel}^{-1}(\underline{q},\lambda)$. One notices that $\chi_{\perp}^{-1}(\underline{q},\lambda)$ has a maximum as before at q_o. Its temperature dependence has to be determined self-consistently from (9.111,112). The numerical calculations of [9.78] show that $\chi_{\perp}(\underline{q},\lambda)$ no longer has a pole in the temperature range where the superconducting phase is stable and hence no second-order transition into a cryptoferromagnetic state appears. In fact, in this approximation, $\chi(\underline{q},\lambda)$ does not diverge at any temperature [9.97,98]. The mathematical reason is that $\chi_{\perp}(\underline{q},\lambda)$ has its maximum on a whole sphere of radius q_o instead of only at one point in the *Brillouin* zone. This leads to large fluctuations in the wavelength or the phase of a possible helical structure and does not allow true long-range order to exist.

This case has some analogy with cholesteric liquid crystals [9.99,100]. Here one finds that the phase of the helix is pinned by surfaces. Presumably in magnetic superconductors, similar mechanisms (grain boundaries, anisotropies) will stabilize a helical or other cryptoferromagnetic state and may also cause the hysteresis effects observed in $HoMo_6S_8$ [9.23].

Finally, let us point out that also in the absence of a phase transition to a cryptoferromagnetic state, the neutron scattering cross section may show a maximum at the wave vector q_0 due to fluctuations in the transverse magnetization. Its width, however, will be influenced by the above-mentioned pinning effects.

9.4 Alternative Forms of Pairing States

Up to now the assumption has been tacitly made that the most favorable pairing of electrons is in a spin-singlet state with zero pairing momentum. Indeed, until now no superconductor has been found where that is not the case. Despite this, we shall discuss in the following pairings in other than s-wave states. The reason for this will become clear below.

There are two different types of alternative pairing which have been proposed. These are states with finite pairing momentum and states with p- (or higher angular momentum) pairing. Both types of states have the property that they are very sensitive with respect to a finite electron-mean free path ℓ. In particular, for $\ell \ll \xi_0$ they cannot exist. However, the finite-momentum paired states as well as p-wave paired states are less sensitive to a uniform exchange field than a ground state with s-wave pairing, except when the latter has a short spin-orbit mean free path.

Assume that one has a magnetic superconductor with mean free path $\ell > \xi_0$. In case the magnetic system prefers ferromagnetic order, one would have to investigate whether states with finite pairing momentum or with p-wave pairing become energetically more favorable than s-wave pairing states with zero pairing momentum. In some ternary systems, the Fermi velocity and hence the intrinsic coherence length ξ_0 are very small [9.54]. Therefore the condition $\ell > \xi_0$ does not seem to be impossible to fulfill. From this point of view the search for states with non-BCS type of pairing deserves some attention.

9.4.1 Ground States with Finite Pairing Momentum

Consider again a superconductor in a uniform exchange field with an interaction Hamiltonian given by (9.28). Furthermore, let us assume that the electron-mean free path is sufficiently long ($\ell \gg \xi_0$) that its effects can be neglected. Then it is found that with increasing h and $T < 0.56\ T_{co}$, the system first goes over from the BCS ground state into a state with finite pairing momentum $\underline{q}_0$ via a first-order phase transition into the normal state [9.18,19]. The reason for the occurrence of a finite pairing momentum state is its finite spin susceptibility. The state can therefore lower its energy in the uniform exchange field. The finite susceptibility results from unpaired electrons at the Fermi surface which also ensures that the *total* momentum of the electronic system is zero in the ground state (Bloch's theorem). The details of the distribution of the unpaired electrons can be found in [9.18]. The region in values of h, for which the ground state has $q_o \neq 0$, depends on T and on the shape of the Fermi surface. For a spherical surface and T = 0 it is given by $0.707 < h < 0.755$. The transition from the superconducting ground state with $q_o \neq 0$ to the normal state is obtained by setting the coefficient of the $|\Delta(\underline{r})|^2$ term in a Ginzburg-Landau expansion equal to zero under the assumption that $\Delta(\underline{r}) = \Delta_o \exp(i\underline{q}_o r)$. This results in the following equation for the determination of the maximum field h as a function of q_o at a given temperature T [9.79]:

$$\ln T/T_{co} = \pi T \,\mathrm{Re}\left\{\sum_n \left[\frac{1}{q_o v_F}\left(\mathrm{arc\,tanh}\,\frac{q_o v_F - 2h}{2|\omega_n|} + \mathrm{arc\,tanh}\,\frac{q_o v_F + h}{2|\omega_n|}\right) - \frac{1}{|\omega_n|}\right]\right\} \tag{9.113}$$

where, as before, $\omega_n = 2\pi T(n + 1/2)$. For $q_o = 0$ this equation reduces to (9.30,31), but for $T/T_{co} < 0.56$, larger values of h are obtained by choosing $q_o \neq 0$.

The stability of a superconducting ground state with $\underline{q}_o \neq 0$ can be considerably extended for strong-coupling superconductors or when the Fermi surface deviates from spherical symmetry [9.80-82]. In particular, when it has flat portions the range of h values can become arbitarily large. The sensitivity of a ground state with $q_o \neq 0$ to a finite electron-mean free path has been pointed out above [9.83].

9.4.2 P-Wave Pairing

The idea of constructing a superconducting ground state from electrons paired in p-waves instead of in s-waves followed the original BCS theory almost immediately and gained particular interest when it was realized that He^3 is a Fermi liquid. For an account of those early developments, we refer to [9.84,86]. Generally, we expects that p-wave pairing may become energetically more favorable than s-wave pairing when the electron-electron interaction is repulsive for large-angle scattering and attractive for small-angle scattering. In this case, the wave function of an electron pair will have a small amplitude for short distances which disfavors s-wave binding (e.g., in He^3 it is the hard-core repulsion which prevents pairing of He atoms in s states). When generalizing the BCS pairing to p-waves, one introduces the anomalous pair amplitudes

$$\psi_{\alpha\beta} = \langle c_{\underline{k}\alpha} c_{-\underline{k}\beta} \rangle \quad . \tag{9.114}$$

The order parameter is then written in terms of the three possible substates of an $\ell = 1$ pair $[|\uparrow\uparrow\rangle, |\downarrow\downarrow\rangle, 2^{-1/2}(|\uparrow\downarrow\rangle + |\downarrow\uparrow\rangle)]$ as

$$\psi(\hat{k}) = \psi_{\uparrow\uparrow}(\hat{\underline{k}})|\uparrow\uparrow\rangle + \psi_{\downarrow\downarrow}(\hat{\underline{k}})|\downarrow\downarrow\rangle + \psi_{\uparrow\downarrow}(\hat{\underline{k}})(|\uparrow\downarrow\rangle + |\downarrow\uparrow\rangle) \quad . \tag{9.115}$$

Since in (9.115) the values of $\underline{k}$ are restricted to $|\underline{k}|$ close to the Fermi surface, we have denoted with $\hat{\underline{k}}$ a unit vector specifying a point at the Fermi surface. The relative size of the different components $\psi_{\alpha\beta}(\hat{\underline{k}})$ depends on the specific features of the interaction. For example, in He^3 the A phase (Anderson-Brinkman-Morel state) is characterized by $\psi_{\uparrow\downarrow}(\hat{\underline{k}}) = 0$ while in the B phase (Balian-Werthamer state), all $\psi_{\alpha\beta}(\hat{\underline{k}}) \neq 0$. In the absence of an external magnetic field, the latter has the same thermodynamic properties as a BCS (s-wave) superconductor.

For a detailed study of the implications and consequences of these two types of p-wave pairing, we refer to a number of excellent reviews [9.84-88]. Here we are interested in the question: under what circumstances might one expect p-wave pairing to occur in a superconductor. First we would speculate that a p-wave superconductor, if there is one, is likely to be in a Balian-Werthamer state. This is so since in He^3, the Anderson-Brinkman-Morel state is stabilized only by an attractive interaction of the fermions via spin fluctuations. This seems to be a special feature of the He^3 system. When one decomposes the effective electron-electron interaction potential $V(\underline{k},\underline{k}')$ into its angular components

$$V(\underline{k}\underline{k}') = \sum_{\ell} (2\ell + 1)V_{\ell}(k,k')P_{\ell}(\hat{\underline{k}}\hat{\underline{k}}') \quad , \tag{9.116}$$

where the $P_{\ell}(\hat{\underline{k}}\hat{\underline{k}}')$ denote the Legendre polynomials, one generally expects $V_{\ell=0}$ to be the largest component. Thus, the tendency to form s-wave pairs will generally be stronger than that to form p-wave pairs. Therefore one can hope to realize the latter only if there are (external) interactions present which suppress s-wave pairing and do not, or have little influence on p-wave pairing. Even in that case one is facing the difficulty that an electron-mean free path ℓ of the order of or less than the coherence length of the p-wave paired state is strongly suppressing this pairing. Therefore, only systems with a small Fermi velocity and hence a relatively short intrinsic coherence length would seem to be appropriate candidates for p-wave pairing. It is known that the Chevrel phases in the BCS ground state have a coherence length of only a few tens of Angstroms [9.54]. In order not to have a much larger coherence length in a p-wave paired ground state, $V_{\ell=1}$ would have to be of the same order of magnitude as $V_{\ell=0}$. There are not yet any estimates available of the actual size of $V_{\ell=1}$ in the Chevrel phases as there are, e.g., for Pd [9.89-93].

Let us assume for the moment that the problem of a sufficiently long mean-free path could be handled. Furthermore, let us assume that an external magnetic field is applied. Since the Chevrel phases have a short intrinsic coherence length, their critical fields are sufficiently high so that the effect of the field on the electron spins is very important. The latter is different for an s-wave and a p-wave paired superconductor. In particular, for the Balian-Werthamer ground state the opposite spin component reduces the spin susceptibility at T = 0 to 2/3 χ_n. Thus, a large external field favors p-wave pairing as compared with s-wave pairing. It is, therefore, tempting to speculate whether in the Chevrel phases in high magnetic fields there will be a phase transition from an s-wave to a p-wave paired ground state. The problem remains as to how one might possibly detect such a transition. This problem has not yet been studied in full detail. There are calculations available for the tunneling density of states in a p-wave superconductor which differs substantially from that in a superconductor with s-wave pairing [9.94]. However, one might also think of detecting other characteristic features such as the modifications in the Josephson current between a p-wave and an s-wave superconductor [9.95]. A superconducting ring consisting of one half of an s-wave and half of a p-wave superconductor should not be able to support a persistent current. The Meissner effect in p-wave superconductors has been discussed in [9.96].

Acknowledgment. We would like to thank Dr. R.A. Ferrell, Dr. U. Krey, Dr. D. Rainer, Dr. W. Rietschel, Dr. C.M. Varma and Dr. G. Zwicknagl for a number of helpful discussions and suggestions.

References

9.1 H. Suhl, B.T. Matthias, E. Corenzwit: Phys. Chem. Solids *11*, 346 (1959); N.E. Philipps, B.T. Matthias: Phys. Rev. *121*, 105 (1961); B.T. Matthias, H. Suhl: Phys. Rev. Lett. *4*, 51 (1960)
9.2 D.K. Finnemore, F.H. Spedding: Bull. Am. Phys. Soc. [2] *10*, No.3, 358 (HCI) (1965); D.K. Finnemore, D.L. Johnson, J.E. Ostenson, F.H. Spedding, B.J. Beaudry: Phys. Rev. *137*, A 550 (1965)
9.3 K. Yoshida: Phys. Rev. *110*, 769 (1958)
9.4 F. Reif: Phys. Rev. *106*, 208 (1957); C.M. Androes, W.D. Knight: Phys. Rev. *121*, 779 (1961)
9.5 R.A. Ferrell: Phys. Rev. Lett. *3*, 262 (1959)
9.6 P.W. Anderson: Phys. Rev. Lett. *3*, 325 (1959)
9.7 W. Baltensperger, S. Strässler: Phys. Kondens. Materie *1*, 20 (1963)
9.8 A.A. Abrikosov, L.P. Gor'kov: Zh. Eksperim. i Teor. Fiz. *39*, 1781 (1960) [English transl.: Soviet Phys.-JETP *12*, 1243 (1961)]
9.9 F. Reif, M.A. Woolf: Phys. Rev. Lett. *9*, 315 (1962)
9.10 H. Suhl: In *Low Temperature Physics*, ed. by C. De Witt, B. Dreyfus, P.G. de Gennes (Gordon and Breach, New York 1962)
9.11 K. Maki: In *Superconductivity*, Vol.2, ed. by R.D. Parks (Marcel Dekker, Inc. New York 1969)
9.12 M.A. Jensen, H. Suhl: In *Magnetism*, Vol.II B, ed. by G.T. Rado, H. Suhl (Academic, New York, London 1966)
9.13 Ø. Fischer, M. Peter: In *Magnetism*, Vol.V, ed. by G.T. Rado, H. Suhl (Academic, New York, London 1973)
9.14 S. Roth: Appl. Phys. *15*, 1 (1978)

9.15 M.B. Maple: In *Magnetism*, Vol.V, ed. by G.T. Rado, H. Suhl (Academic, New York, London 1973)
9.16 E. Müller-Hartmann: In *Magnetism*, Vol.V, ed. by G.T. Rado, H. Suhl (Academic, New York. London 1973)
9.17 V. Jaccarino, M. Peter: Phys. Rev. Lett. *9*, 290 (1962)
9.18 P. Fulde, R.A. Ferrell: Phys. Rev. A*135*, 550 (1964)
9.19 A.I. Larkin, Y.N. Ovchinnikov: Zh. Eksperim. i. Teor. Fiz. *47*, 1136 (1964) [English transl.: Sov. Phys.-JETP *20*, 762 (1965)]
9.20 W.A. Fertig, D.C. Johnston, L.E. De Long, R.W. Mc Callum, M.B. Maple, B.T. Matthias: Phys. Rev. Lett. *38*, 987 (1977)
9.21 M. Ishikawa, Ø. Fischer: Solid State Commun. *23*, 37 (1977)
9.22 A summary is given by W. Thomlinson, G. Shirane, J.W. Lynn, D.E. Moncton: In Chap.8 of this volume
9.23 J.W. Lynn, A. Raggazoni, R. Pynn, J. Joffrin: J. de Phys. Lett. (France) *42*, L-45 (1981)
9.24 P.W. Anderson, H. Suhl: Phys. Rev. *116*, 898 (1959)
9.25 E.I. Blount, C.M. Varma: Phys. Rev. Lett. *42*, 1079 (1979)
9.26 R.A. Ferrell, J.K. Bhattacharjee, A. Bagchi: Phys. Rev. Lett. *43*, 154 (1979)
9.27 U. Krey: Intern. J. Magnetism *3*, 65 (1972); Intern. J. Magnetism *4*, 153 (1973)
9.28 C.A. Kuper, M. Revzen, A. Ron: Phys. Rev. Lett. *44*, 1545 (1980)
9.29 H. Matsumoto, H. Umezawa, M. Tachiki: Solid State Commun. *31*, 157 (1979); see also M. Tachiki, A. Kotani, H. Matsumoto, H. Umezawa: Solid State Comm. *31*, 927 (1979);
M. Tachiki: Proc. Intern. Conf. Ternary Superconductors, ed. by G.K. Shenoy, B.D. Dunlap, F.Y. Fradin (North-Holland, Amsterdam 1981)
9.30 R.J. Noer, W.D. Knight: Rev. Mod. Phys. *36*, 177 (1964)
9.31 P. Gor'kov: Zh. Eksperim. i. Teor. Fiz. *48*, 1772 (1965) [English transl.: Sov. Phys.-JETP *21*, 1186 (1965)]
9.32 K. Maki, P. Fulde: Phys. Rev. *140*, A 1586 (1965)
9.33 P. Fulde, K. Maki: Phys. Rev. *139*, A 788 (1965)
9.34 A.A. Abrikosov, L.P. Gor'kov: Zh. Eskperim. i. Teor. Fiz. *42*, 1088 (1962) [English transl.: Sov. Phys.-JETP *15*, 752 (1962)]
9.35 L.P. Gor'kov, A.I. Rusinov: Zh. Eksperim. i. Teor. Fiz. *46*, 1363 (1964) [English transl.: Sov. Phys.-JETP *19*, 922 (1964)]
9.36 D. Rainer: Private communication
9.37 K.H. Bennemann: Phys. Rev. Lett. *17*, 438 (1966): for corrections see [9.48]
9.38 D. Rainer: Z. Physik *252*, 174 (1972)
9.39 P. Entel, W. Klose: J. Low Temp. Phys. *17*, 529 (1974)
9.40 C.M. Soukoulis, G.S. Grest: Phys. Rev. B*21*, 5119 (1980)
9.41 A. Sakurai: Solid State Commun. *25*, 867 (1978)
9.42 Ting-Kuo Lee: Solid State Commun. *34*, 9 (1980)
9.43 S. Maekawa, M. Tachiki: Phys. Rev. B*18*, 4688 (1978)
9.44 Y. Machida, D. Youngner: J. Low Temp. Phys. *35*, 449 (1979); J. Low Temp. Phys. *36*, 617 (1979)
9.45 J. Keller: Proc. Intern. Conf. Ternary Superconductors, ed. by G.K. Shenoy, B.D. Dunlap, F.Y. Fradin (North-Holland, Amsterdam 1981)
9.46 R. Avenhaus, Ø. Fischer, B. Giovannini, M. Peter: Helv. Phys. Acta *42*, 649 (1969)
9.47 P. Fulde, K. Maki: Phys. Rev. *141*, 275 (1966)
9.48 J. Keller, R. Benda: J. Low Temp. Phys. *2*, 141 (1970)
9.49 G. Sarma: J. Phys. Chem. Solids *24*, 1029 (1963)
9.50 K. Maki, T. Tsuneto: Prog. Theor. Phys. (Kyoto) *31*, 945 (1964)
9.51 P. Fulde: Adv. in Phys. *22*, 668 (1973) [note the difference in the definition of b by a factor of $\Delta(0)$]
9.52 N.R. Werthamer, E. Helfand, P.C. Hohenberg: Phys. Rev. *147*, 295 (1966)
9.53 K. Maki: Phys. Rev. *148*, 362 (1966)
9.54 Ø. Fischer: In *Proc. 14th Intern. Conf. on Low Temperature Physics*, Vol.5, ed. by M. Krusius, M. Vuori (North-Holland, Amsterdam 1975)
9.55 H. Engler, P. Fulde: Phys. kondens. Materie *7*, 150 (1968)
9.56 G. Zwicknagl, P. Fulde: Z. Physik *43*, 23 (1981)

9.57 A.A. Abrikosov, G.P. Gor'kov, I.Y. Dzyaloshinskii: *Quantum Field Theoretical Methods in Statistical Physics*, 2nd ed. (Pergamon Press, Oxford 1965)
9.58 G. Zwicknagl: Private communication
9.59 K. Machida, K. Nokura, T. Masubara: Phys. Rev. Lett. *44*, 821 (1980); Phys. Rev. Lett. *22*, 2307 (1980)
9.60 T. Jarlborg, A.J. Freeman: Phys. Rev. Lett. *44*, 178 (1980)
9.61 M.J. Nass, K. Levin, G.S. Grest: Phys. Rev. Lett. *46*, 614 (1981)
9.62 T.V. Ramakrishnan, C.M. Varma: Phys. Rev. B*24*, 137 (1981)
9.63 J. Keller: J. Magn. Magn. Mater., in print
9.64 H. Suhl: J. Less-Comm. Met. *62*, 225 (1978)
9.65 L.N. Bulaevskii, A.I. Rusinov, M. Kulič: J. Low Temp. Phys. *39*, 255 (1980)
9.66 C.M. Varma: In *Superconductivity in d- and f-Band Metals*, ed. by H. Suhl, M.B. Maple (Academic, New York 1980) p.391
9.67 P.G. de Gennes: *Superconductivity of Metals and Alloys* (Benjamin, New York, Amsterdam 1966)
9.68 M.V. Jarič: Phys. Rev. B*22*, 3503 (1980);
E.I. Blount, C.M. Varma: Phys. Rev. B*22*, 3507 (1980)
9.69 J. Kasperczyk, P. Tekiel: Acta physica polonica A*57*, 11 (1980);
J. Kasperczyk, G. Kozlowski, H. Romejko, P. Tekiel: Acta physica polonica A*57*, 17 (1980)
9.70 M. Tachiki, A. Kotani, H. Matsumoto, H. Umezawa: Solid State Commun. *32*, 599 (1979)
9.71 M. Tachiki, H. Matsumoto, T. Koyama, H. Umezawa: Solid State Commun. *34*, 19 (1979)
9.72 C.G. Kuper, M. Revzen, A. Ron: Solid State Commun. (in press)
9.73 M.V. Jarič, M. Belič: Phys. Rev. Lett. *42*, 1015 (1979);
M.V. Jarič: Phys. Rev. B*20*, 4486 (1979); Phys. Rev. B*22*, 463 (1980)
9.74 H.S. Greenside, E.I. Blount, C.M. Varma: Phys. Rev. Lett. *46*, 49 (1981)
9.75 M. Tachiki, H. Matsumoto, H. Umezawa: Phys. Rev. B*20*, 1915 (1979)
9.76 R.E. Glover: In *Progress in Low Temperature Physics*, Vol.6, ed. by C.J. Gorter (North-Holland, Amsterdam 1970);
W.J. Skocpol, M. Tinkham: Rpts. Prog. Phys. *38*, 1049 (1975)
9.77 B.I. Halperin, T.C. Lubensky, Shang-Keng Ma: Phys. Rev. Lett. *32*, 292 (1974)
9.78 N. Grewe, B. Schuh: Phys. Rev. B*22*, 3183 (1980)
9.79 D. Saint-James, G. Sarma, E.J. Thomas: *Type II Superconductivity* (Pergamon Press, Oxford 1969) Sect.6.1.3
9.80 S. Takada, T. Izuyama: Progr. Theor. Phys. (Kyoto) *41*, 635 (1969)
9.81 L.N. Bulaevskii: Zh. Eksperim. i. Teor. Fiz. *65*, 1278 (1973) English transl.: Sov. Phys.-JETP *37*, 1113 (1974)
9.82 K. Aoi, W. Dieterich, P. Fulde: Z. Physik *267*, 223 (1974)
9.83 L.W. Gruenberg, L. Gunther: Phys. Rev. Lett. *16*, 996 (1966)
9.84 A.J. Leggett: Rev. Mod. Phys. *47*, 331 (1975)
9.85 J.C. Wheatley: Rev. Mod. Phys. *47*, 415 (1975)
9.86 P.W. Anderson, W.F. Brinkman: In *The Physics of Liquid and Solid Helium*, ed. by K.H. Bennemann, J.B. Ketterson (Wiley, New York 1977)
9.87 D.M. Lee, R.C. Richardson: In *The Physics of Liquid and Solid Helium*, ed. by K.H. Bennemann, J.B. Ketterson (Wiley, New York 1977)
9.88 P. Wölfle: Rpts. Prog. Phys. *42*, 269 (1979)
9.89 D. Fay, J. Appel: Phys. Rev. B*16*, 2325 (1977)
9.90 J. Appel, D. Fay: Solid State Commun. *28*, 157 (1978)
9.91 F.J. Pinki, P.B. Allen, W.H. Butler: Phys. Rev. Lett. *41*, 431 (1978)
9.92 I.F. Foulkes, B.L. Gyoffry: Phys. Rev. B*15*, 1395 (1977)
9.93 M. Grodzicki, J. Appel: Phys. Rev. B*20*, 3659 (1979)
9.94 L.J. Buchholtz, G. Zwicknagl: Phys. Rev. B*23*, 5788 (1981)
9.95 J.A. Pals: Phys. Lett. *56*A, 414 (1976)
9.96 K. Scharnberg, R.A. Klemm: Phys. Rev. B*22*, 5233 (1980)
9.97 We thank H. Kleinert for bringing this problem to our attention
9.98 H. Kleinert: Phys. Lett. *83*A, 294 (1981)
9.99 T. Lubensky: Phys. Rev. Lett. *29*, 206 (1972)
9.100 M.J. Stephen, J.P. Straley: Rev. Mod. Phys. *46*, 617 (1974)
9.101 H. Rietschel, H. Winter: Phys. Rev. Lett. *43*, 1256 (1979)
9.102 W. Klose, P. Entel, M. Peter: Z. Physik *264*, 51 (1973)

Additional References with Titles

Chapter 2

F. Acker, H.C. Ku: Time-dependent magnetization and resistivity in $DyRh_4B_4$. J. Magn. Magn. Mater. *24*, 47-53 (1981)

F. Acker, H.C. Ku: Magnetic order and superconductivity in $Ho(Ir_xRh_{1-x})_4B_4$ compounds. Phys. Rev. B*25*, 5692-5697 (1982)

N.S. Bilonizhko, I.B. Krib, Yu.B. Kuz'ma: Neodymium-nickel-boron system. Dopov. Akad. Nauk Ukr. RSR, Ser. B*4*, 21-23 (1982)

C. Boekems, R.H. Heffner, R.L. Hutson, M. Leon, M.E. Schillaci, J.L. Smith, S.A. Dodds, D.E. MacLaughlin: μSR measurement of rare-earth moment dynamics in the $Ho_xLu_{1-x}Rh_4B_4$ alloy system. J. Appl. Phys. *53*, 2625-2627 (1982)

H.F. Braun, J. Zábrácszky: "Interaction of Superconductivity and Magnetism in Solid Solutions $(R_{1-x}R'_x)_5Ir_4Si_{10}$ (R = Sc, Lu; R' = Er, Dy)", in Proc. Europhysics Conf. on Solid State Phys., March 22-25, 1982, Manchester, UL (to be published)

H.F. Braun, M. Pelizzone, K. Yvon: "Ferromagnetic Borides with Incommensurate Rare Earth (R) and Iron Sublattices: $R_{1+\varepsilon}Fe_4B_4$", in Proc. 7th Int. Conf. on Solid Compounds of Transition Elements, June 21-26, 1982, Grenoble (to be published)

H.F. Braun, M. Pelizzone: "Coexistence of Superconductivity and Antiferromagnetic Ordering in $(Sc_{1-x}Dy_x)_5Ir_4Si_{10}$ Solid Solutions", in Proc. 4th Conf. on Superconductivity in d- and f-Band Metals, June 28-30, 1982, Karlsruhe (to be published)

B. Chevalier, A. Cole, P. Lejay, J. Etourneau: New ternary silicides in rare earth-rhodium-silicon systems. Crystal structure and magnetic properties. Mater. Res. Bull. *16*, 1067-1075 (1981)

B. Chevalier, A. Cole, P. Lejay, M. Vlasse, J. Etourneau, P. Hagenmuller, R. Georges: Crystal structure and magnetic properties of new rare earth ternary equiatomic silicides RERhSi. Mater. Res. Bull. *17*, 251-258 (1982)

B. Chevalier, P. Lejay, A. Cole, M. Vlasse, J. Etourneau: Crystal structure, superconducting and magnetic properties of new ternary silicides LaRhSi, LaIrSi and NdIrSi. Solid State Commun. *41*, 801-804 (1982)

I. Felner, I. Nowik: Mixed valency of europium in $EuPd_6B_4$. Solid State Commun. *39*, 61-63 (1981)

Z. Fisk, S.E. Lambert, M.B. Maple, J.P. Remeika, G.P. Espinosa, A.S. Cooper, H. Barz, S. Oseroff: Magnetic and superconducting properties of rare earth osmium stannides. Solid State Commun. *41*, 63-67 (1982)

H.C. Hamaker, H.B. MacKay, M.S. Torikachvili, L.D. Woolf, M.B. Maple, W. Odoni, H.R. Ott: Observation of the coexistence of superconductivity and long-range magnetic order in $TmRh_4B_4$. J. Low Temp. Phys. *44*, 553-568 (1981)

K. Hiebl, M.J. Sienko, P. Rogl: Magnetic behavior and structural chemistry of $RE(Os,Ir)_4B_4$ borides (RE: rare earth). J. Less-Common Metals *82*, 21-28 (1981)

K. Hiebl, P. Rogl, M.J. Sienko: Superconductivity in the pseudoternary system YRh_4B_4-$LuRh_4B_4$-$ThRh_4B_4$. J. Less-Common Metals *82*, 201-209 (1981)

K. Hiebl, P. Rogl, M.J. Sienko: Structural chemistry and magnetic properties of the compounds $EuOs_4B_4$ and $EuIr_4B_4$ and of the solid solutions $(RE)Os_4B_4$-$(RE)Ir_4B_4$ (RE: Ce, Pr, Sm). Inorg. Chem. *21*, 1128-1133 (1982)

J.L. Hodeau, M. Marezio, J.P. Remeika, C.H. Chen: Structural distortion in the primitive cubic phase of the superconducting/magnetic ternary rare-earth rhodium stannides. Solid State Commun. *42*, 97-102 (1982)

J.L. Hodeau, M. Marezio, J.P. Remeika: "The Crystal Structure of $Sn(1)_xEr(1)_{1-x}Er(2)_4Rh_6Sn(2)_4Sn(3)_{12}Sn(4)_2$, a Re-entrant Superconductor", in Proc. 7th Int. Conf. on Solid Compounds of Transition Elements, June 21-26, 1982, Grenobe (to be published)
A. Jayaraman, J.P. Remeika, G.P. Espinosa, A.S. Cooper, H. Barz, R.G. Maines, Z. Fisk: High pressure synthesis of the mixed valent and nonsuperconducting ternary $YbRh_xSn_y$ compound. Solid State Commun. *39*, 1049-1051 (1981)
K. Klepp, E. Parthé: RPtSi phases (R = La, Ce, Pr, Nd, Sm and Gd) with an ordered $ThSi_2$ derivative structure. Acta Crystallogr. B *38*, 1105-1108 (1982)
K. Klepp, E. Parthé: LaIrSi with an ordered $SrSi_2$ derivative structure". Acta Crystallogr. B *38*, 1541-1544 (1982)
Yu.B. Kuz'ma, N.S. Bilonizhko, N.F. Chaben, G.V. Chernjak: X-ray investigation of the rare earth metal-iron triad metal-B systems. J. Less-Common Metals *82*, 364 (1981)
A.R. Moodenbaugh, D.E. Cox, H.F. Braun: Neutron-diffraction study of magnetically ordered $Tb_2Fe_3Si_5$. Phys. Rev. B *25*, 4702-4710 (1982)
S.I. Mykhalenko, N.F. Chaban, Yu.B. Kuz'ma: About the interaction of boron with the rare earth metals and molybdenum, tungsten and rhenium. J. Less-Common Metals *82*, 365 (1981)
M. Pelizzone, H.F. Braun: "Magnetism of Ternary Rare Earth-Cobalt-Silicides", in Proc. Europhysics Conf. on Solid State Phys., March 22-25, 1982, Manchester, UK (to be published)
M. Pelizzone, H.F. Braun, J. Muller: Magnetic properties of $RCoSi_2$ compounds (R = rare earth). J. Magn. Magn. Mater. (in press)
P. Rogl, K. Hiebl, M.J. Sienko: "Structural Chemistry and Magnetic Behavior of RM_4B_4-Borides", in Proc. 7th Int. Conf. on Solid Compounds of Transition Metals, June 21-25, 1982, Grenoble (to be published)
K.B. Schwartz, C.T. Prewitt, R.D. Shannon, L.M. Corliss, J.M. Hastings, B.L. Chamberland: Neutron powder diffraction study of two sodium platimum oxides: $Na_{1.0}Pt_3O_4$ and $Na_{0.73}Pt_3O_4$. Acta Crystallogr. B *38*, 363-368 (1982)
G.K. Shenoy, D.R. Noakes, D.G. Hinks: Determination of crystalline electric fields in $RERh_4B_4$ compounds. Solid State Commun. *42*, 411-414 (1982)
G.K. Shenoy, D.R. Noakes, G.P. Meisner: Mössbauer study of superconducting $LaFe_4P_{12}$. J. Appl. Phys. *53*, 2628-2630 (1982)
S.K. Sinha, G.W. Crabtree, D.G. Hinks, H. Mook: Study of coexistence of ferromagnetism and superconductivity in single-crystal $ErRh_4B_4$. Phys. Rev. Lett. *48*, 950-953 (1982)
A.M. Stewart: Absence of anomalously large conduction-electron polarization in superconducting rare-earth ternary compounds. Phys. Rev. B *24*, 4080-4081 (1981)
C.U. Segre: "Superconductivity and Magnetism in Compounds with the $Sc_2Fe_3Si_5$-Type Structurce"; Ph.D. Thesis, University of California, San Diego (1981)
G. Venturini, B. Malaman, J. Steinmetz, A. Courtois, B. Roques: Distribution des atomes metalliques dans les structures apparentees des trois composes ternaires equiatomiques: RhMnSi, isotope de Co_2P, RhMnGe, isotope de TiFeSi et PdMnGe, isotype de Fe_2P. Mater. Res. Bull. *17*, 259-267 (1982)

Chapter 4

F. Acker, H.C. Ku: Magnetic order and superconductivity in $Ho(Ir_xRh_{1-x})_4B_4$ compounds. Phys. Rev. B *25*, 5692 (1982)
F. Acker, L. Schellenberg, H.C.Ku: "Magnetism and Superconductivity in Tb $(Ir_xRh_{1-x})_4B_4$ and $Ho(Ir_xRh_{1-x})_4B_4$", in Proc. 4th Conf. Superconductivity in d- and f-Band Metals, June 28-30, 1982, Karlsruhe, FRG (to be published)
F. Acker, H.C. Ku: Time dependent magnetization and resistivity in $DyRh_4B_4$. J. Magn. Magn. Mater. *24*, 47 (1981)
F. Behroozi, G.W. Crabtree, S.A. Campbell, M. Levy, D.R. Snider, D.C. Johnston, B.T. Matthias: Experimental determination of magnetic Landau parameters in $ErRh_4B_4$. Solid State Commun. *39*, 1041 (1981)
C. Boekema, R.H. Heffner, R.L. Hutson, M. Leon, M.E. Schillaci, J.L. Smith, S.A. Dodds, D.E. MacLaughlin: μSR measurement of rare earth moment dynamics in the $(Ho_xLu_{1-x})Rh_4B_4$ ternary alloy system. J. Appl. Phys. *53*, 2625 (1982)

G.L. Christner, B. Bradford, L.E. Toth, R. Cantor, E.D. Dahlberg, A.M. Goldman, C.Y. Huang: Sputter deposition of thin films of superconducting $Er(RhE)_4$. J. Appl. Phys. *50*, 5820 (1979)
B.D. Dunlap, G.K. Shenoy: Effects of magnetic atoms on the properties of ternary superconductors. Hyperfine Interact. *10*, 903 (1981)
F.Y. Fradin, K. Kumagai: "^{11}B and ^{103}Rh Study of Spin Dynamics in $Y_{1-x}Re_xRh_4B_4$", in Proc. 4th Conf. Superconductivity in d- and f-Band Metals, June 28-30, 1982, Karlsruhe, FRG (to be published)
K. Hiebl, M.J. Sienko, P. Rogl: Magnetic behavior and structural chemistry of $RE(Os,Ir)_4B_4$ borides (RE= rare earth). J. Less Common Metals *82*, 21 (1981)
K. Hiebl, P. Rogl, M.J. Sienko: Structural chemistry and magnetic properties of the compounds $EuOs_4B_4$ and $EuIr_4B_4$ and of the solid solutions $(RE)Os_4B_4$ - $(RE)Ir_4B_4$ (RE = Ce, Pr, Sm). Inorg. Chem. *21*, 1128 (1982)
K. Hiebl, P. Rogl, M.J. Sienko: Superconductivity in the pseudoternary system YRh_4B_4-$LuRh_4B_4$-$ThRh_4B_4$. J. Less Common Metals *82*, 201 (1981)
A. Jayaraman, J.P. Remekia, G.P. Espinosa, A.S. Cooper, H. Barz, R.G. Maines, Z. Fisk: High pressure synthesis of mixed valent and nonsuperconducting ternary $YbRh_xSn_y$ compounds. Solid State Commun. *39*, 1049 (1981)
D.C. Johnston, B.G. Silbernagel: ^{11}B NMR studies of the electronic and magnetic properties of $LuRh_4B_4$ and $GdRh_4B_4$. Phys. Rev. B *21*, 4996 (1980)
H.C. Ku, S.E. Lambert, M.B. Maple: "The Occurrence of Superconductivity in the Magnetic System $Tb(Rh_{1-x}Ir_x)_4B_4$", in Proc. 4th Conf. Superconductivity in d- and f-Band Metals, June 28-30, Karlsruhe 1982, FRG (to be published)
B. Lachal, M. Ishikawa, A. Junod, J. Muller: Direct evidence for the first-order phase transition at the lower critical temperature in $Er_{1-x}Ho_xRh_4B_4$. J. Low Temp. Phys. *46*, 467 (1982)
L.-J. Lin, C.P. Umbach, A.M. Goldman: "Tunneling into Magnetic Superconductors", in Proc. 4th Conf. Superconductivity in d- and f-Band Metals, June 28-30, Karlsruhe, 1982, FRG (to be published)
H.A. Mook, S.K. Sinha, G.W. Crabtree, D.G. Hinks, M.B. Maple, Z. Fisk, D.C. Johnston, L.D. Woolf, H.C. Hamaker: "Neutron Scattering Studies of the Magnetic Ordering in Ternary Rare Earth Compounds", in Proc. 4th Conf. Superconductivity in d- and f-Band Metals, June 28-30, Karlruhe, 1982, FRG (to be published)
H.A. Mook, W.C. Koehler, S.K. Sinha, G.W. Crabtree, D.G. Hinks, M.B. Maple, Z. Fisk, D.C. Johnston, L.D. Woolf, H.C. Hamaker: Superconductivity and magnetism in the $Ho_{1-x}Er_xRh_4B_4$ system. J. Appl. Phys. *53*, 2614 (1982)
H.R. Ott, W. Odoni, H.C. Hamaker, M.B. Maple, L.D. Woolf: "Transport and Magnetic Properties in the Coexistence Region of Magnetic Superconductors", in Proc. 4th Conf. Superconductivity in d- and f-Band Metals, June 28-30, Karlsruhe, 1982, FRG (to be published)
G.K. Shenoy, D.R. Noakes, D.G. Hinks: Determination of crystalline electric fields in $RERh_4B_4$ compounds. Solid State Commun. *42*, 411 (1982)
S.K. Sinha, G.W. Crabtree, D.G. Hinks, H.A. Mook: Study of coexistence of ferromagnetism and superconductivity in single-crystal $ErRh_4B_4$. Phys. Rev. Lett. *48*, 950 (1982)
C.P. Umbach, A.M. Goldman, L.E. Toth: Rare-earth oxides as artificial barriers in superconducting tunneling junctions. Appl. Phys. Lett. *40*, 81 (1982)
C.P. Umbach, A.M. Goldman: Pair tunneling in $ErRh_4B_4$ films. Phys. Rev. Lett. *48*, 1433 (1982)
K.N. Yang, S.E. Lambert, H.C. Hamaker, M.B. Maple, H.A. Mook, H.C. Ku: "Low Temperature Investigation of Magnetic Order and Superconductivity in the Pseudoternary System $Ho(Rh_{1-x}Ir_x)_4B_4$", in Proc. 4th Conf. Superconductivity in d- and f-Band Metals, June 28-30, Karlruhe, 1982, FRG (to be published)

Chapter 5

R.C. Lacoe, S.A. Wolf, P.M. Chaikin, C.Y. Huang, H.L. Luo: Partial gapping of the Fermi surface and superconductivity in $Eu_xMo_6S_8$. Phys. Rev. Lett. *48*, 1212 (1982)

M.S. Torikachvili, M.B. Maple, R.P. Guertin, S. Foner: Superconducting and magnetic interactions in $La_{1.2-x}Eu_xMo_6S_8$ pseudoternary compounds. J. Appl. Phys. *53*, 2619 (1982)

M.S. Torikachvili, M.B. Maple: Enhancement of the upper critical magnetic fields in $La_{1.2-x}Eu_xMo_6S_8$ compounds. Solid State Commun. *40*, 1 (1981)

S.A. Wolf, W.W. Fuller, C.Y. Huang, D.W. Harrison, H.L. Luo, S. Maekawa: Magnetic-field induced superconductivity. Phys. Rev. B *25*, 1990 (1982)

Chapter 7

P. Bonville, R. Chevrel, J.A. Hodges, P. Imbert, G. Jehanno, M. Sergent: "Low Temperature Properties of $TmMo_6S_8$ and $YbMo_6S_8$", in Proc. Int. Conf. Applications of the Mössbauer Effect. Dec. 14-18, 1981, Jaipur, India

P. Bonville, J.A. Hodges, P. Imbert, G. Jehanno, R. Chevrel, R. Sergent: Low temperature Mössbauer study and magnetic susceptibility measurements on $YbMo_6S_8$ and $TmMo_6S_8$. Rev. Phys. Appl. *15*, 1139 (1980)

B.D. Dunlap, G.K. Shenoy: Effect of magnetic atoms on the properties of ternary superconductors. Hyperfine Interact. *10*, 903 (1981)

B.D. Dunlap, D. Niarchos: Crystal field effects in $RERh_4B_4$ compounds. Solid State Commun. (in press)

M.M. Abd Elmeguid, H. Micklitz: Valence of Eu in $EuMo_6S_8$ under pressure. J. Phys. C *15*, 1479 (1982)

F.Y. Fradin, P.K. Tse, J.W. Downey: "Low Field ^{11}B NMR in YRh_4B_4", in Proc. 16th Int. Conf. Low Temp. Physics, Physica *107*, 753 (1981)

J.D. Jorgensen, D.G. Hinks, F.J. Rotella, G.K. Shenoy, D.R. Noakes: Fe defect sites in $YbFe_xMo_6S_8$. Bull. Am. Phys. Soc. *27*, 320 (1982)

J.D. Jorgensen, D.G. Hinks, D.R. Noakes, P.J. Viccaro, G.K. Shenoy: Valence and delocalization of Yb in the Chevrel phase $YbMo_6S_8$. Phys. Rev. (to be published)

K. Kumagai, F.Y. Fradin: "^{11}B Study of Spin Dynamics in $Y_{1-x}Re_xRh_4B_4$", in Proc. 4th Conf. Superconductivity in d- and f-Band Metals, June 28-30, 1982, Karlsruhe, FRG (to be published)

K. Kumagai, F.Y. Fradin: "Effect of Superconductivity on Spin Dynamics in $Y_{1-x}Re_xRh_4B_4$", in Proc. Int. Conf. Magnetism, Sept. 6-10, 1982, Kyoto, Japan (to be published)

R.W. McCallum, F. Pobell, R.N. Shelton: ^{125}Te Mössbauer effect in Chevrel phase compounds $Mo_6(Se_{1-x}Te_x)_8$. Phys. Lett. A *89*, 316 (1982)

D.R. Noakes, G.K. Shenoy: The effect of crystalline electric fields on the magnetic transition temperatures in rare-earth rhodium borides. Phys. Lett. (in press)

G.K. Shenoy, D.R. Noakes, D.G. Hinks: Determination of crystalline electric fields in $RERh_4B_4$ compounds. Solid State Comm. *42*, 411 (1982)

G.K. Shenoy, B.D. Dunlap, D. Niarchos: "Magnetic Interactions in Ternary Superconductors", in Proc. Int. Conf. Applications of the Mössbauer Effect, Dec. 14-18, 1981, Jaipur, India

G.K. Shenoy, D.R. Noakes, G.P. Meisner: Mössbauer study of superconducting $LaFe_4P_{12}$. J. Appl. Phys. *53*, 2628 (1982)

G.K. Shenoy, G.W. Crabtree, D. Niarchos, F. Behroozi, B.D. Dunlap, D.G. Hinks, D.R. Noakes: "Magnetism and Crystal Fields in Ternary Superconductors", in *Crystalline Electric Fields in f-electron Systems* (Plenum, New York 1982) p. 431

G.K. Shenoy, F. Probst, J.D. Cashion, P.J. Vaccaro, D. Niarchos, B.D. Dunlap, J.P. Remeika: ^{166}Er and ^{119}Sn Mössbauer studies of the re-entrant ternary superconductor $ErRh_{1.1}Sn_{3.6}$. Solid State Comm. *31*, 53 (1981)

G.K. Shenoy, B.D. Dunlap, F.Y. Fradin, S.K. Sinha, C.W. Kimball, W. Potzel, F. Probst, G.M. Kalvius: Magnetic dilemma in superconducting $ErRh_4B_4$. Phys. Rev. B *21*, 3386 (1980)

Subject Index